BRIEF CONTENTS

VIDEO PRODUCTION HANDBOOK

Leonard Shyles

Villanova University

Houghton Mifflin Company   Boston   New York

Sponsoring Editor: George Hoffman
Associate Production Editor: Christina Lillios
Senior Production/Design Coordinator: Carol Merrigan
Manufacturing Manager: Florence Cadran
Editorial Assistant: Joy Park

Cover design: Linda Manly Wade
Cover Images © 1997 PhotoDisc, Inc.

Photo Credits: Fig. 2.8, property of AT&T archives. Reprinted with permission of AT&T; Fig. 4.8a, copyright by Tektronix, 1997; Fig. 4.9, copyright by Tektronix, 1997; Fig. 4.12, Chesterwood, a National Trust Historical Site, Stockbridge, MA. Photographer: De Witt Ward; Fig. 5.3, photo courtesy of Vinten Inc.; Fig. 5.4, photo courtesy of Arriflex Corporation; Fig. 5.5, photo courtesy of Shotmaker Inc.; Fig. 5.6, photo courtesy of Miller Fluid Heads (USA) Inc.; Fig. 5.7a, photo courtesy of Cinema Products Corp.; Fig. 5.9, photo courtesy of Ikegami Electronic (USA), Inc.; Fig. 6.19, photo courtesy of Shure Brothers Inc.; Fig. 6.20, photo courtesy of Sennheiser and Whisper Technologies; Fig. 6.22, photo courtesy of Shure Brothers Inc.; Fig. 6.23, © Spooner/ Gamma Liaison Network; Fig. 7.3, photo courtesy of Steinberg Inc.; Fig. 7.4, photo courtesy of Steinberg Inc.; Fig. 7.9, used with permission from TEAC America Inc.; Fig. 8.1, Dick Van Halsema/Knight-Ridder News; Fig. 8.4, photo courtesy of QUANTEL; Fig. 8.5, photo courtesy of QUANTEL; Fig. 9.12, reprinted by permission of GVG Product. Bill Santo © 1991; Fig. 9.15, photo courtesy of QUANTEL; Fig. 10.1, photo courtesy of Ikegami Electronics (USA), Inc.; Fig. 11.14, photo courtesy of Photofest; Fig. 14.3, Ken Staniforth/ Fox/The Kobal Collection; Fig. C5, photo courtesy of QUANTEL; Fig. C6, photo courtesy of QUANTEL.

The following photos were taken by the author: Introduction photo, Fig. 3.5, Fig. 4.1, Fig. 4.2, Fig. 4.5, Fig. 4.7, Fig. 4.8(right), Fig. 4.14, Fig. 5.1, Fig. 5.2, Fig. 5.7(right), Fig. 5.8, Fig. 5.10, Fig. 5.11, Fig. 5.12, Fig. 5.13, Fig. 5.14, Fig. 5.15, Fig. 5.16, Fig. 5.17, Fig. 5.18, Fig. 5.19, Fig. 5.20, Fig. 5.22, Fig. 5.25, Fig. 6.9, Fig. 6.10, Fig. 6.11, Fig. 6.12, Fig. 6.14, Fig. 6.15, Fig. 6.16, Fig. 6.18, Fig. 7.7, Fig. 8.2, Fig. 8.3, Fig. 8.12, Fig. 8.13, Fig. 8.14, Fig. 8.15, Fig. 8.16, Fig. 8.17, Fig. 8.18, Fig. 8.27, Fig. 9.7, Fig. 10.2, Fig. 10.7, Fig. 10.13, Fig. 11.15, Fig. 14.2, Fig. C4, Fig. C7.

Printed in the U.S.A.

ISBN: 0-395-71488-5

123456789-CRS-02 01 00 99 98

CONTENTS

PART ONE

COMMUNICATING WITH VIDEO   1

1 Video as Communication   2

2 How Television Works   17

7 Audio Processing and Aesthetics

8 Graphic and Set Design

9 Video Processing   204

The Video Switcher   207

More Complex Effects and Transitions   221

Digital Video Effects Generators   225

Key Terms   227

Questions for Review   227

10 Field Production   229

The Gulf War: A Case Study in Field Production   231

Three Basic Concepts   232

Electronic News Gathering   234

Electronic Field Production   253

Multicamera Remote Production   258

Key Terms   263

Questions for Review   263

11 Editing: Aesthetics and Techniques   265

Editing Aesthetics   267

12 Writing and Script Formats   300

13 Producing and Directing   326

14 Performing   355

Why a new book on video production? First, recent changes in the production field demand a fresh look at the process. Second, many of the books currently available emphasize the latest technology but fail to explore underlying principles in sufficient depth. This book reverses that trend.

For example, Chapter 1, "Video as Communication," outlines the importance of audience analysis in production planning as well as understanding the program's topic, content, and purpose. This book stresses the idea that audience analysis determines program goals. Communication concepts from both ancient and modern traditions clarify how the production process is simplified and made more manageable when understood in these terms.

Chapter 2, "How Television Works," continues to emphasize conceptual foundations. Using simple language, the chapter describes in greater depth than other production texts the physical nature of radio energy and how television transmission is possible. This information demystifies technology, making it easier to assess the impact of new technologies on the production process.

Chapter 3, "Light and Lenses," explains how light interacts with lenses and why lenses and aperture settings produce their effects. It does not simply list a set of rules to follow about which lens to use to achieve what effect. Therefore, you will better understand what lenses to use in *any* production situation.

In short, this book follows the philosophy that if you give a person a fish, you feed him or her for a day, but if you *teach* a person to fish, you feed that person for a lifetime. Put even more simply, a guiding principle of this book is that *there is nothing quite as practical as a good theory*, and in today's rapidly changing world, understanding is the best tool for adapting to change.

Although this book is more in-depth than most in that it deals with both studio and field production, it is also briefer than many comprehensive texts. This is because the book emphasizes enduring production principles rather than information that will surely become outdated in a year or two. For example, instead of presenting pages of photos and descriptions of the latest models of microphones, the book explains how the three types of microphone elements used in *all* microphones work. In this way, enduring principles are featured in place of passing trends, thus saving you a great deal of time while deepening your understanding of the production process.

One of the book's strongest features is its treatment of the aesthetic aspects of video production. For example, Chapter 4, "Lighting Equipment and Design," presents a unified rationale for lighting design beginning with a philosophy of *naturalism*. From this starting point, the chapter presents a reasoned approach to lighting in *any* production situation.

Similarly, Chapter 5, "Using the Camera," aims to give you an aesthetic understanding of how and why cameras and camera mounts may be selected and positioned for different programs. The chapter considers elements of both static and motion visuals, and gives attention to how different images shape meaning for the audience.

Chapter 6, "Understanding Sound and Microphones," and Chapter 7, "Audio Processing and Aesthetics," explain the physical and aesthetic nature of sound and the proper use of microphones and other sound sources to shape program content effectively for the audience.

PREFACE

Chapter 8, "Graphic and Set Design," comprehensively tackles one of the fastest-changing parts of the production field. Although it covers many of the latest computer-based tools, including some discussion of the brave new world of computer-generated virtual sets, it emphasizes underlying principles rather than showcasing the latest line of new equipment. This emphasis affords you the best means for coping with rapid change. For example, in the near future, when high-definition television (HDTV) begins to enter the American home, the current screen standard 4:3 *aspect ratio* of width to height will change to a new standard of 16:9, more like that used in film. Your TV screen will get wider. However, after this conversion, understanding the impact of the television system on graphic production will still be essentially the same for the television producer. It will remain critical to know how brightness and contrast, scanning area and frame rate, color compatibility, pace of presentation, and a number of other aesthetic factors influence the audience and their perception of the graphics put on screen. These are perennial concerns for the television producer, and their coverage is one of this book's strong points.

Chapter 9, "Video Processing," and Chapter 10, "Field Production," introduce their subject matter in a way that simplifies and organizes understanding of two major areas of rapid change in the production field. Chapter 9 does so by building a video switcher from the ground up, starting with the simplest transitions among video sources and advancing logically to the most complex. The chapter also discusses some key differences between studio and postproduction switchers, preparing you for the discussion of on-line and off-line approaches to video editing. Rather than differentiating equipment in terms of "high-tech" and "low-tech"—descriptions destined for obsolescence even before the ink is dry—Chapter 10 presents the more enduring concepts of *reach, range,* and *interactivity* in explaining how electronic news gathering (ENG), electronic field production (EFP), and large-scale remotes are accomplished.

Chapter 11, "Editing: Aesthetics and Techniques," presents both editing theory and method. First, using a fictional sitcom treatment as a starting point, the chapter applies editing concepts to an actual example, thus making theory concrete. Then it presents the technical side of editing, including discussions of analog versus digital, linear versus nonlinear, and on-line versus off-line editing. This approach makes the chapter a comprehensive editing primer for the 1990s.

Chapter 12, "Writing and Script Formats," covers the basic principles of television writing, emphasizing how characteristics of the television medium and market affect the writer's task. In particular, it explores the relationship of pictures to words, as well as the need to have a strong sense of the audience to make your writing successful. The second half of the chapter presents clear examples of different script formats, each appropriate to different types of programs. Few comprehensive production texts cover basic writing principles for television in this much depth; some neglect the topic entirely.

Chapter 13, "Producing and Directing," explains the responsibilities, roles, and tasks of the producer-director in making video programs. The chapter stresses the need to view the producer's role in terms of five key factors: *space, time, material, money,* and *personnel.* Presenting the producer's job in these terms simplifies your understanding of what television producers do. The chapter then explains the job of the director, describing why and how the director gets involved in various production functions. The chapter discusses the director's job in terms of different production phases, with special emphasis on how to direct in the production phase, when critical decisions must be made, often with little time to spare. The chapter offers directing techniques to make your directing clear and easy to understand by your crew and therefore more successful.

Finally Chapter 14, "Performing," teaches you how to put your best foot (and face) forward as a television performer. The chapter describes how technical, social, and aesthetic aspects of the television medium influence the way you look and sound on television and tells you how to prepare the best possible performance. In addition, the chapter presents a clear rationale for selecting your wardrobe and using makeup.

Several other features of the book deserve mention. First, this is the only comprehensive production text that features an Industry Voices section, interviews with industry professionals, from free-lancers to network veterans, giving their views on how things have changed, what the current state of the art is in their fields, what the latest trends are, and where things seem to be headed in the future. The industry voices interviewed for this book include a producer for a major network's sports remotes, a graphic designer for "The Late Show with David Letterman," a successful free-lance writer, a computer graphics specialist, an independent producer making corporate videos and promotional pieces, a technical network field operations manager, a makeup artist for daytime and prime-time television, a lighting designer, and an audio expert. These practitioners also tell what they look for in an intern. For any student who wants to work in the field of broadcasting and electronic media, the Industry Voices feature is invaluable.

Second, each chapter contains a Professional Pointers section of boxed material clarifying, where necessary, some of the finer points of technical information. Each chapter also provides a list of key terms, terms that practitioners expect you to master to be able to work alongside them in the industry.

This book also provides the latest information on the recent FCC decision to adopt a digital standard for television broadcasting. Located at the end of Chapter 2, this section describes how the new digital standard will affect the broadcasting industry, especially as digital television (DTV) becomes more compatible with the computer and film industries.

In general, the fourteen chapters of this book cover all aspects of video production in both studio and field, offering a brief, clear, and conceptual approach to the craft of television production.

Len Shyles, May 1997
Villanova, PA

ACKNOWLEDGMENTS

I thank Villanova University for providing the atmosphere essential for the completion of this book: the library staff, the administration of the College of Arts and Sciences, colleagues, and students. I especially wish to thank

Dr. Marguerite Farley, my chairperson at Villanova, who gave me her support, trust, and encouragement to complete this project. Marguerite passed away in February 1996 and is profoundly missed. Thanks also to Patty Trenchak for reading early drafts and for being a great student.

I thank Michael Young, principal, Young Design, Inc., Falls Church, VA, friend and talented electronics engineer, for checking technical information in the audio and television technology chapters for accuracy. Any mistakes are mine. I also thank Morgan Besson of the Villanova University physics department, Shelden Radin of the Lehigh University physics department, and Erwin Cohen, friend and demanding editor, who helped with the chapter on lenses.

Thanks also to Thomas McCain, Director of the Center for Advanced Study of Telecommunications (CAST) and Professor of Communication and Journalism at The Ohio State University, for locating material on the digital television standard recently adopted by the FCC, which appears in Chapter 2.

At CBS, I thank Joseph Flaherty, Senior VP of Technology, who granted me access to the CBS studios; Andy Barry, VP of Production Services; and Bill Naeder of Remote Operations for their insights about production techniques.

I thank Louis Shore, Executive Director of Labor Relations, Paramount Pictures, for putting me in touch with many industry voices who graciously provided me with insights and information about the state of the field.

At Houghton Mifflin, I thank Margaret Seawell, Sponsoring Editor, who trusted me to do this book and had faith that it would make a contribution to teaching and education; George Hoffman, Sponsoring Editor, who continued vigorous support for this project after it was under way; and Kara Maltzahn, Editorial Assistant, who helped me complete the project in a timely fashion.

Special thanks go to Doug Gordon, developmental editor from P.M. Gordon Associates in Philadelphia, for his writing knowledge, humor, and human touch, who made the editing phase of the project a pleasure.

I also thank Scott Brady, my assistant in television production classes; colleague Terry Nance; student Alan Watt; and Al Petrasko, ENG producer from CBS local television, New York, who modeled for photos featured in the book; makeup artist Rob Rosiello and model Sarah Schmittinger, actors from the Villanova Theatre Department, who demonstrated makeup application techniques; and Wesley Truitt of the Villanova Theatre Department for providing scale models and set sketches.

I also thank one of my oldest and best friends, Ernest Scaglione, Jr.—construction worker, bartender, schoolteacher, and hilariously irreverent comedian—who graciously hosted me on numerous occasions in Brooklyn when I was conducting interviews with production personnel in the New York area.

I thank Janice Hillman, my wife, best friend, and artist, who helped develop early sketches for many of the images that appear in these pages; my parents, Judith and Milton, for making this project possible; and my children, Rebecca and Daniel, for keeping it possible and making it all worthwhile. All of these people have made this writing experience a rewarding, often pleasurable journey.

Finally, I thank the following reviewers for their help on this project:

Gary C. Dreibelbis, Solano Community College
Ron Osgood, Indiana University
Ann D. Jabro, Penn State University
Janice D. Tanaka, University of Florida

Bob Lochte, Murray State University
Rick Houlberg, San Francisco State University
David Baule, Milwaukee Area Technical College
David Vest, Colorado State University
John Fryman, Texas Tech University
Bob Harris, Ithaca College
Frances K. Gateward, The American University
Michael Lasater, Western Kentucky University
Robert K. Avery, University of Utah

DEDICATION

To Janice, Rebecca, and Daniel

INTRODUCTION: THE VIDEO PRODUCTION PROCESS

Why do you watch television and video? Depending on your needs and wants, television can tell you what's new, whether to take an umbrella when you go out, or where to get the best deal on a car. It can entertain you with old movies, the latest music video, or live drama. It tries to persuade you to buy products and services and even to elect political candidates. You can find travel programs to spur interest in distant places you have never visited; you can take college classes or learn about microscopic animals or distant galaxies. Moon landings have been televised, as well as concerts to raise money for people starving in underdeveloped countries.

Television and video enable you to see things in the comfort of your living room that would otherwise cost a lot of time, money, and effort to see firsthand. For example, with multiple cameras, telephoto lenses, and instant replays, you can watch sporting events in even greater detail than you could if you were there in person. With videotape, moreover, you can watch them time and time again if you choose.

Nowadays you can also *do* things through television. Home shopping networks offer you the chance to buy products after seeing them demonstrated on television. You can order pay-per-view programs and, in some cases, even conduct business transactions. Business organizations increasingly use video to communicate with both their clients and their own staffs.

No wonder, then, that many people are interested in learning video production. But the complex production environment of a modern television studio or field facility can be a daunting sight for students. Where does one start? How can anyone begin to master so much complicated equipment? I remember my own impressions when I entered a television studio and control room for the first time. By contemporary standards, the technology I saw there was ancient, but I felt as though I had just discovered an incredibly advanced civilization. My attention was instantly focused not on people (classmates and future crew members) but on the many strange, new machines. In addition to the television cameras on heavy tripods, there were banks of recording and editing machines, monitors, consoles, and complicated switching devices for manipulating sounds and images. Heavily insulated cables hung on walls in

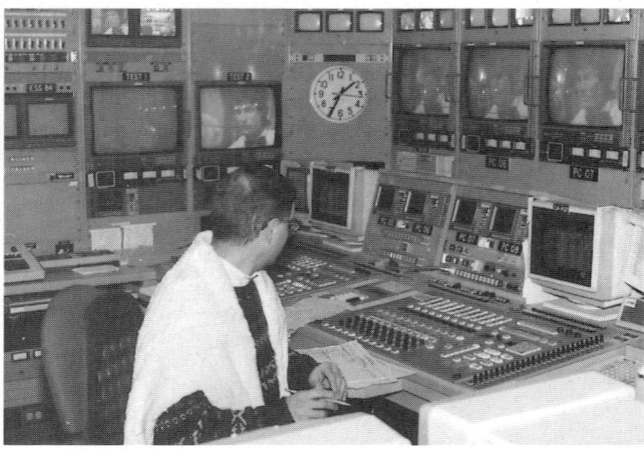

A contemporary television studio, full of fascinating and complex electronic equipment.

coils or were strung from lights on metal grids fastened to the studio ceiling. Other cables were slung haphazardly across the studio floor.

As I was, many production students are initially dazzled by the elaborate machinery. At first blush, it may seem that learning the craft of television production primarily involves mastering this wonderful technology. In fact, however, learning about the equipment is only a small part of learning video production.

On that first day I spent in a studio, my attention eventually shifted from the various fascinating contraptions to the crew members. I was witnessing the up-tempo activity of a production in progress. Camera operators were following directions received through their headsets. An assistant director was fielding questions from the control room. The audio operator was routing sounds from microphones and music tapes. Various other people I couldn't identify were busy at other tasks. Yet somehow the director was coordinating all of this frenetic activity. I began to realize that the director was orchestrating the actions of the crew in much the same way that a musical conductor leads a symphony.

Comparing a television director and crew to a musical conductor and orchestra is a useful way to begin understanding the production process. Just as a symphony conductor communicates his or her goals to musicians through the special language of music during rehearsals, so the television director uses jargon unique to the craft of television production to get the desired performance from the crew. Just as musicians must understand musical notation and be able to perform with special technical skill on cue, so the television crew must understand the language of video production and stand ready to execute the director's commands. When everything is coordinated properly, the crew functions as a graceful and cohesive unit, like an orchestra performing a musical composition. The product is a smooth, polished, well-produced, "air-quality" video.

This success cannot occur without communication, collaboration, and understanding. Video is not just a technology-centered activity; it is also a cooperative social activity. Though the making of a program surely requires technical knowledge and skill, it also demands great people skills. Just as a musical conductor cannot perform an orchestral composition single-handed but must rely on successful communication with a talented group of well-trained professionals, so a television director *must* successfully communicate with a crew to produce a quality television program. Lone artists need not apply.

We can push the analogy further. Just as conductors must first learn a musical score and then communicate their concept of how it should be played to the musicians, the television director must first get the program concept clear in his or her own mind, then communicate it successfully to the crew and talent on whom the project depends. Only then can the program be successfully produced. For this reason, for the director, *the crew is the first audience*. Hence, even prior to production, the director must find some means of clearly articulating the program concept to the crew.

To understand this process of communication and collaboration more clearly, let's examine the basic stages of television production and the major roles played by the different crew members.

THE STAGES OF VIDEO PRODUCTION

Video production is commonly discussed in terms of three principal stages: preproduction, production, and postproduction.

Preproduction

The **preproduction** stage of video production is the planning and development phase. It involves considering everything that needs to be arranged before entering the studio or field to shoot a television show. These concerns include formulating a budget, a program concept, and an overall treatment for the program. Preproduction also involves scriptwriting, revision, and approval; acquisition of facilities, studio space, and materials; scheduling and holding production meetings; and selecting cast and crew. In addition, agreements, contracts, permits, and legal issues are dealt with at this stage, including union matters. Contingency plans to handle last-minute changes are also addressed.

Preproduction cannot be taken lightly. It is in this phase that the program concept is first articulated and clarified. During this phase, the director begins to formulate the concept, considers the potential audience, and meets with the crew to work out details, refine ideas, and answer crucial questions. If those questions are not answered satisfactorily, the project may be canceled before it ever reaches the next stage, production.

Production

The **production** stage includes all of the activities designed to produce the actual program, including setup, rehearsals, and shooting. Rehearsals can

---

P R O F E S S I O N A L   P O I N T E R S

Important Questions to Answer During Preproduction

- Is this program a worthy project on which to expend time, money, and effort?
- What is the program's purpose?
- Does the program have cultural or other value?
- Will there be an audience for the program? Can I define the audience in specific terms?
- Can this project be done with available facilities and materials?
- What will the script be like?
- How long will the program take to make?
- Do I have enough competent personnel to crew the program?
- What talent and facilities will I need?
- What will I do if some components I am now counting on fall through?
- What flexibility do I have to deal with unforeseen problems?
- What audio and video sources must I have?
- What rights, agreements, permissions, and release forms must I clear to avoid lawsuits?
- What ethical concerns should be considered in shaping the content of the production? (For instance, how will the editing be done to ensure that people's views are accurately represented?)
- How long will the finished program be?

---

include run-throughs and coordinating the use of all technical equipment. Sets are constructed, graphics prepared, props and costumes made ready. Lights are hung, aimed, and focused. Cameras and all other material items are gathered and used as needed to produce the actual show. If the production is a field shoot, facilities are deployed at the production site, and the program is either produced live or recorded for later editing into the final product. The overall quality of the show is monitored at this stage, and necessary changes are made. Reshoots are done if needed. After the show is finished, the *strike* or *teardown* is done, meaning all the equipment is put away or packed up and the production space is cleared.

In the production stage, the trick is to remain calm, flexible, efficient, and effective. One must cope with adversity. It helps to keep the program concept clearly in mind so that you can make last-minute changes if needed and still meet the goals you set when you first articulated the project. All of this involves not only careful thought and technical mastery but also coordinated group activity—a team effort.

Postproduction

The **postproduction** stage is devoted to evaluating the program and editing it if necessary. Then the marketing, publicity, distribution, and exhibition of the finished product are undertaken. Any unpaid salaries and bills are paid, and personnel involved in the production are thanked, including cast, crew, and peripheral agents who helped make the production run smoothly. Often a post mortem meeting is held to discuss problems that occurred and determine how the production process might be improved in the future.

WHO DOES WHAT?

Everyone has heard of directors and producers, but their precise responsibilities in video production are not always understood. The roles of other team members, such as the technical director and the floor manager, may be even more mysterious. What, then, does each team member actually do during the three stages of video production? The following sections outline a number of the major roles.

Writer   During preproduction, the **writer** works with the producer and director to develop the script and format of the show. The writer writes, edits, and revises the script until it is approved. Then, during the production stage, the writer rewrites as needed. Even in postproduction the writer may be needed to smooth out transitions and provide supplemental material.

Producer   In preproduction, the **producer** develops the program concept, outlines the budget, and selects the crew. The producer approves the script and the director's treatment of it, and supervises and coordinates personnel, scheduling, and payroll. The producer also handles all contracts and other legal matters, including the necessary permits and releases. During the production phase, the producer supervises all production activities, monitors the rehearsals, and suggests needed changes. The producer also maintains the production schedule and keeps it within budget. In postproduction, the producer supervises and approves the editing, runs the marketing and publicity effort, and pays any remaining unpaid bills.

Director   In preproduction the **director** is involved in all meetings of the creative team. The director has input into the program format, the content, and the treatment, including the script, lighting, sets, and audio. In consultation with the producer, the director also works out the assignments of talent and crew. During production, the director runs the rehearsals, works with crew and talent, practices camera shots, and (as the title implies) directs the actual shooting of the program. In postproduction, the director works with the producer to supervise the editing.

Assistant Director   In preproduction, the **assistant director (AD)** helps the director plan and develop the program, and generally provides any assistance the director asks for. During production, the AD acts as a human clock, timing segments and previewing and readying upcoming camera shots before they

I N D U S T R Y
voices

Darcy Antonellis

Vice President of Technical and Olympic Operations, CBS

Q: You're vice president of technical operations. But what is "operations"?

A: Operations essentially deals with both the production aspects and the technical aspects of a show and those details needed to support a show. Our clients include the Sports Division, the Entertainment Division, and the News Division at CBS, and also non-CBS clients. For example, we tape "As the World Turns" and "Geraldo" here. We tape all of our news magazines with the exception of "Face the Nation," which is done in Washington. For our clients, we provide them with people and facilities to help them do their show so they can concentrate on the editorial or creative content.

Q: When a client comes to you, what happens?

A: Let's say they have a remote to do. We technically manage the broadcast. I'll use the example of the president's *State of the Union Address*, which we just finished. We sit down at a meeting with the producers, director, lighting director, and other support people, and they lay out how it should be produced. They tell us the site. They tell us the talent—in this case, Connie Chung and Dan Rather. They tell us the show is going to air for an hour during prime time. They tell us the production facilities they'll require, e.g., number of cameras, types of lenses, audio components.

We then go on a survey, look at the facility, look at access to the building, and decide how we will be able to transmit. Can we use a production mobile unit as the control room, or do we have to do what is called a "carry in," where you literally set up a control room inside? How will we be able to pull cables? Is there enough power, or do we have to bring our own? Do we need special credentials to get people in and out?

Q: Tell me about your work history. How did you get here?

A: I got my electronics engineering degree from Temple University, and I started at CBS as an engineering intern. I learned broadcast engineering from the standpoint of designing systems. I worked as a technician for about one year at WCAU in Philadelphia to get hands-on experience in the building and construction of facilities. Then I came back to New York, back to a management role in the engineering department, and from there was promoted to operations.

The engineering function here tends to be more "off-line" because you are building facilities; you're not dealing with the day-to-day air requirements. When I stepped into operations, that was a big transition because all of a sudden there were firm deadlines. It's a matter of airtimes and seconds. I started working a lot more closely with production personnel, the producers, the directors, the editorial people who create the broadcast. I enjoy this aspect of my job much more than production design work.

Q: All jobs have good and bad aspects. What are yours?

A: The pull for me is the desire to be thorough as an engineer. But with firm deadlines, that's not always possible. The business is demanding by trade. You have an air date and time, and there's no pushing that back and saying "I'll do it tomorrow." Frustration for me is juggling priorities. You have to recognize that your department's sole reason for being is to provide the resources to allow the production folks to excel in their broadcasts. We're supposed to help them be successful. That makes CBS successful, and we all win. Because we service News, Entertainment, and Sports, as well as outside clients, everything comes first, and you can't say no to anyone. So we're spinning a lot of plates.

Q: What do you look for in a worker or intern?

A: A few things. One is *energy*. Another is *optimism*. There are plenty of people who will tell you why something can't be done. I find very little use for that. There's a difference between saying "It can't be done" and saying "If we do it, here are some of the roadblocks we need to figure out how to get over."

You also have to *love the business*. The hours are horrible. You start out in entry-level jobs and get paid very little—in engineering somewhere in the $30,000 range, in production even lower, depending on the market.

But I would encourage people to start out in small markets because you get to do everything. That's a big plus. It's hard to do that on the network level. It's important to get a good foundation and learn all phases of the business. You get to take that with you to the network level. And that's good, because there's just a certain amount of background that's expected when you arrive here.

go to air. The AD also helps cue other crew members to deliver needed program elements, such as music. In postproduction, the AD helps with editing, timing, and keeping track of segments.

Technical Director   In preproduction, the **technical director (TD)** works with the producer and director to select technical facilities (often abbreviated *fax*). During production, the TD operates the video switcher, the machine that permits transitions among different video sources. If the switcher is needed during postproduction editing, the TD operates it then as well. During all phases, the TD is responsible for the technical quality of the program.

Audio Technician   In the preproduction phase, the **audio technician,** along with the producer and director, plans the audio treatment and approach the program will use. He or she prepares all audio sources and selects microphones. During production, the audio technician assigns the audio crew and deploys all the audio materials for the program. At this stage, the audio board that routes all audio signals (from microphones as well as other audio sources) is set up. During the shooting of the program, the audio technician executes the audio cues of the director. In postproduction, the audio technician operates the audio console to help with the editing process. Additional audio sources, such as music and narration, may be added at this time.

Lighting Director   During preproduction, the **lighting director (LD)** consults with the entire creative team (including the producer, director, costume designer, makeup artist, and scenic designer) to develop the best lighting approach. Since set and costume colors affect and are affected by the lighting, it is crucial to plan the lighting design *holistically*, with input from all involved about how the set and talent should look. A light plot is then developed accordingly. During production, the LD hangs, aims, and focuses the lights. Adjustments are made, and light levels are balanced. Then, during shooting, the LD executes any lighting changes and operates the dimmer board on cue.

Scenic Designer   Like the LD, the **scenic designer** (also known as the *set designer* or the *art director*) works with the creative team in preproduction to develop the sets and set dressings. Scale drawings and plots are prepared illustrating where all set pieces will go. During production, the scenic designer supervises set construction and makes changes as needed.

Graphic Artist   The **graphic artist** can have a significant effect on the visual impact of the program. In preproduction, the graphic artist consults with the creative team about the graphics that will be used and then prepares them, either electronically or on hard copy. Graphics can include textual material, pictorial illustrations, and complex, computer-generated graphic effects. During production, the graphic artist makes any necessary changes and operates the graphic equipment on cue from the director or the AD. In postproduction, the graphic artist supplies any additional materials needed during the editing.

Floor Manager   During the production stage, the **floor manager** coordinates activities in the studio or in the talent area in the field, feeding cues from the director to the talent. The floor manager is the director's main connection to the operation. The floor manager also supervises set and costume changes during production.

Camera Operator   During both rehearsal and shooting, the **camera operator** prepares and operates the camera, following the cues of the director received over a headset.

Video Technicians   During the production phase, the **video technicians** help maintain picture and transmission quality. They adjust cameras to reduce color and brightness variations. They also help with special effects if needed.

Talent   As you are probably aware by now, the term **talent** in television and video refers to the performers and actors. This category includes reporters, newscasters, hosts of game shows or interview shows, "guests" on those shows, singers, and comedians—in other words, all of the people who appear on camera.

In Part III of this book, we return to the responsibilities of the writer, producer, director, and talent to examine each of these crucial roles in more detail. In Part II, as we delve into the technical aspects of video production—lighting, camera operation, audio, graphics, set design, video processing, and editing—the tasks of the other crew members will become clearer. First, though, Part I of the text focuses on a crucial, overarching concept that we cannot emphasize enough: the idea that video is, above all, a form of communication.

KEY TERMS

preproduction *(xxi)*

production *(xxi)*

postproduction *(xxiii)*

writer *(xxiii)*

producer *(xxiii)*

director *(xxiii)*

assistant director (AD) *(xxiii)*

technical director (TD) *(xxv)*

audio technician *(xxv)*

lighting director (LD) *(xxv)*

scenic designer *(xxv)*

graphic artist *(xxv)*

floor manager *(xxvi)*

camera operator *(xxvi)*

video technicians *(xxvi)*

talent *(xxvi)*

QUESTIONS FOR REVIEW

1. Why is video production dependent not only on technical expertise but also on communication and collaboration?

2. What key tasks are carried out in each of the three stages of video production?

3. Name the major personnel involved in video production, and describe each person's main responsibilities.

PART ONE

COMMUNICATING WITH VIDEO

1 Video as Communication

T hink of the various means television and video offer to communicate your ideas. Television can broadcast events to virtually every household in the country, and to much of the rest of the world, live—as they occur. Television can also present animation, music, and sound effects. Like theater, television shows actors performing in costume and makeup, with lighting that can be altered in an instant to reflect changing moods. Like film, video offers audiovisual effects that the stage cannot easily duplicate, including flashbacks featuring instant changes of sets, actors, and costumes. In short, the communication options available to the television director are formidable. By combining live transmission with audiovisual capability, television has become both powerful and pervasive. It is a potent communication medium not only because of what it can do but because of how it can do it.

The potential of television and video to reach and influence audiences should challenge you to learn the communication principles that can enhance your programs' effectiveness. This chapter describes some of those principles. The topics covered include:

THE PROGRAM CONCEPT
identifying the concept • justifying the concept
AUDIENCES, MEDIA, AND MESSAGES

our changing media • tailoring programs to audiences • practical advice from the past • modern communication research • information theory • linear and nonlinear communication models • message effects • the universal audience

THE PROGRAM CONCEPT

Identifying the Concept

Before shooting any video program, you need a clear idea of the program's topic, purpose, and intended audience. You should also try to determine the context in which your program will be shown, as well as the genre or program category (if any) into which it fits.

Context A moment's reflection suggests that all television programs, and all messages in general, are made by some source for some audience for some purpose. If we accept this premise, it is plain that program content is by nature situated in some social or cultural *context;* it is never free-floating outside the matrix of human activity. Of course, although programs are directed toward a *target* or intended audience, they may also be seen by unintended viewers.

Understanding the social and cultural context of a television program helps you decide how it should be produced. Part of understanding a program's context includes knowing the source's reasons for making it and the audience's motivations for watching it. Knowing these things makes your program objectives clearer and simplifies all your production decisions. By grounding your program in the needs, values, and expectations of the audience, you can make it more engaging, perhaps even compelling, and therefore more successful.

In addition, knowing about other programs that deal with similar subject matter and how other program types or genres have treated similar topics is a distinct advantage. Of course, it is possible to develop a program that defies comparison to anything produced before. However, producing a new program that is similar to past successes can increase your audience's ability to identify and accept the new material.

Topic A program's *topic* can influence whether the program treatment should be serious or humorous, passionate or cold, light or heavy. Topics can include politics, history, science, literature, news, current affairs, sports and fitness, humor, cuisine, and innumerable other aspects of the human condition. The content or subject matter that a television program can feature is virtually limitless. In most cases, however, it is possible to compare your proposed program to past successful ones to get clues about its production.

Purpose Identifying the program's *purpose* involves specifying the goals and effects the producer wishes to achieve with an audience. For example, you can produce a program about AIDS whose purpose might be to inform or educate viewers about how to avoid contracting the disease. A different purpose might be to persuade viewers that more funding is needed for research, with the objective of getting them to write checks for a worthy cause. For other programs you might consider entirely different purposes, including sheer entertainment. Depending on the objective(s) of the program you are producing, your decisions regarding content and treatment can vary.

Genre *Genre* means the basic type or category of the program: fiction or non-fiction, comedy or tragedy, period or contemporary, and so on. Many catego-

PROFESSIONAL **POINTERS**

Questions to Use in Analyzing Your Potential Audience

- Whom do you wish to reach?
- Can you describe the audience in terms of demographic, lifestyle, and psychographic variables?
- What benefits will the program offer the audience?
- Is anything known about what audience members do with their time?
- What have audience members seen before? What did they think of it?
- Can the audience attend to, accept, and understand the program, as well as enjoy it, be moved by it, and profit from it in some way?
- At what level should the script be written to hold the audience?
- What references and associations will advance the program's main idea?
- How hard can you work the audience members without losing them?
- What message(s) will the audience reject?

rizations can be applied on increasingly detailed levels. For example, within the comedy genre, there may be slapstick, screwball, and high-brow, to name but a few. In the drama category, we can find westerns, cop and crime shows, and medical shows, among others. Knowing where your proposed program fits within the various program types or genres can help you plan your production.

Audience   Audience-related concerns include the demographic characteristics of the intended or target audience (such as age, gender, educational level, occupation, race, religion, geographical location), the audience's interests and hobbies (sometimes referred to as *lifestyle* variables), and cognitive and emotional qualities (sometimes called *psychographic* variables). The audience's experience, motivation, and abilities may also be considerations. For example, if the intended audience is exclusively children ages six to twelve, the script will obviously be pitched at a different level than one suited to a sophisticated adult audience.

Justifying the Concept

In most cases, the effort used to produce a quality television program is great, even for short programs. In addition, costs for services of talent and crew, as well as for rental, use, and maintenance of facilities, can be high. Therefore, winning support for your project depends on developing a proposal that justifies the concept in the eyes of a funding agent. *The proposal is a persuasive document, designed to convince a funding agent to finance your idea.*

Checklist for Developing Your Program Concept

The more you know about the funding agent, audience, topic, context, genre, and purpose of the program you are producing, the better you can determine how to assemble all production elements. All these factors do overlap, but each one should be considered separately. As you develop your program concept, make sure you have a clear idea of each of the following elements:

_____ *What* the message is

_____ *Who* is making the program and *why*

_____ *How* the message is to be delivered

_____ *For whom* the program is being made

_____ *For what* purpose

In many cases, agents fund video projects that win favorable review from among competing proposals. In the broadcasting industry, new pilots for network series are produced after considering numerous ideas that compete for resources from programming executives. In the corporate world, decisions to produce a program to motivate the sales force or to introduce a new product into the marketplace are made after carefully considering needs and weighing a number of alternatives. In educational settings, video teaching modules may be produced based on an analysis of what constitutes the latest knowledge and information about a special topic.

How, then, can programmers write proposals that justify the program concept? Although we deal with this question in detail in Chapter 13, one point should be made here. The proposal should demonstrate a significant need for the program, both from the audience's perspective and from the perspective of the funding agent. In commercial television, desirable programs are those believed capable of attracting audiences for advertisers. In the corporate world, approved projects are those believed to offer beneficial results in generating profits, solving some problem, or satisfying a need as defined by company executives. In government, top priority usually goes to videos perceived as serving the general welfare in problem areas such as public health, the environment, and safety. In every case, success hinges on convincing the funding agent that the program will attract the desired audience.

The budget, of course, is always important to those who control the funds, and later chapters of this book will refer often to budget concerns. But it is important not to let budget issues diminish your creativity. At the beginning of the brainstorming process, imagine what the show would look and sound like if you were to have unlimited budget, facilities, personnel, and time; that is, eliminate the practical side of the equation and conceptualize the project exclusively from a communication perspective. Later you will have plenty of time to modify your plan in terms of practical limitations and bring yourself back to earth. *In short, in the beginning, budget issues should be secondary to concept considerations.*

Our Changing Media

Audience analysis in the television field is becoming more complicated as media choices and functions increase. It was once reasonable to describe the television audience as a large, undifferentiated mass of passive viewers. But just as the magazine industry, which once featured general-interest magazines such as the *Saturday Evening Post* and *Life,* has evolved into a plethora of special-interest periodicals serving more highly segmented audiences, so has the video marketplace changed, and this trend continues right before our eyes. Today's video environment commonly offers dozens and perhaps hundreds of broadcast and cable channels featuring highly specialized programs. Video shops offer thousands of more selections. Business, education, and government all use videos aimed at specific groups of people. This trend has resulted in more fractionalized or segmented video audiences.

Interactivity   In addition to fragmentation of the audience, **interactivity** has entered the marketplace, enabling viewers to not only receive video information but also to send messages back to programmers during broadcasts. For years, thanks to the telephone, call-in shows have offered interaction with the hosts and guests. Today, on-line computer systems enable home shopping networks to gauge whether to continue displaying a product or move on to a different one based on the level of purchases made during the show. Viewers are increasingly able to use television interactively to satisfy personal needs and desires in such areas as ordering pay-per-view programs, initiating banking and consumer transactions, and even engaging in distance learning classes. Clearly, the trend toward a more interactive relationship between television and viewers has modified the traditional functions of television to entertain, inform, educate, and persuade.

Convergence   Related to interactivity is the increasing **convergence** of communication media. Various other communication technologies, such as telephones, computers, and satellite and cable services, are being integrated with the television medium to create products and services unimaginable a decade ago. This convergence of technologies makes greater interactivity possible.

The Key to New Technology: User Appeal   To succeed, new media technologies that offer programming alternatives must appeal to potential users. That is, they must come to be perceived as offering significant advantages over what they replace. An engineer's opinion of a new device is secondary, perhaps even irrelevant, to whether the item will succeed in the marketplace. Doubters of this view need only recall past media inventions that offered improved programming options but failed as consumer products: eight-track tape, quadraphonic stereo, and video laser disks, for example. Therefore, the most powerful computer, or any other program package or communication device, is not the one with the greatest fidelity, storage capacity, or speed of delivery but *the one that people will use*.

Tailoring Programs to Audiences

Despite recent technological changes, such as increased channel capacity and greater interactivity, we human beings are still essentially the same two-eyed, one-brained communicators we have been for thousands of years. That is, greater programming potential by itself is no guarantee of programming success. Television content must still command attention from some audience to succeed. Put simply, the toughest step in communication is still the last few inches from the eyes and ears to the brain. Therefore, increased program availability underscores the need for programmers to know more, not less, about audiences than ever before to compete in an increasingly complex marketplace.

Analyzing the linkage between audience characteristics and program appeals has been an ongoing enterprise in broadcast research for decades. Program makers, especially those in commercial television, avidly seek to determine what features of program content attract audiences in the hope of finding out how to maximize ratings and justify the sale of higher-priced commercial time to advertisers. One reason spin-offs of successful television shows and film sequels are so popular with producers is that they offer the chance to exploit formulas that have been successful in the past. When a program is a hit, the maxim is: do not mess with success, but make more like it.

In addition to attracting large audiences, programmers and advertisers want to attract certain *types* of audiences according to desired demographic, lifestyle, and psychographic categories. For example, the advertiser of an acne cream will want to know which types of program content attract teenagers. Similarly, the manufacturer of a denture adhesive will look for program characteristics that appeal to senior citizens. Broadcast audience research is a full-time and lucrative business for such companies as Arbitron and Nielsen, Inc., which provide research data to help advertisers and programmers measure audience preferences as well as learn the *uses* and *gratifications* specific programs offer their viewers. Knowing what motivates people to watch, recall, and act on the programs they see is also important to corporations, public-interest groups, political parties, public relations agencies, market research firms, and other businesses that depend on message effectiveness for their success.

Practical Advice from the Past

Arbitron and Nielsen are not the first to research audience preferences and methods of appeal. Since the time of the ancient Greeks, specialists in rhetoric and communication have analyzed the best ways to reach audiences. In fact, many ancient principles and techniques of persuasion are still relevant to modern forms of communication.

Down through the centuries, communication experts have suggested that the best communicators are those who know history and current affairs as well as the caprices of the human heart. Good communicators understand human emotion, memory, imagination, hope, fear, and human will. Further, successful messages invariably appeal to the *rational, emotional,* and/or *ethical* dimensions of human nature. The following sections examine these three basic types of appeal.

Ethos, or Source Credibility   Ethical appeals are often the most powerful means of persuading an audience. Audience perception of **ethos** or *source credibility*—that is, the ethical character or trustworthiness of a source—is strongly related to the audience's knowledge of a source's past experience and behavior. Depending on the level of expertise a source displays on a given topic and the audience's knowledge of the source's past behavior, habits, and character, an audience may judge a source to have either high or low credibility. In modern terms, ethos is also thought of as the degree of *charisma* a speaker is perceived to possess.

High levels of admiration and respect for speakers can help make their statements seem worthy of belief. This may explain why advertisers often cast doctors, or actors who have portrayed doctors, as spokespersons in commercials for headache remedies and other medicines. Since doctors can be reasonably trusted to know how to treat headaches, a product endorsement is believed to be credible if it appears to come from a medical expert. Frequently, the performer wears a lab coat and appears in a set resembling a doctor's office.

Other highly credible sources frequently seen endorsing headache remedies include people performing high-stress jobs, such as jackhammer operators and kindergarten teachers. Similarly, sports figures such as football great Joe Namath are often hired to endorse muscle liniments and other pain relievers. The appeal to credibility in these cases lies not in the sources' credentials but in their experience. (Notice how the terms *credible* and *expert* relate to the words *credentials* and *experience*.) Beyond the selection of talent and the roles they are cast in, other production elements (props, makeup, script, and so forth) are often fashioned to reflect the advertiser's desire to use high-ethos appeals to sell products.

Logos, or Reasoning Ability   Knowledge of the audience's reasoning ability, or **logos** (pronounced with an *s* rather than a *z* sound at the end), is also critical in fashioning messages. To engage the audience, television advertising often uses rational appeals featuring tightly structured arguments, sometimes with one premise left out. Presenting an argument with a key part missing invites audience members to fill in the blank or make a desired inference, encouraging them to aid and abet the source in the persuasion process.

A concrete example of such an appeal to logos is seen in the insurance company commercial in which the central character has just become aware of a new insurance plan offering more comprehensive coverage than her current one. The character muses in a voice-over while shown looking over some insurance brochures: "People should have the best insurance coverage they can afford. This new plan is more comprehensive than my current plan, yet costs less than I'm paying now. . . ." These lines invite the audience to complete the argument: "It's a good idea to switch plans."

Structuring video messages to engage the audience's reasoning ability is a powerful means of encouraging active participation in the communication process. It increases the audience's involvement and presumably influences buying behavior.

Pathos, or Emotional Appeal   Appeals to **pathos,** or the emotional nature of the audience, work by arousing the audience's passions, feelings, and sympathies. Though whipping up strong emotions can warp the audience's reasoning ability and judgment, such appeals aid in the persuasion process.

The emotional lives of human beings include feelings of love, hate, fear, confidence, shame, pity, benevolence, indignation, and envy, among others. Notice how this litany describes the frequently lurid subject matter on such programs as "Hard Copy," "A Current Affair," and tabloid talk shows, which capitalize on emotionally charged topics to get audience attention. Consider too the many high-budget commercials that rely on raw, sensuous audiovisual stimulation, including rhythmic cutting, quick camera movements, music, color, and arresting and provocative visuals. Often such stimuli are used with the goal of having a direct effect on our bodies, not our minds. They arouse us to get our attention. Provocative music videos routinely use the same techniques to arouse audience attention.

Age and Message Receptivity   Understanding human experience as a function of age provides further insight into message receptivity. We know that certain types of programs—cartoons, to use an obvious example—succeed more with younger audiences, whereas more sedate programs attract older audiences. Thus, video programmers need a clear idea of the age-related characteristics of the potential audience.

More than two thousand years ago, the Greek philosopher Aristotle discussed human life in terms of three ages, which we can roughly label youth, the prime of life, and old age. Youth, compared to other stages of life, is a time of strong desires, powerful sexual urges, and a relative lack of self-control. The longings of youth, said Aristotle, are "keen but not deep." Young people are passionate, quick to anger, and highly resentful if slighted. They love "honor" and victory more than money. They hold a strong belief in human goodness, and they are trusting and hopeful but easily deceived. They are high-minded, not yet humbled by life. They live lives of impulse rather than calculation. Finally, they are devoted to their friends and tend to make mistakes on the side of intensity and excess. When they injure others, they do so not through malice but through insolence.

Of course, no existing description of youth or any other age is complete, nor does it apply equally to every individual in the group. It is interesting, though, to consider whether Aristotle's ancient description of youth fits popular television characters such as Bart Simpson, Beavis and Butthead, and others watched avidly by youth audiences. What other programs or characters might fit this mold?

In contrast, according to Aristotle, elderly individuals often possess qualities of extreme moderation. They "think" but never "know." All is doubtful; nothing is firm. Elderly people are also cynical and tend to interpret everything pessimistically. They are suspicious and mistrustful as a result of sad experience. Relative to other ages, elderly individuals have been humbled by life. They crave the mere necessities, are not generous, and tend to be cold and calculating, with less regard for honor than for expediency. Finally, unlike youth, elderly people are less concerned with what people think of them and live mainly in memory rather than anticipation. In light of this profile, it is understandable that programs for the elderly population are very different from programs for the youth market.

People in the so-called prime of life, intermediate between youth and old age, tend to be both confident and cautious. They judge each case by the merits of the facts, says Aristotle, and they pursue both honor and expediency. They manage to balance self-control with valor. If this is an accurate characterization, perhaps it can help us in designing a program for, say, forty-year-olds.

Message Construction and Style   In addition to presenting insights about the nature of the audience, communication theorists have advanced notions about style—in other words, how the material is presented. Traditionally, style considerations have involved language, diction, order, and arrangement. In television production, style extends to the many visual aspects of the set, props, costumes, makeup, graphics, and nonverbal sound effects.

Good style also means delivering a *lively* message with *clarity* and *appropriateness*. The liveliness aspect of style (or what contemporary analysts call *dynamism*) is especially relevant to television, since one of television's greatest strengths lies in its ability to cover events such as sports and on-the-spot news live—as they happen. The liveliness aspect of video also offers the chance to demonstrate complex processes and techniques in real time, such as when infomercials demonstrate how to use the latest cooking utensil or exercise equipment.

Since television is preeminent in its ability to broadcast live coverage of real events to audiences, programmers should consider how they can imbue every appropriate element of a production with liveliness. This may mean deciding when to cut to an extreme close-up of the main character's grimacing face as he struggles to escape harm or how to sequence and edit a scene in a lively and engaging way to maximize audience involvement.

Also, since video programs incorporate a great many audio and visual elements into a unified whole, it is helpful to consider each element in terms of how it contributes to the overall style of the program. Even decorative considerations should be tempered by the totality of what you know about a particular program's content, purpose, and likely audience reaction and comprehension.

Modern Communication Research

Building on the insights of past centuries, modern communication research offers detailed findings about the nature of audiences and messages. Theories dealing with message construction, meaning, information transmission, message impact, audience characteristics, and persuasion can help you produce more effective video programs.

For example, most modern theorists agree that defining communication as simply the transmission of information is inadequate, since a message sent may not always be the same as a message received. This is because people do not communicate messages directly to receivers but must mediate them through the use of symbols or images. In other words, messages must be *encoded* before they can be transmitted, and sometimes the encoding process itself gets in the way of successful meaning transmission. The simple parlor game of post office, where one person whispers a message to another, who then whispers it to the next player, and so on, resulting in a garbled final product, illustrates this point clearly.

In the case of video, meanings may at times be lost because messages are sent to viewers who in most cases have never even met the sender. To overcome this additional limitation, producers often use devices such as redundancy and repetition to ensure that messages and their intended meanings are successfully received. Advertisers have learned well the lessons of repetition in brand name advertising, often repeating their central theme over and over again throughout an advertising campaign or even within a single commercial. They use redundancy techniques when they both show *and* tell about some key aspect of a new product they wish to sell. Of course, advertisers' commitment to audience research, which includes finding out as much as possible about their target audience before constructing their appeals, allows them to maximize the fit of their messages to their audience.

The following sections offer some modern communication theories and models useful for crafting video programs.

Information Theory

Information as Reduction of Uncertainty   Information theorists have defined *information* as a measure of the degree of randomness in a situation. For example, if you are watching television and the set suddenly stops working, you may not know whether the problem is with the receiver or the electrical outlet (or some other cause, such as an unpaid utility bill). To find out, you may disconnect the receiver and plug in a working lamp to test the outlet. Or you may plug the TV set into a different socket to troubleshoot the problem. By conducting these tests, you are reducing uncertainty in the situation, thereby gaining information.

The utility of understanding information this way is that it can help you decide what elements to include in a production based on the level of familiarity your audience is likely to have for a given subject. For instance, if you are producing a cooking show, depending on the knowledge base of your audience, it may be unnecessary to explain what "beat three large eggs" means but quite necessary to explain what "fold in three egg whites" means. If you assess your audience correctly, you can avoid the dual problems of telling them what they already know, thus boring them, or neglecting to tell them what they might not yet know, leaving them confused and frustrated. Of course, knowing about information theory is no guarantee that you will correctly determine how the program should proceed, but it can alert you to consider your audience from an information perspective and help you plan your program more effectively.

Noise as an Unintended Signal Component   Information theorists define **noise** as any part of a signal that is unintended by the source. Noise inhibits the reception of a message by a receiver. Thus, noise impedes communication.

Noise can take many forms. Common examples include static on the radio and whispering in a crowded movie theater. Even dirt on the television screen can constitute noise, though the dirt is perfectly silent. Noise can also include a speaker's distracting hairstyle or clothing. If a speaker's odd mannerisms distract the audience from the speech, they may be considered noise from the perspective of information theory.

Linear and Nonlinear Communication Models

A *model* is a descriptive, simplified representation of some phenomenon or process. For example, maps generally model some aspect of geography. A road map may model the highway system in some part of the world, while a precipitation map may model the annual rainfall in the same area. Depending on what the map is intended to do, it can model any number of aspects of what it represents.

In the communication field, theorists use models to describe, explain, and test propositions about the way messages and meanings are transmitted between senders and receivers. Communication models usually include a *channel* component to account for how much information can be put through a communication system and still be delivered accurately.

Linear models of communication emphasize the one-way flow of information from a sender to a receiver. Early broadcast media systems such as network television and local radio were linear in nature because they sent messages from a single source to a large number of receivers. The components of the communication model used to represent network television included a sender or transmitter, a message, a channel, and a number of receivers.

However, not all communication phenomena can be adequately rendered with such a one-way flow model. For example, a letter to the editor, a call-in television show, and a cable touchpad that permits viewers to communicate with a program host demand that the one-way flow model be expanded to accommodate **feedback,** or a two-way component. Such a *nonlinear* or *transactional* model better captures the process of message exchange brought about by interactive media technologies, especially systems that permit near or actual real-time, two-way communication among users.

Video programmers can benefit from considering both linear and nonlinear communication models. For example, if you already possess a great deal of information about how an audience is likely to respond to a particular program, it may be unnecessary to solicit feedback from audience members. However, if the program is relatively novel and unproven, or if certain major changes are being contemplated for an ongoing production, it may be wise to collect some audience feedback or pilot-test the new idea to determine where strengths and weaknesses lie so that improvements can be made in line with audience suggestions. Focus group research is one method of collecting audience feedback to avoid disaster. Transactional communication models recognize the give-and-take or play aspect of communication: you don't play *at* people, you play *with* them.

Message Effects

Communication theorists recognize that messages sent are not always messages received and that meanings are in people, not in messages. This is so in part because people bring to the communication setting a welter of opinions, attitudes, values, beliefs, and life experiences that uniquely color their interpretations of messages.

Selective Perception and Retention   One stable research finding is that people tend to positively assess and accept those parts of messages they perceive to be consistent with their values, beliefs, attitudes, and opinions, and they tend to negatively assess and reject those parts of messages viewed as inconsistent. *Selective perception* and *retention,* as cognitive psychologists call these tendencies, are well documented in political campaign settings, where research has shown the tendency for partisan voters to assess a candidate's campaign performance in line with their political ideology or party membership.

Behavioral Aspects of Message Effects　Message effects may also influence behavior and action, not just cognitive perceptions. The behavioral aspect is especially important in persuasive settings, where the desired objective is to move audience members to do something after exposing them to a message. For television programmers, persuasive appeals that influence buying and voting behavior are particularly important.

Modern studies have applied scientific research methods to build on Aristotle's notions of ethos, logos, and pathos. Much of this work has focused on experiments attempting to locate the origins of source credibility. Among the variables believed to influence source credibility are message context, source attractiveness, the audience's knowledge of the source's agenda, audience incentive and level of need for attending to a particular message, and the nature of the audience's existing belief system regarding the topic of discussion.

Consistency and Balance Theories of Persuasion　As mentioned earlier, audience members' predispositions, including their attitudes, opinions, beliefs, values, and life experiences, may explain why some messages succeed in influencing audiences whereas others fail. A number of consistency and balance theories have been advanced to account for these outcomes. For example, *congruity theory* asserts that people naturally seek consistency in the world around them in order to make sense of things. When they encounter messages that square well with other things they have learned and accepted and that do not contradict their basic beliefs about the nature of things, people are more likely to accept such messages. Conversely, when people encounter messages that go against the lessons they have learned over time, they perceive those messages to be dissonant and confusing, and tend to ignore or reject them.

Similarly, *cognitive dissonance theory* asserts that people try to reduce uncertainty and tension in their interactions with the world by either making their behavior consistent with their belief systems or making their belief systems consistent with their behavior. By achieving consistency and balance between their behavior and their belief systems, people maximize their feelings of comfort with the way they view and act in the world. For example, a cigarette smoker may shun messages that describe the hazards of smoking to avoid the tension between cognition and behavior that such messages might cause. Or a car buyer who has just spent twenty thousand dollars on a new car may ignore an advertisement for the same car priced at sixteen thousand dollars to avoid the tension such a message might bring. Cognitive dissonance theory suggests that when messages create tension between the cognitive and behavioral aspects of a person's life, the individual will seek a way to avoid either the behavior or the messages that generate discomfort.

Consistency and balance theories suggest that when predispositions toward certain topics or stimulus objects are strong, it is extremely difficult to change audience attitudes. This may be why it is probably fruitless in most cases to try to convince diehard political party members to vote against their parties' candidates in political elections or to persuade Ku Klux Klan members to support minority rights legislation. In such cases, it may be wise to try to determine which audience members are less committed to such deeply ingrained values and then direct your contrary persuasive appeals to them.

Long-Term Public Communication Campaigns   While changing an audience's deeply held values through the use of persuasive communication may be difficult, must we deny the possibility that an amazing conversion will result from time to time? Is it not possible that a change of heart can occur even in an individual who holds extreme views on some topic or whose behavior has been deeply habituated over time? Although genuine conversions may be rare, it is only reasonable that we allow for them at least in principle.

The types of persuasive campaigns that try to change deeply ingrained views among audiences tend to be those that take place over long periods of time, and they resemble comprehensive training and education programs. Such programs tend to use *internal change agents,* or highly respected spokespersons from the ranks of the target audience, to deliver the campaign message most persuasively. In addition to extending over a long period of time, the persuasive appeal has both cognitive and behavioral elements.

Examples of such long-term public communication campaigns using television include campaigns to get teenagers to stay off drugs ("Just say no"), drivers to stay sober and use seat belts, campers to practice fire safety, volunteers to join the army ("Be all that you can be"), and heart patients to alter their eating habits to lower their blood cholesterol levels. On a broader scale, China has launched a national campaign to get families to have only one child. This long-term effort includes rewards for families in each neighborhood that fulfill the behavioral goals of the program, as well as disincentives for those who do not.

For video programmers, the value of audience analysis, balance and consistency theories, and the lessons from long-term public communication campaigns lies in recognizing how messages can be structured to gain compliance from an audience. Since television is concerned largely with convincing audiences through the use of symbolic appeals to think, feel, and act in line with programmers' messages, it is in your interest as a television programmer to understand as completely as possible the nature of the audience in planning your productions.

The Universal Audience

In practical terms, audience measurement is always incomplete, for a number of reasons. First, the actual audience for a particular program is almost never completely known, since a comprehensive study normally is not done to determine, for every potential viewer, who has actually watched the program and with what level of interest. Further, when a program is crafted for a specific audience, actual viewers may differ from the target audience profile in significant ways. Among intended audience members, some may watch for a moment and then tune away, whereas others may either choose not to watch at all or remain unaware of the program's existence.

However incomplete audience data are for a given program, some programs seem to have a more universal appeal than others, consistently attracting a wider audience across demographic, psychographic, and lifestyle categories and even across generations. One such program is the "I Love Lucy" television series of the 1950s. This program has garnered the praise of critics and the loyalty of huge, diverse, and enthusiastic audiences since its debut, and has continued to enjoy both critical acclaim and rerun success in syndication for nearly half a century.

When we say that a program has "universal appeal," we seem to rely on an ideal concept of a universal audience. The notion of the universal audience has value for program makers in that it encourages them to consider those program qualities that have enduring aesthetic appeal. At the same time, the idea of the universal audience seems to run counter to all we have said thus far about audience analysis, differentiation, and segmentation. Recall that most of our discussion of audiences has emphasized the *particularity* of audience characteristics and how message appeals should be adapted to key differences among audience segments to maximize effectiveness. We pointed out that today's rapid proliferation of program choices makes it necessary to target more highly segmented audiences.

Of course, not all television programs need to pass an ideal litmus test of appealing to a universal audience to be considered worthy. Those that do seem to capture some enduring quality about culture or to objectify some ultimate truth about human nature may deserve to be elevated to the status of art (for instance, Lucille Ball's classic performances or Dustin Hoffman's portrayal of Willy Loman in the made-for-TV presentation of *Death of a Salesman*). When such projects gain a broad or more universal appeal, they usually do so by revealing some truths or foibles of human nature in their themes in an aesthetically pleasing, unified way.

Many programs never achieve such status yet may still be worthwhile. Many serve practical goals: to entertain, report news, teach language or cooking, sell bath soap, or help people improve their tennis game or first aid skills. How should we evaluate such programs? It is important to remember that although practical programs may not be intended to transmit culture and transcend time, they may still be made poorly or well. Therefore, program makers should continue to strive to produce them according to standards of excellence for what these programs say and how they say it.

Here is the main point: in planning any program, it is essential to think through carefully what your program is intended to do and then decide how best to produce it so that it communicates its message clearly to its audience. Try to keep this point in mind as you read later chapters of this book, which delve into the more technical aspects of how video conveys its messages.

KEY TERMS

interactivity *(6)*	pathos *(8)*
convergence *(6)*	noise *(11)*
ethos *(8)*	feedback *(12)*
logos *(8)*	

QUESTIONS FOR REVIEW

1. What questions do you need to answer in developing a program concept?

2. What attributes of the audience are especially important to video producers?

3. How have recent changes in technology affected the way video communicates with an audience?

4. What three types of audience appeal did Aristotle identify, and how are they relevant to video?

5. What do the modern communication concepts of *information, noise, feedback, selective perception and retention, consistency,* and *balance* have to tell us about effective video production?

2 How Television Works

The desire to communicate from afar is part of human nature. Before radio and television made it possible for people to broadcast sounds and images across oceans and continents at the speed of light, people invented less powerful means to communicate to distant places. For instance, they created the megaphone, which extends the reach of the voice (though not greatly). Other methods of sending messages long distance have included beacons, semaphore flags, drums, smoke signals, and telegraphy.

One feature these alternative means of communicating share is that they all rely on prior agreements (or *codes*) between senders and receivers about what various signals will mean. For example, if someone does not know the sounds or letters associated with flag positions in semaphore, the message will be lost even if the flags can be clearly seen. One must be initiated into the code to be able to interpret the information sent.

Successful communication relies on both unimpeded reception of a message's physical component and accurate decoding of the pattern of information (or *intelligence*) it contains. Of course, it is still possible to misunderstand or misinterpret the meaning of a message after it is successfully received, but cultural issues of meaning are not even considered until an encoded message is received and decoded.

How is intelligence carried in a message system? A common feature of the systems just mentioned is the need to *vary* some aspect of the signal to encode information. If the sender of semaphore flag signals fails to move the flags, no message is transmitted (except, perhaps, that something may be wrong with the sender). Similarly, if a Morse code telegrapher sends nothing but dots at regular intervals, there will be no information to decode because there is nothing in Morse code associated with an endless sequence of dots. In the technical language of radio and television broadcasting, the term for creating a pattern through variation is *modulation*. In this chapter, *modulation* is used as a synonym for imposing a pattern, or a change or variation, on a transmitted signal.

Among the most advanced and successful systems ever developed for communicating at a distance is television. This chapter introduces some key concepts concerning the physical nature of radio energy, which makes television possible. These concepts include how an audio signal and a televised scene are encoded, transmitted, received, and decoded. The chapter also describes some of the capabilities and constraints that shape and limit the video medium. Finally, the chapter addresses recent developments in expanded video systems, such as high-definition television and digital television, which will soon provide a cornucopia of new video services, increasing television's potential for both producers and consumers.

The chapter's topics include:

**THE BASICS OF
ELECTROMAGNETIC RADIATION**

propagating radio waves • converting sound into electrical energy • modulating radio waves with amplified audio signals

**FROM RADIO TO
TELEVISION TRANSMISSION**

channel space • converting light into electrical energy

**PRINCIPLES OF
A MONOCHROME SYSTEM**

the basics of scanning • interlaced scanning • receiver operation

**COLOR TRANSMISSION
AND RECEPTION**

chrominance, luminance, and saturation • signal transmission • reception

**PRODUCTION IMPLICATIONS OF
VIDEO TECHNOLOGY**

**THE EXPANDED TELEVISION
SYSTEM**

the adoption of a digital television standard • characteristics of the new standard

THE BASICS OF ELECTROMAGNETIC RADIATION

Among the natural phenomena that make broadcasting possible is the passage (*propagation*) of radio waves, or electromagnetic radiation, through space. At the simplest level, radio energy can be generated by rotating a loop of copper wire in a magnetic field. Such rotation induces an electric current in the wire. As the wire passes through each full rotation, the intensity and direction of the flow of electrons vary in an orderly pattern called a *sine wave*. As Figure 2.1 indicates, sine waves produced by continuous rotation feature several characteristics, including a **frequency** (the number of cycles per second, or cps), a **period** (the time it takes for one cycle to occur), an **amplitude** (the magnitude of the voltage in the wire), a **wavelength** (the length of one cycle in meters), and a **phase** (the difference between the same points on different waves). Radio waves travel at the speed of light, about 186,000 miles per second, or 300 million meters per second.

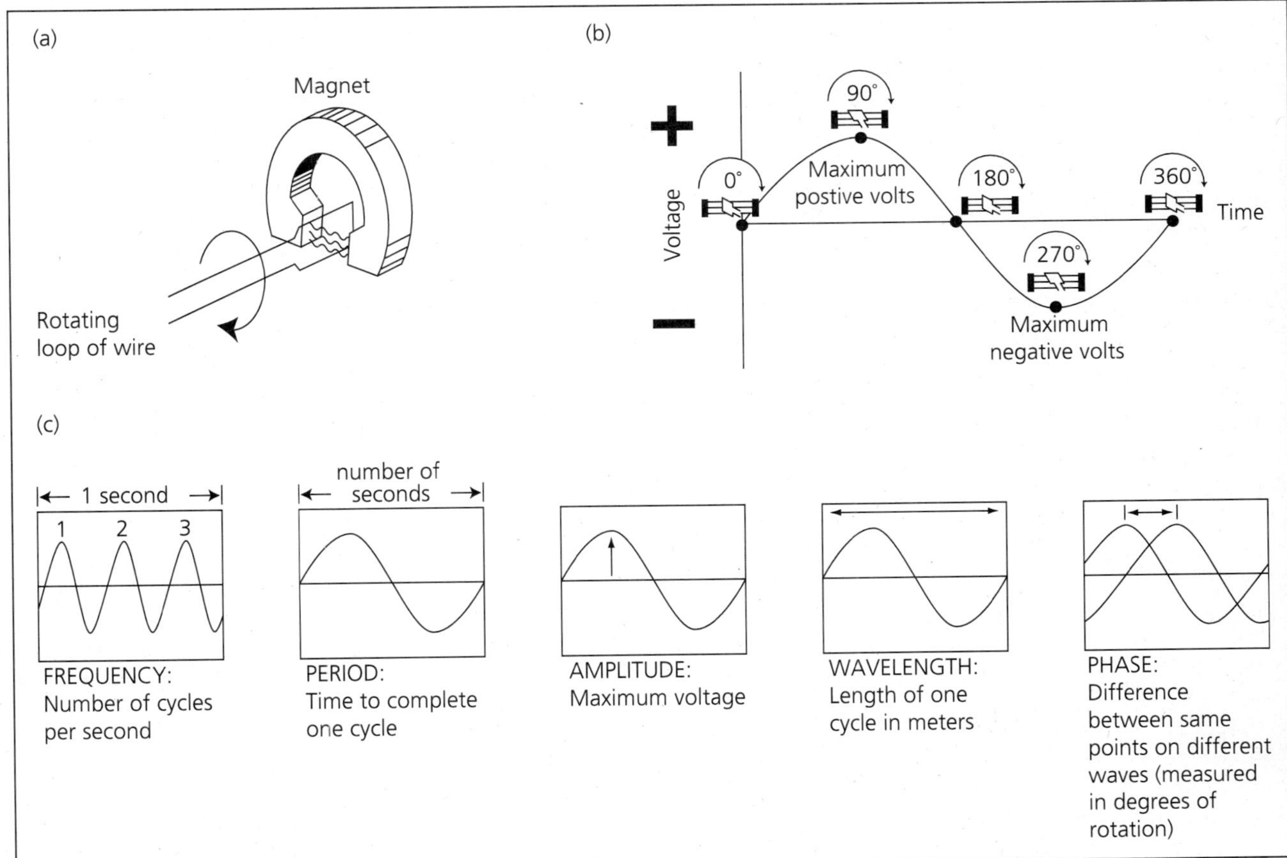

Figure 2.1   *The basic sine wave of radio energy. (a) The wave is produced by a loop of wire rotating in a magnetic field. (b) One cycle of a sine wave, as the loop goes through a full (360 degree) rotation. (c) Properties of the sine wave.*

Propagating Radio Waves

In 1819 the Danish scientist Hans Oersted, while experimenting with electrical effects of magnetic fields on certain materials, learned that magnetism and electricity are related. Then, in 1831, Michael Faraday discovered *induction*, the ability of an electric current in a wire to create a similar current in a nearby wire with no physical connection between them. Based on Faraday's discovery, Joseph Henry developed the first efficient electromagnet.

In 1837 Samuel F. B. Morse made practical use of Henry's electromagnet by patenting a long-distance telegraph system. As telegraphy developed, it was observed that some "leakage" of current from telegraph wires appeared to magnetize some nearby metallic objects. This phenomenon was explained in 1865 by the English physicist James Clerk Maxwell, who presented evidence that the electrical impulses emitted from wires travel through space in a manner similar in form and speed to light waves. Maxwell called these impulses **electromagnetic waves**.

Thomas Edison tried to capitalize on this leakage phenomenon to send telegrams to people aboard moving trains. Unfortunately, the waves the telegraph wires sent into the atmosphere were a chaotic mixture of signals leaking from other wires in the area, making the patterned dots and dashes from any particular message unintelligible. The problem of how to separate one electromagnetic wave from many was solved by the German scientist Heinrich Hertz. In 1887 Hertz demonstrated that an electromagnetic wave could be propagated and then detected amid other waves using an *oscillating circuit,* which produces an electric current that changes direction at a stable frequency. An example of an oscillating circuit (albeit a relatively slow one compared to radio frequencies) is the one found in a typical American household electrical outlet, which supplies alternating current (AC) at 60 cycles per second. In honor of Hertz's discovery, the unit called a *hertz* (abbreviated Hz) was adopted in the 1960s to mean "cycles per second."

It was soon confirmed that a radio wave, when propagated *at a stable frequency,* will not mix with waves of other frequencies. By 1895 the Italian scientist Guglielmo Marconi had sent the first wireless telegraph message. These early wireless messages were in the form of Morse code, using the simplest modulation technique, namely an *interrupted continuous wave (ICW).* In this method of radio transmission, a continuous, alternating current, made up of a succession of identical sine waves, is broken into a series of pulses corresponding to the dots and dashes of Morse code. This is done simply by opening and closing the circuit for relatively short or long durations to turn the radio wave on or off.

Although this type of signal is still widely used, it is limited in that it does not vary enough to carry sounds, such as music or speech. Eventually advances in digital technology would make it possible to store enough pulses of information in binary patterns of 0s and 1s to render sounds and images on CDs, video disks, and computers. In Marconi's day, however, further advances were needed to permit broadcasting of audio signals.

Converting Sound into Electrical Energy

The advance from Morse code to the sending of an electrical replica of the human voice (*voice modulation*) was made possible by Alexander Graham Bell. In

1876 Bell invented the telephone, which makes a current of electricity vary with changing sound waves generated by the human voice. The telephone transmits a pattern of electricity that faithfully matches a pattern of sound waves made by speech. How does this happen?

The microphone in a telephone converts sound waves (vibrations in the air) into a matching pattern of electric current. To do this, sound waves created by the voice are directed onto a diaphragm of thin metal in the mouthpiece of a telephone handset, which vibrates according to the pattern of sound waves imposed on it.

The telephone mouthpiece contains a metal diaphragm (typically a circular piece of thin aluminum) that forms the top of a cylinder containing carbon particles. The carbon particles can conduct electricity. When sound waves enter the mouthpiece, they cause the aluminum to vibrate so that the carbon particles are alternately squeezed and loosened. When electricity is made to flow through the cylinder, the current increases and decreases as the carbon particles are squeezed and released. Loud sounds cause sound waves to press hard on the diaphragm, compressing the carbon particles tightly, thus making it easier for electric current to flow. This permits a large amount of electricity to pass through the circuit. When the sound is low, less pressure is exerted on the carbon particles, allowing them to remain more loosely packed and making it harder for current to pass, resulting in a smaller current.

In this way, the current passing through the telephone circuit copies the pattern of sound waves striking the diaphragm. If it is a close copy—an accurate reproduction—we call it **high fidelity**. *Fidelity* means "faithfulness to the original." This process of changing sound into a pattern of electrical energy is termed *transduction,* and the telephone is therefore a **transducer**.

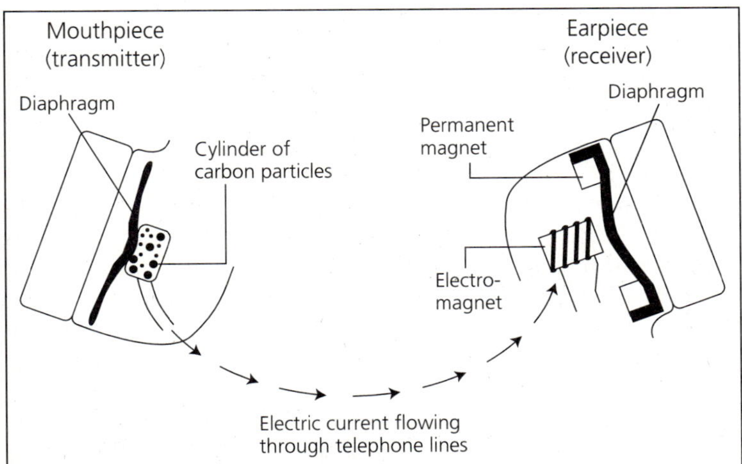

Figure 2.2   *Operation of the telephone, a simple transducer. Sound entering the mouthpiece vibrates a metal diaphragm atop a cylinder of carbon particles through which an electric current is passing. This vibration produces a pattern of electric current that replicates the pattern of the sound waves. At the receiver end, the incoming current creates variations in the strength of the earpiece's electromagnet. These variations cause the receiver diaphragm to vibrate, reproducing the original sound.*

At the receiving end, the electrical pattern is transformed back into sound. The telephone earpiece has a diaphragm that can freely vibrate in and out. In the center of the diaphragm is a coil of wire used to form an electromagnet. A permanent magnet surrounds the electromagnet, supplying a force against which the electromagnet pulls. As the incoming current varies in strength, so does the electromagnet, and the diaphragm vibrates at the same rate, vibrating the air around the diaphragm. The sound waves generated by this motion reproduce the original sound. Figure 2.2 diagrams this process.

The audio coding and decoding process in video microphones and loudspeakers works in essentially the same way as in the telephone. Standard video microphones, though, are sensitive to a wider range of the audio spectrum and have higher fidelity than the small microphones found in telephone mouthpieces. Likewise, television speakers have more power and fidelity than a telephone earpiece.

Modulating Radio Waves with Amplified Audio Signals

The telephone made it possible to project an electrical version of the human voice through long distances over wires and then recover a replica of the original sound from the transmitted electricity. It soon became possible to voice-modulate radio waves in a similar manner without wires. This change resulted from the work of two electrical engineers, England's Sir John Ambrose Fleming and America's Lee De Forest.

Attenuation and Amplification   Like sound waves, radio waves naturally tend to dissipate as they move farther away from their source. As distance increases, the strength of the waves decreases. This phenomenon is called *attenuation*. One way to picture this process is to imagine the effect of dropping a stone into a pond of still water. The stone causes circular waves of water to move away from the point where it hits, and as the waves move outward, their strength diminishes. At some distance, the original disturbance attenuates to such a degree that the water remains undisturbed by the original splash.

In Fleming's day, it was already well known that electron movement produces current in a closed circuit. In the language of electrical theory, Fleming knew that a *voltage* applied to a metal wire conducts electrons toward the positive end. What Fleming discovered, however, was that an electrode inside an evacuated heated filament lamp (a glass *vacuum tube*) also conducts an electric current. Fleming noticed a one-directional current between the heated filament (the *cathode*) and the positive electrode, known as the *anode* or the *plate*. Because it contained two elements, Fleming called the device a *diode*.

Lee De Forest extended Fleming's work by interposing a thin, open-mesh grid between the heated filament and the anode. When a separate voltage was fed to the grid, De Forest could control the magnitude of electricity flowing from the cathode to the plate. With a grid, De Forest obtained a large voltage change at the plate from just a small voltage change on the grid. Thus, by introducing a third element to Fleming's diode, De Forest's *triode* made it possible to amplify weak radio signals received from distant radio transmitters. Figure 2.3 diagrams the triode vacuum tube, the original heart of radio amplifiers. Since the 1950s, successive generations of solid-state technologies (tran-

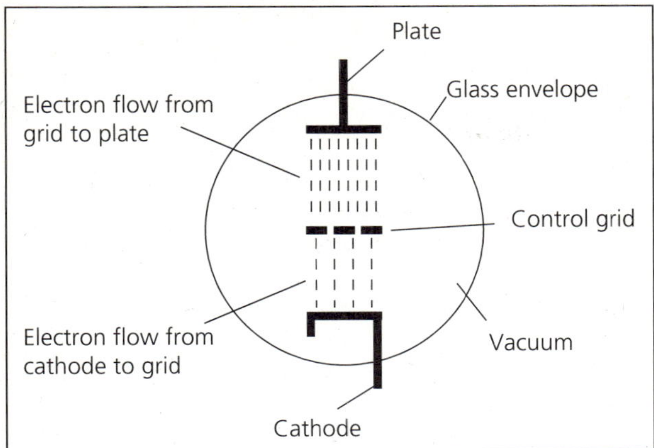

Figure 2.3   *A triode vacuum tube solved the problem of amplifying radio signals. Small voltage changes on the control grid modify the electrical flow from the cathode to the plate.*

sistors, semiconductors, integrated circuits, and microprocessors) have replaced vacuum tubes, but the concept of amplification is still the same.

Modulating the Carrier   By feeding an electrical signal converted from a pattern of sound waves (such as a voice over a microphone) to the grid of a triode, relatively weak audio signals could be amplified enough to be used for radio transmission. However, before sound waves could be transmitted to distant points without wires, the amplified audio signal had to be superimposed onto a *radio frequency carrier* (or *RF carrier*). This was because sound waves are pressure waves and do not propagate across space at the speed of light as electromagnetic radio waves do.

The radio frequency carrier is created with an *oscillator,* an electronic circuit that produces a sine wave at a specific frequency. The RF carrier is made to vary by an audio signal (voice or other information) superimposed on it. In other words, the pattern imposed on the RF carrier is sound, converted into an electrical signal, supplied by a microphone or some other audio source (for example, a CD or cassette tape).

The two most common techniques for modulating a radio wave are amplitude and frequency modulation. When an audio signal modulates the amplitude of a carrier, the process is called **amplitude modulation (AM)**. When the audio signal modulates the frequency of a carrier, we speak of **frequency modulation (FM)**. In AM radio, the carrier consists of a sine wave whose amplitude copies the variations of the audio signal. In FM radio, it is the frequency of the carrier wave that the audio signal changes. Figure 2.4 illustrates these two common types of voice modulation in radio broadcasting.

FM is a far superior modulating technique than AM because it produces better fidelity with much higher noise immunity. For example, auto ignition noises and high-tension lines cause hum and static on AM reception because those disturbances affect the amplitude of the received carrier. In contrast, since FM detects changes in frequency only, it generally is not affected by impulse noise from the atmosphere and manmade sources.

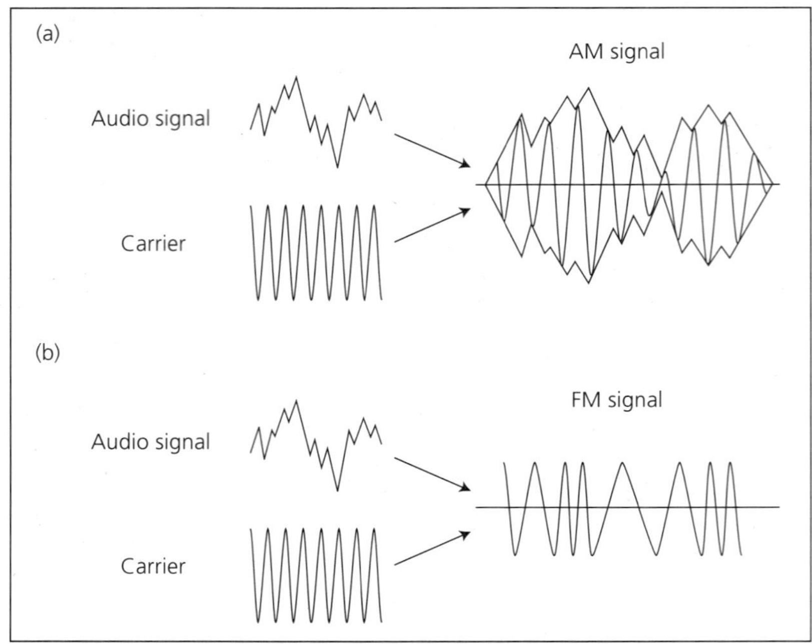

Figure 2.4   *AM and FM signals. (a) In AM transmission, the audio signal modulates (varies) the amplitude of the carrier wave. (b) In FM transmission, the audio signal modulates the frequency of the carrier wave.*

Transmitting the Carrier   After the audio signal is imposed on the RF carrier, it may be further amplified. Finally, it is fed from a transmitter to an antenna for propagation. In standard AM transmission, the range of frequencies used for radio carriers is between 535 and 1,705 kilohertz (abbreviated kHz, meaning thousands of hertz). Each channel is allocated a space or **bandwidth** of 9 kHz. This means enough spectrum space is allotted to provide for a maximum of 130 standard AM radio channels in any given area.

A radio wave propagates in all directions unless it is intentionally altered from this pattern. The effective coverage area can extend many miles surrounding the transmitting antenna. It is possible for millions of radio sets in a coverage area to receive the signal being transmitted at a particular frequency. However, because the signal attenuates as distance increases, it must once again be amplified at the receiver to make it strong enough to drive a speaker.

Demodulating the Carrier   The function of the radio receiver is to tune in a particular frequency, detect the modulated carrier operating at that frequency, and remove the audio signal from the carrier. This part of the process is known as **demodulation**. The isolated audio signal is then amplified and directed to a speaker so that the original audio information can be heard. Figure 2.5 presents a block diagram of the demodulation process.

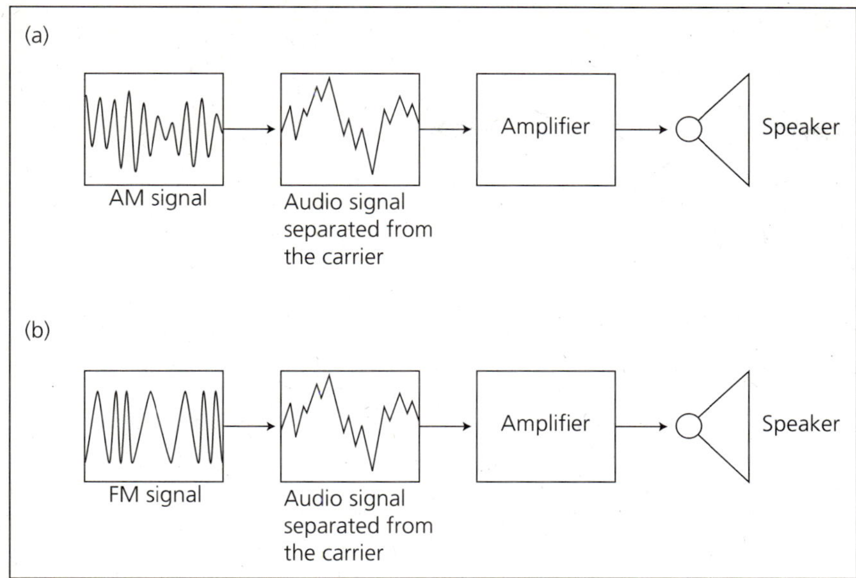

Figure 2.5 *Demodulation of (a) AM signals and (b) FM signals.*

FROM RADIO TO TELEVISION TRANSMISSION

So far, this chapter has provided a basic model of how audio information is transmitted via radio energy to distant points and then recovered. How does radio energy enable us to do the same with pictures? Before we examine the process of television transmission, let's look at some preliminary facts.

Channel Space

In using radio energy to transmit sound plus full-motion visual images, a greater bandwidth is needed than for sound alone. As a result, in American broadcasting, whereas the bandwidth allocated for standard AM radio is only 9 kHz per channel, the bandwidth for each television channel is more than 660 times larger, or 6 MHz (6,000 kHz). This means one television channel contains enough channel space to accommodate more than 600 AM radio stations!

How much of the radio spectrum to allocate for each television station was determined after much technical debate and testing by the National Television System Committee (NTSC). The first objective of the NTSC was to suggest technical standards that would permit an acceptable level of picture definition or **resolution**. With enough resolution, the television image would be clear and aesthetically pleasing. However, the NTSC also wanted to conserve spectrum space, using no more than necessary for each channel assignment. The NTSC's job was tricky, because as picture resolution increases, bandwidth for each channel must increase to accommodate finer levels of detail, thus reducing the total number of possible channels. The standard bandwidth of 6 MHz for each television channel, adopted in 1941, allowed 4.5

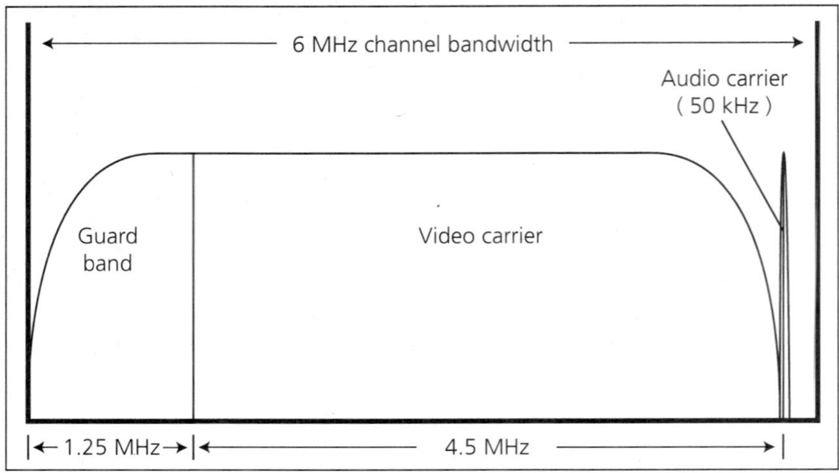

Figure 2.6   *Audio and video portions of a standard 6 MHz television channel.*

MHz for the AM-modulated video signal, a complex video waveform that would include synchronization, scanning, blanking, and, eventually, color information. The remaining 1.5 MHz provided two things: (1) a *guard band* or buffer between adjacent channels operating in the same geographical area to reduce interference of one channel with another and (2) space for transmitting the FM-modulated audio portion of the television signal. Figure 2.6 diagrams these features of the television channel.

Why is amplitude modulation used for the video portion and frequency modulation for the audio component of the television signal? First, as mentioned earlier, FM is less susceptible to noise and interference than AM, making it less subject to static and therefore more suitable for audio reception. Further, AM is better suited for video transmission because it exhibits fewer problems caused by multipath *reception* of the signal. *Multipath reception* occurs when the same signal, due to reflections from obstacles such as buildings and bridges, reaches a receiving antenna from more than one path. Since the distance these multipath signals travel usually varies, different parts of the signal arrive at the antenna at the same time. For AM signals, this causes less severe interference at the television receiver than would occur with FM signals.

Converting Light into Electrical Energy

Just as telephone and radio technologies harness natural qualities of electricity and electromagnetic radiation to transmit voice-modulated audio signals, television relies on the natural phenomena of photoelectric effects, including photoconductivity and the photoemissive effect, to convert light into and back from electrical energy.

Photoconductivity   To change light into electricity, video depends on **photoconductivity,** which occurs when light on some metals increases the flow of electricity in the metal. One of the earliest examples of photoconductivity was observed in 1873 with the metal selenium. When selenium was used in an electrical circuit, the current passing through it increased during exposure to light. For television, selenium responds too slowly to light to be useful for replicating natural motions. But cesium silver and other silver-based materials are excellent for such applications.

Photoemissive Effect   In the **photoemissive effect,** discovered by Heinrich Hertz in 1887, visible light results from some materials' exposure to energy that may not be visible to the eye. Sources of such energy include streams of electrical energy or photons of higher-than-visible light energy, such as ultraviolet rays or x-rays.

In the picture tube of a television receiver, the inside of the screen is coated with fluorescent material. When the screen is struck with a stream of electrons from an electron gun, the screen glows as a result of the photoemissive effect. As the stream of electrons is strengthened, the portion of the screen struck by the electrons glows more brightly. When the stream is weakened, the glow decreases. If the stream of electrons can be modulated in accordance with the darker and brighter portions of a scene focused by the lens of a television camera, that scene can be rendered on the screen. If the recreation process can be done quickly enough, smooth motion can be rendered convincingly.

In black-and-white (*monochrome*) television receivers, the fluorescent material needs to glow only with a range of brightness from dark to light roughly proportional to the intensity of the stream of electrons hitting it. Color is not important; only the brightness information matters. However, in color television reception, it is necessary to use different materials that glow with different colors when streams of electrons hit them. To understand this process, let's first look at the major components of the monochrome television system.

PRINCIPLES OF A MONOCHROME SYSTEM

Television cameras (Figure 2.7) use a lens system to focus light from a scene to be televised into the camera's **pickup tube** or, in integrated-circuit or microprocessor systems, into a **charge-coupled device (CCD).** The pickup tube or the CCD is the element in the camera at which light reflected from a scene is converted into an electrical signal. The output is then amplified and fed to external circuits for recording, routing to closed-circuit locations, broadcast from a transmitter, or transmission via cable or satellite to some distant point.

Within a studio complex featuring more than one camera, each camera is connected to a **camera-control unit (CCU).** The CCU enables a technician to adjust and match operation for all cameras to eliminate jarring differences in how they render the same scene. The video signal is also routed to the **viewfinder** of each camera for immediate viewing by each camera operator.

The Basics of Scanning

It is not possible to transmit the many details of a given picture simultaneously over the same circuit, since combining picture information in this way would lead to a chaotic mixing of the signal in a single output. Instead, small

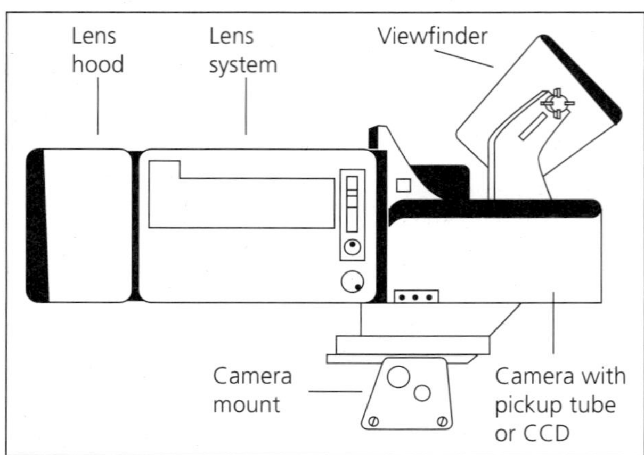

Figure 2.7 *The most basic parts of a video camera.*

areas of the picture must be converted into separate magnitudes of electric current that match the brightness information present in each portion; then each portion is sent out discretely. This is done so that each bit of picture information can be received and converted into light without being confused with any others.

In theory, we could create a separate circuit for each area of the screen and then send all of the information at once, but such a method is impractical since it would require more than two hundred thousand separate circuits for just one station's program (see Figure 2.8). Instead, a *scanning* method is used to transmit the brightness information for each area in turn. Scanning makes it possible to use just one channel for each program.

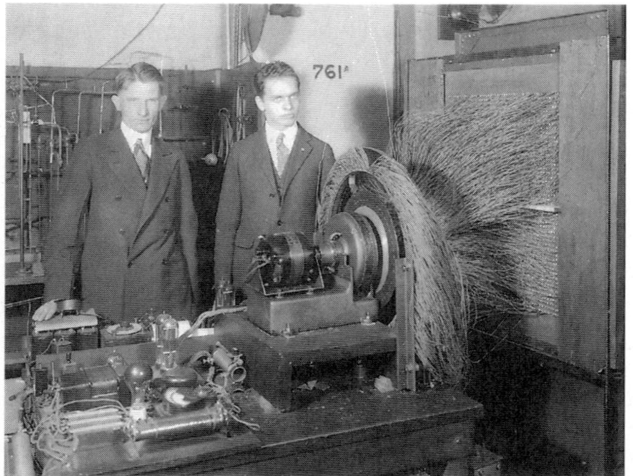

Figure 2.8 *The Bell Telephone television receiver of 1927, which used thousands of separate circuits to compose a picture. What a nightmare! The impracticality of such a device prompted development of the electronic scanning method shown in Figure 2.9.*

The original monochrome television camera tube converted picture information into electrical signals by focusing light onto a mosaic of cesium silver *picture el*ements, or **pixels**. In such a system, when a scene to be televised is focused on the mosaic, electrons become stored in each pixel in direct proportion to the intensity of light focused on each one. Stored electrons are instantly attracted by an anode in the camera tube, leaving the mosaic with a copy of the original scene in the form of varying amounts of electrical charge. This pattern is analogous to an exposed photographic image.

In American broadcasting, the mosaic currently consists of 525 horizontal lines containing about 211,000 pixels. An electron gun is used to scan each line from left to right and top to bottom in an orderly fashion. As the electron beam passes each pixel, it replaces electrons lost to the anode, enabling the video signal to exist in an external circuit. This signal may be coupled to video amplifiers for immediate transmission.

Interlaced Scanning

The human visual system detects *flicker,* a source of severe eye fatigue, below about 45 image presentations per second. To defeat flicker problems, the film industry has adopted a standard of twenty-four frames per second, each illuminated twice, for a rate of forty-eight presentations per second. For television, a system called *interlaced scanning* is used to avoid flicker problems (Figure 2.9).

Interlaced scanning takes advantage of **persistence of vision,** the tendency for an image to persist in our sight for a short period of time after the stimulus is no longer physically present to our eyes. Instead of having 525 successive sweeps of the screen, interlaced scanning uses two separate scans of 262.5 lines. The electron beam alternately scans the odd-numbered lines of the 525 lines and then the even-numbered lines, thus creating the illusion of covering the entire field twice.

Each successive scan of 262.5 lines is called a **field**. Since line frequency (normal wall current, or AC power) in the United States is 60 Hz, it is convenient to scan each field in 1/60 of a second. As a result, sixty fields per second are televised, a rate fast enough to eliminate the flicker problem. The complete scanning of all 525 lines, or two successive fields, is called a **frame**. This means thirty frames per second are televised.

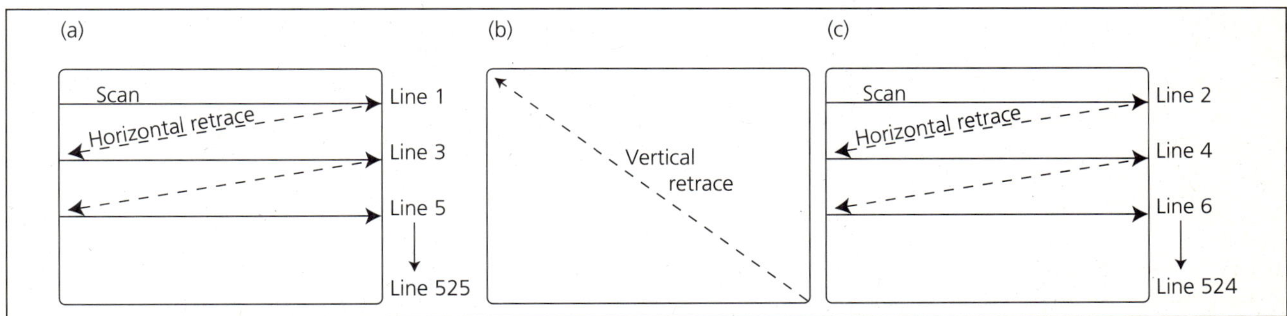

Figure 2.9   *Interlaced scanning. (a) The electron beam scans the odd-numbered lines of the screen. (b) The beam then retraces vertically to the top left starting point. (c) Finally, the beam scans the even-numbered lines, completing the total 525-line frame.*

The movement of the electron beam across each scanning line is accomplished by the use of electromagnet coils surrounding the neck of the camera's cathode ray tube (CRT). As the fixed electron gun generates a stream of electrons at the tube face, varying magnetic forces, generated within the coils surrounding the CRT, bend the stream of electrons along its path. In this way, the camera performs its work without using any mechanical parts. This makes the scanning process extremely reliable.

Each time the beam finishes a line, it returns to the extreme left position, but shifted downward to the next odd or even line, to begin scanning again. This move back to the left is called *horizontal retrace*. When the beam finishes scanning the last line, it returns once again to the top left position to begin the entire process over again. This move back is called *vertical retrace*. During each retrace, the electron beam is turned off to eliminate spurious illuminations. The signal to turn off the electron beam is called the *blanking signal*. The blanking and retrace signals, along with the synchronization information needed to keep the receiver precisely in step with the transmitter, are embedded in the overall television signal.

Receiver Operation

A television set receives video, audio, and all ancillary signals to replicate coherently the original televised scene and audio information. It has a loudspeaker, a phosphor-coated picture tube, an electron gun, and circuits for synchronization and scanning purposes. Regardless of tube size, the standard proportion of tube height to width, called the **aspect ratio,** is three units by four units. As with the television camera, the neck of the picture tube is fitted with magnetic deflection coils that control the direction of the electron beam produced by the electron gun. A beam of electrons scans horizontal paths across the picture tube's phosphor coating.

When a television signal is received, the sound component (the FM part of the signal) is routed to audio circuits, where it is demodulated and sent to the loudspeaker. The video signal (the AM portion of the signal) is routed to the picture tube, where it directs the electron beam to emit electrons in amounts that vary in concert with the brightness levels of the original scene captured by the camera. As the electron beam sweeps across the face of the picture tube, its varying intensities cause variations in the brightness of the phosphors, replicating the original scene.

To synchronize the video signal so that pixels can be reassembled without mixing them up, deflection coils around the neck of the picture tube are fed horizontal and vertical sync pulses from the original video signal. These pulses control the deflection of the electron beam across the screen, thus keeping the receiver in step with the original signal from the television camera.

COLOR TRANSMISSION AND RECEPTION

Color television broadcasting began after the monochrome system was already in place and millions of black-and-white sets were in use. This made it desirable to find a color system compatible with monochrome technology.

Color information had to be added without changing the 6 MHz bandwidth set aside for each TV channel. Also, black-and-white and color receivers had to be capable of receiving both monochrome and color signals. To accomplish this, the transmission of information had to be virtually identical for both monochrome and color systems.

Chrominance, Luminance, and Saturation

To transmit **chrominance** (color or hue) information, the color camera's optical system separates the light entering it into three primary colors: red, blue, and green. It is a fortunate characteristic of human vision that virtually any color can be reproduced from these additive primary colors. Further, any colored light can be specified with only two additional qualities: **luminance,** or brightness, and **saturation,** or vividness. Saturation can be thought of as the degree to which a color is free of impurities, such as dilution by white light. Low-saturation colors are paler, whereas highly saturated colors are purer and more vivid. (See Figure C1 in the section of color plates.)

In early color systems, light entering the lens was broken into its primary color components using filters and a set of dichroic mirrors. A *dichroic mirror* passes light at one wavelength while reflecting light at other wavelengths. Today most color cameras use a prism block (called a *beam splitter*) instead of dichroic mirrors to break light into its primary colors. Once the light has been split, the separate light beams are directed into three camera pickup tubes for processing into video signals. When a CCD microprocessor is used, different wavelengths of light are absorbed by a silicon lattice at different depths to distinguish colors. In either case, the patterns of electrical voltage generated in an external circuit match the levels of the original pattern of light received by the camera.

Additive Versus Subtractive Color Systems

In *additive* color systems such as video, a complete, natural-color image is produced by mixing light from three or more separate images, each of which emits light of a primary color. In other words, as the name implies, colors are added together to produce the color we see.

In many *subtractive* systems, light from a white source falls on a surface that absorbs light selectively so that light of certain wavelengths is not reflected from areas where it is not wanted. In other subtractive systems, the white light is transmitted through a succession of dye images that absorb light of certain wavelengths in specific areas of the image. In either case, the subtractive system starts with a white light and subtracts from it to create the color we see.

All practical color television systems are based on additive principles, whereas painting and nearly all forms of color photography may be classified as subtractive processes. Color printing from halftone plates is a complex process involving both additive and subtractive principles.

In light-based systems such as television, the additive primary colors are *red, blue,* and *green.* In pigment systems such as painting and color photography, the subtractive primary colors are *red, blue,* and *yellow.*

Some cameras use a single imaging element with a filter to separate incoming light into its component values. Others use filters to separate incoming light into only two colors and additional microprocessors to assign values to the third color needed to reproduce the colors the camera is "seeing."

Signal Transmission

In the color camera, video signals from the three pickup tubes (or the CCD) are combined to produce a signal containing all of the picture information to be transmitted. Signals are combined using a phase-shifting technique so that they can be transmitted in one video channel and then retrieved without confusion. The overall signal contains the audio and picture information as well as blanking and synchronization pulses. This *colorplexed* video signal modulates the video carrier for transmission to receivers.

Black-and-white television sets treat the color portion of the colorplexed video signal as though it were part of the intended monochrome transmission. To avoid degraded reception, the chrominance signal is masked by tying it to the scanning motions. This way, any pixels brightened by interference during one line scan are made to darken by an equal amount on the next line scan. The net effect of chrominance signal interference over successive scans is thus virtually zero.

Reception

The tube in the color television receiver contains three electron guns that project separate electron beams, which deflect simultaneously in the standard interlaced scanning pattern over the face of the picture tube. One gun projects the red color signal, one projects the blue, and the third projects the green.

The screen of the color television receiver is coated with phosphor dots that glow either red, blue, or green when struck by a stream of electrons. The three color phosphors are uniformly distributed over the face of the picture tube, arranged in adjacent groups of three dots that form tiny triangles, each containing a phosphor dot for each primary color. The dots are so small that a single one cannot be distinguished by the viewer's eye. The color of any one triangle is the additive function of the varying intensities with which each dot in the triangle is made to glow by the strength of the electron beam hitting it. Virtually any color may be rendered with this method. If electrons from all three guns strike their respective dots in a triangle with the right intensity, the color of that triangle will appear white. If no electrons strike a trio of dots in a triangle, the color of that triangle will be black. In this way, black-and-white images are possible on a color receiver.

To ensure that the electron beams from the red, blue, and green guns hit only phosphor dots that glow red, blue, and green, respectively, a metal plate is inserted close to the phosphor coating between the electron guns and the screen (see Figure 2.10). This plate, called a *shadow mask,* is pierced with more than 200,000 holes. The shadow mask is carefully positioned so that it shields two of the dots in each triangle from being hit by unwanted electrons. In this way, the electron beams are confined to the phosphor dots of the proper color.

PRODUCTION IMPLICATIONS OF VIDEO TECHNOLOGY

Knowing that cars need oil to operate can help make driving easier, more successful, and fun. Similarly, knowing the capabilities and limits of video technology can help you make smart production decisions. For example, by now it should be apparent that television or video, as it is presently constituted, is a "low-resolution" medium. This means picture detail is limited by the 525-line standard. The current line standard provides a level of picture detail that is frequently more degraded than that provided by either film or photogra-

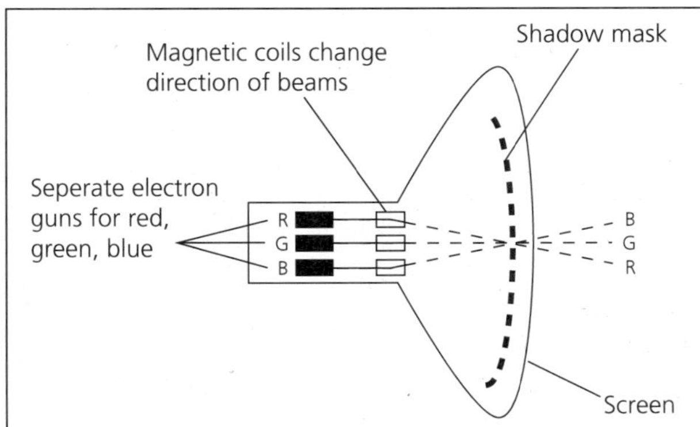

Figure 2.10 *Diagram of the television receiver, showing how the shadow mask keeps the separate electron beams targeted at the proper points on the screen.*

phy. The pixel composition of the television image can be easily seen by looking at the screen from a distance of a few inches through a droplet of water or with the aid of a magnifying glass.

What does this technical limitation mean for production strategy? For one thing, when small objects (or processes) are featured in a program, you may need to arrange tight close-ups to convey key information to the audience. But sometimes even moving in for an extreme close-up may not be enough. For example, if you are demonstrating a knitting stitch with very thin thread, you may need to provide larger-than-life graphics. Similarly, written text must be big and bold to be seen clearly on the screen.

Since cameras use light to create a video signal, it should be clear that controlling light is critical to the rendering of objects. As an obvious example, lighting schemes that send incident light directly into the camera from a light source can wash out the image. Chapters 3 and 4 show how lighting considerations based on the physical nature of television can make sharp differences in picture quality.

Another production implication of television technology is the way a monochrome system sees colors compared to the way the color system sees them. On a black-and-white monitor, two different colors with the same luminance value will look identical. For home viewing, where color TV sets are now the norm, this problem may seem irrelevant. However, camera viewfinders and control room monitors are still often monochrome. Imagine the confusion that would occur during a broadcast of a football game in which the teams had different-colored uniforms with the same luminance values. To the production crew watching monochrome monitors, the two teams might look identical. Video production people need to keep such considerations in mind to do their jobs efficiently.

THE EXPANDED TELEVISION SYSTEM

In the past few decades, the development and growth of microwave, cable, and satellite technologies and facilities have made live video transmission and reception possible from almost any location on earth. The camcorder has put production capability into the hands of the audience. Videotape and recording hardware now enable receivers to store programs for later use, giving audiences the luxury of watching programs at their convenience. Editing, special-effects machines, and video paintboxes permit virtually endless enhancement and manipulation of video images. High-quality audio and multiple-speaker configurations offer stereo and surround sound for consumers' home entertainment systems. Interactive multimedia developments have increased users' ability to engage in dialogue with program providers and with one another. Digital technology is hastening the convergence of televisions and personal computers. Projection and big-screen video have begun to influence the thinking of home builders, who frequently include entertainment theater space as a selling point. Even the remote control has changed the way we watch television, as well as influenced the way programmers think about how to capture and hold our attention.

From these facts, it is clear that technical innovation has made video more complicated than it was in its infancy. Yet the core of the system still uses radio energy to transmit television signals, and it will continue to do so even in the coming age of digital television.

The Adoption of a Digital Television Standard

Although all these developments have had a significant impact on the production, content, and use of television, perhaps the greatest change was set in motion by the Federal Communications Commission (FCC) on December 24, 1996. On that date, culminating years of research and discussion, the FCC announced its decision to adopt a new standard for **digital television (DTV),** an advanced television system based on digital technology.[1]

Originally the main goal of developing an advanced television system was to provide the American public with a sophisticated method of sending and receiving higher-quality video images. The projected high-resolution system was known as **high-definition television (HDTV).** However, the rapid development of digital technologies expanded the objectives of public-interest groups, computer and television receiver manufacturers, telecommunications providers, cable and satellite television producers, filmmakers, broadcasters, and other parties interested in how to deliver enhanced video and audio services to the American public. Experts began to believe the new video system should include movies on demand, telephone ("video dial tone") and computer services, interactive programs, distance learning, paging systems, home shopping, data transmission, and other services as yet unimagined. Ironically, with all the possibilities created by the development of DTV, some experts have begun questioning whether the long-promised HDTV will remain a major use of the new system.

To see the capabilities of the new system, it helps to understand the process by which it was developed. The FCC originally had several goals in mind. It wanted an advanced television technology that would

1. Put more choices in video programming in the hands of the American consumer.
2. Feature dramatically better aural and visual resolution than the old NTSC system.
3. Provide innovative services through a data transmission capability.
4. Offer compatibility with computers to spur innovation and competition.

To achieve these goals, the FCC began inquiries in 1987 into the potential for advanced television services. At that time, industry research teams suggested more than twenty systems. By February 1993, after a decision that any new system must be fully digital, the field had been narrowed to four potential systems. In May 1993, seven companies and institutions representing the

[1]See Federal Communications Commission, Fourth Report and Order, *Advanced Television Systems and Their Impact upon the Existing Television Broadcast Service,* M.M. Docket No. 87–268 (December 24, 1996). This document is available on the World Wide Web at <www.fcc.gov/Bureaus/Mass_Media/Orders/fcc96493.txt>.

four remaining systems formed the so-called *Grand Alliance*[2] to develop a "best of the best" system to present to the FCC for approval. Over the following years, a digital system was developed, tested, documented, and eventually recommended to the FCC.

The Advanced Television System Committee (ATSC), a fifty-four member group including television workers, television and film producers, trade associations, television and equipment manufacturers, and segments of the academic community, endorsed the Grand Alliance's proposal as representing "the best digital broadcast television system in the world," with unmatched flexibility and ability to incorporate future improvements. Nevertheless, some industry members objected to having the government impose the standard. Others thought it would be better to allow market forces to dictate standards. Some suggested having the government issue standards only for spectrum allocation, transmission, and reception, leaving all other matters (such as frame rates, number of scanning lines, and aspect ratio) to be settled by the free market.

Ultimately the FCC decided that relying on market forces would be a mistake. It might make the conversion process from the existing analog system more difficult. It might lead to the development of incompatible systems that would prove too costly to consumers, who might have to invest in several different receivers to gain access to different programs. In addition, the FCC reasoned that a government-mandated standard would be the best way to guarantee universal access to broadcasting services for all Americans. Because Americans rely on television as a primary source of information and entertainment, the FCC decided the goals of certainty and reliability took on a special significance. For all these reasons, the FCC rejected the idea of relying on market forces and adopted the Grand Alliance proposal by the end of 1996.

Since the FCC had allowed industry members to draft crucial parts of the new standard, it was able to characterize the arrangement as "voluntary." Moreover, the new standard was designed to be readily adapted to many different uses. Let's look now at the specifics of the standard and the myriad new possibilities it offers.

Characteristics of the New Standard

Like the NTSC television format that came before it, the new DTV standard calls for each television channel to occupy a 6 MHz bandwidth. To fit the more complex digital signal demands of DTV (at times with twice the picture resolution of the current NTSC format) into the same space used for current analog signals, digital compression techniques are used. However, unlike the NTSC format, the DTV standard remains relatively flexible in such areas as scanning technique and line format.

For example, to promote compatibility ("interoperability") among various services, the DTV system can broadcast and receive both interlaced-scanned programs and those produced in a new, noninterlaced scanning format called

---

[2]The members were AT&T, General Instrument Corporation, Massachusetts Institute of Technology, Philips Electronics North America Corporation, Thomson Consumer Electronics, the David Sarnoff Research Center, and Zenith Electronics Corporation.

progressive scanning. In **progressive scanning,** each line of video is scanned in order, with no skipping, at a maximum rate of sixty frames per second (double the current NTSC frame rate).

In addition, whereas the NTSC horizontal line format is fixed at 525 lines, the DTV standard offers two line formats: one with 720 lines per frame and one with 1,080 lines per frame. Both line formats use a 16:9 aspect ratio of width to height rather than the 4:3 ratio of the NTSC standard. Moreover, the DTV pixels are distributed on the screen in a "square" arrangement so that they are equally spaced in horizontal and vertical directions, as opposed to the "rectangular" NTSC arrangement, in which the distance between pixels is greater horizontally than vertically. Together the 16:9 aspect ratio and the square pixel arrangement mean that when 720 horizontal lines are being scanned (not counting those lost to blanking and retrace), 1,280 vertical lines of pixels are used, for a total of 921,600 pixels potentially contributing to the overall video image. Similarly, when 1,080 horizontal lines are used, 1,920 vertical lines of pixels are used, for a total of 2,073,600 pixels potentially contributing to the overall image. These numbers, which are five to ten times higher than those associated with the NTSC format, help to convey the improved picture resolution long desired from the new system.

The 16:9 aspect ratio DTV offers is more compatible than the old NTSC system with the format used in many films produced throughout the world. Currently most films shown on television must be cropped or electronically altered to fit the TV screen's aspect ratio, or else they must be letterboxed. **Letterboxing** is a technique used to preserve the original aspect ratio of a film by blacking out portions of the television screen, usually at the top and bottom. Letterboxing is preferred by film purists but often proves annoying to more casual viewers. With the DTV aspect ratio, letterboxing will black out much less of the screen and therefore should be more acceptable. Further, DTV provides frame rates of twenty-four, thirty, and sixty frames per second, making it compatible with film, NTSC video, and computers.

In addition to these characteristics, the new system can provide

1. Layering of video and audio signals. *Layering* in this sense refers to transport of different programs simultaneously over the same channel. For example, the system allows the simultaneous broadcasting of two HDTV programs, five standard-definition programs (with a visual quality better than that currently available), or dozens of CD-quality audio signals. Transmitting multiple data streams on the same channel is called *multicasting*.
2. Rapid delivery of large amounts of data. The contents of the daily newspaper could be sent in less than two seconds.
3. Interactive transmission of educational materials.
4. Universal closed-captioning for deaf viewers.

Despite these innovations, DTV remains adaptable to programs in the old NTSC format. Using letterboxing or a similar technique, it can readily accommodate the NTSC line standard of 525 horizontal lines. It also retains the capability of RF transmission—that is, traditional television broadcasting, which can reach households that do not have cable or satellite services. In sum, the DTV standard allows the system to adapt to new and expanded uses while retaining the ability to transmit programs in the original television format.

With all of these developments, it is easy to see that television will continue to be a powerful and pervasive communication medium. As new configurations enter the marketplace, it will become increasingly important for message makers and consumers alike to understand how the new devices can change the way we use video.

KEY TERMS

frequency *(19)*	charge-coupled device (CCD) *(27)*
period *(19)*	camera-control unit (CCU) *(27)*
amplitude *(19)*	viewfinder *(27)*
wavelength *(19)*	pixels *(29)*
phase *(19)*	persistence of vision *(29)*
electromagnetic waves *(20)*	field *(29)*
high fidelity *(21)*	frame *(29)*
transducer *(21)*	aspect ratio *(30)*
amplitude modulation (AM) *(23)*	chrominance *(31)*
frequency modulation (FM) *(23)*	luminance *(31)*
bandwidth *(25)*	saturation *(31)*
demodulation *(25)*	digital television (DTV) *(35)*
resolution *(25)*	high-definition television (HDTV) *(35)*
photoconductivity *(26)*	
photoemissive effect *(27)*	progressive scanning *(37)*
pickup tube *(27)*	letterboxing *(37)*

QUESTIONS FOR REVIEW

1. How does modulation of a signal make communication at a distance possible?

2. Define and describe the process of transduction in a simple microphone.

3. How do AM and FM differ? Why are FM signals less susceptible than AM to static and interference?

4. Why did video engineers choose a scanning method of transmitting television signals?

5. What components of the video signal must be sent to reproduce a coherent program at the receiver?

6. What are some production implications of current television technology?

7. What are some differences between the NTSC standard and the new DTV standard?

PART TWO

3 Light and Lenses

Light is the universal currency of television; without it, no pictures would be possible. The way light falls on a scene critically affects the way the scene appears. Videographers need some basic understanding of the nature of light to use it to best advantage in making television programs.

Similarly, since light interacts with various lenses (or lens settings) in different but predictable ways, understanding how lenses work offers powerful advantages in planning lighting designs. For example, with the proper selection of lenses you can determine which objects in an image will be kept in clear view and which will be pushed out of focus, thus increasing your ability to shape the impact and meaning of every shot.

This chapter describes physical aspects of light relevant to television production, especially its interaction with lenses and filters. The topics covered are:

LIGHT AS RADIANT ENERGY
the law of reflection • the inverse square law • the role of lenses

FOCUSING CHARACTERISTICS OF LENSES
focal length • f/number

AESTHETIC EFFECTS OF LENSES
depth of field • perspective

ZOOM LENSES

CONTRAST AND COLOR
white balance • color temperature • filters

LIGHT AS RADIANT ENERGY

Physicists define light as "visible electromagnetic wave radiation" or "radiant energy that can be seen" and locate it in that portion of the electromagnetic spectrum between the infrared and ultraviolet wavelengths (Figure 3.1). Light viewed in this way may be thought of as essentially the same phenomenon as radio waves, with this difference: radio transmissions use wavelengths that may be several meters long, whereas wavelengths of light are only a few hundred-thousandths of a centimeter long. At these wavelengths, the energy transmitted is visible to our eyes.

We can learn much about the nature of light by looking at a simple pinhole camera, similar to the arrangement shown in Figure 3.2. In this device, an image of a stationary object is formed on an image surface by light passing through a small hole in a screen. Rays of light reflecting from the object project toward the screen. Since light travels in straight lines (a quality known as *rectilinear propagation*), the rays of light that do not travel exactly in the direction of the hole are blocked. The rays that do pass through the hole form an inverted image of the object. Those rays that come from the top of the object arrive at the bottom of the image surface; similarly, those that come from the bottom of the object arrive at the top of the image surface. Likewise, rays that come from the right side of the object arrive on the left side of the image surface, and those that come from the left side of the object arrive on the right side. Some rays from all parts of the object get through. In this way, an inverted and reversed image of the entire object is formed.

The light scales the image size according to the distances from the object to the pinhole and from the pinhole to the image surface. As the image surface is moved farther from the hole, the image gets larger because the light from the object spreads out in space. Conversely, as the object itself is moved farther from the hole, the image it casts gets smaller. With a fixed amount of light, the brightness of the image decreases as image size increases.

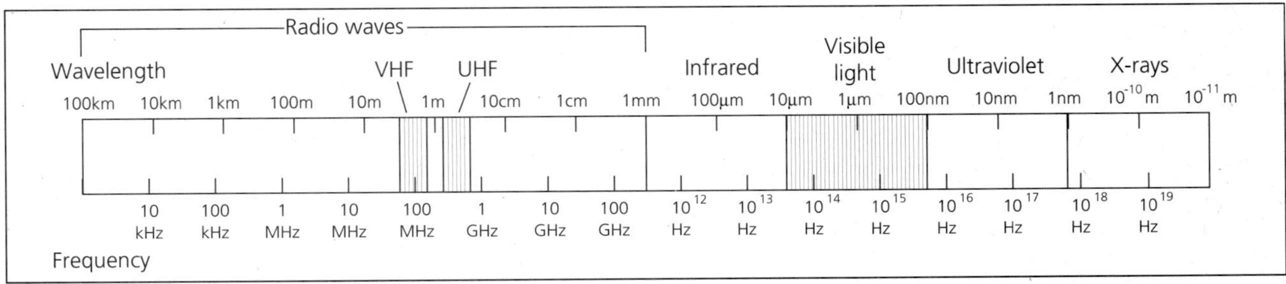

Figure 3.1   *A logarithmic chart of a portion of the electromagnetic spectrum showing the approximate position of visible light with respect to other kinds of waves. Note the areas of the spectrum used by VHF and UHF television.*

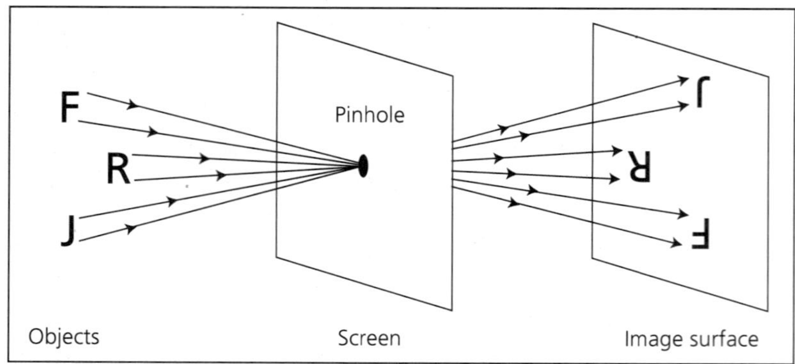

Figure 3.2 *This figure demonstrates the "rectilinear propagation" of light, that is, the fact that light travels in straight lines.*

The Law of Reflection

The reflection of light enables us to see the moon at night and is responsible for most of our visual experience of the world. On smooth surfaces, reflection occurs in a predictable fashion. It follows the **law of reflection,** as illustrated in Figure 3.3.

The figure shows a ray of light striking a mirror. The imaginary line perpendicular to the mirror, at the point where the light strikes, is called the *normal*. The ray approaching the mirror, called the *incident ray,* makes an angle called the *angle of incidence* with the normal. Similarly, the reflected ray makes an *angle of reflection* with the normal. Regardless of how the surface is oriented to the beam of light, *the angles of incidence and reflection are equal.*

The law of reflection is simple but critical in planning lighting designs for television. For example, by paying attention to the law of reflection, you can

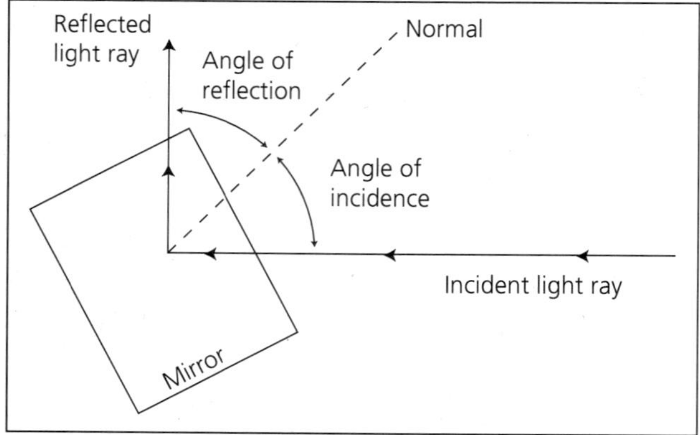

Figure 3.3 *A ray of light striking a mirror demonstrates the law of reflection: the angle of incidence equals the angle of reflection. Both angles are measured from the normal, the imaginary line perpendicular to the mirror at the point where the light strikes.*

direct unwanted light away from the set and the camera lens. In Chapter 4, we will see how crucial this ability can be.

The Inverse Square Law

As light travels away from a source, it spreads out; that is, the area it illuminates increases. At the same time, the intensity of light falling on a given area decreases. In effect, the light beam grows wider but dimmer. To put it scientifically, the area illuminated by a light source increases with the square of the distance from the source. The intensity of illumination on a given area decreases at the same rate. Physicists call this principle the **inverse square law**.

We can see the inverse square law at work when we project an image on a screen with a slide projector (Figure 3.4). If we move the projector farther from the screen and refocus, the image appears less bright. Though the amount of light on the screen remains the same, it is being distributed over a larger area. This means there is now less light per unit area. In fact, the inverse square law tells us that if we double the distance from the projector to the screen, the intensity of light on the screen goes down by a factor of 4; that is, the light for a

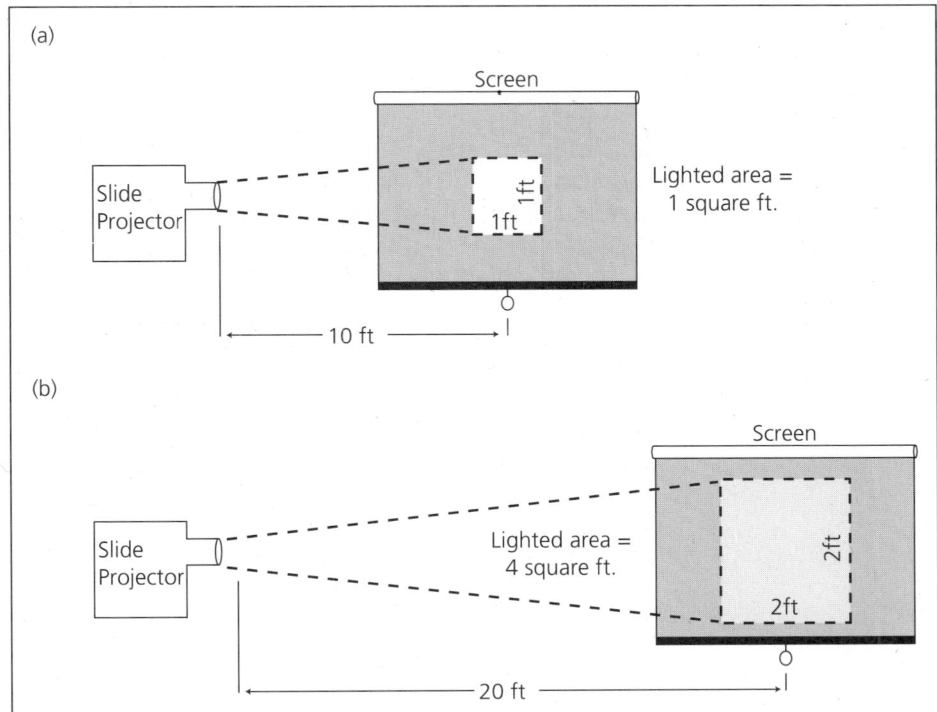

Figure 3.4   *An image from a slide projector illustrates the inverse square law. Comparing the two parts, we see that the screen in part (b) is twice as far from the projector as in part (a). Because the distance to the screen has been multiplied by 2, the area illuminated is four times as large (the square of 2 equals 4), but the intensity of the screen's illumination is only one-quarter as strong (¼ is the inverse of 4).*

given unit area is only one-quarter as bright. Remembering this property of light is important in creating lighting designs that work.

The Role of Lenses

Look again at the pinhole camera in Figure 3.2. As light emerges from the pinhole, it spreads out to cover a wider area, and consequently the sharpness of the image is reduced. To minimize this degradation of the image, we could keep the **aperture** of the camera—the hole through which light enters—extremely small, but that would let in so little light that we would need a much longer exposure time. (With some real-life pinhole cameras, exposure times may be several hours.) Conversely, if we increased aperture size to allow more light in, we would reduce exposure time but suffer an unacceptable loss of image sharpness.

Only by controlling light in some way can the conflicting needs of image sharpness and sufficient light level be met. This is the function of a lens. A lens can converge incoming light rays, making it possible to use larger apertures while still achieving sharp focus.

FOCUSING CHARACTERISTICS OF LENSES

Lenses rely on the fact that light slows down when it passes from air into a denser transparent medium, such as glass. When light passes through a flat sheet of glass, all of the light rays slow down at the same time and then speed up at the same time as they exit the glass, continuing to travel in the same direction. But if the glass is curved or of varying thickness, some light rays take longer than others to get through the glass, and they change direction when they exit; in short, some of the light bends, or *refracts*. Convex (outward-curving) lenses make incoming light rays converge, whereas concave (inward-curving) lenses spread the rays farther apart.

When a convex lens brings light rays to a point, that point is called a **focus**. For imaging purposes, more than one lens is used in almost all cases, and it is essential that the centers of all the lens elements lie on a straight line, called the *axis* of the lens system. If we construct an imaginary plane perpendicular to the axis at the point of focus, we call that plane the *focal plane*. It is here that we place film to receive an image in a traditional still camera.

Focal Length

The **focal length** of a lens is the distance between the center of the lens and the point at which a sharp image of an object is formed, often expressed in millimeters (mm). The factors that influence focal length or bending power of a lens include the density of the glass and the curvature of its surfaces. Lenses of shorter focal length generally have a wider **angle of view**—that is, they capture a larger horizontal and vertical area in front of the lens—whereas

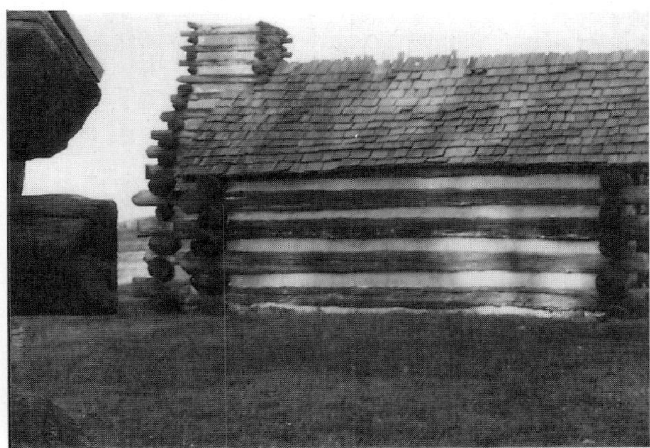 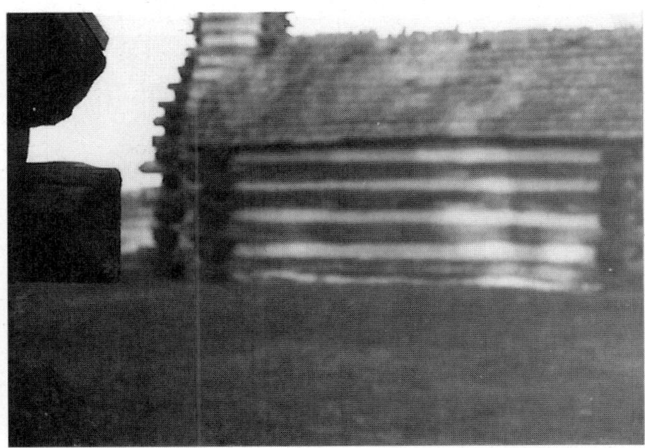

Figure 3.5　*Different depths of field. In the photo on the left, both the foreground and background are in focus. In the photo on the right, shot with a lens that has less depth of field, the foreground remains in focus but the cabin in the background is blurry.*

lenses of longer focal length have a narrower angle of view and provide a larger image size.[1]

Usually a lens receives light from various objects at a variety of distances from the lens. But the lens will focus sharply on the image plane only rays within a certain range of distance. Therefore, some parts of the image will be in focus and some will not. The region of space for which images are acceptably sharp is called **depth of field** (Figure 3.5).

f/number

The brightness of an image is largely determined by the amount of light permitted to go through the lens. To control the amount of light entering a lens, an adjustable aperture is used. In television and photography, an iris diaphragm serves this purpose (Figure 3.6). The intensity of light collected is also influenced by the magnifying properties of the lens. Lenses of different focal lengths have different magnifying properties.

Because of these factors, the ratio of focal length to aperture diameter is a useful way to express information about the amount of light entering the camera. The ratio of focal length to aperture diameter is called the **f/number** or, in the term often used in the video industry, **f-stop number**.

As an example, if a lens has a focal length of 50 mm and an aperture or lens opening of 5 mm in diameter, the f/number or f-stop number is 10. For the same lens, if the aperture diameter is 10 mm, the f/number is 5. For any given lens, as the f/number increases, aperture diameter decreases.

---

[1]Technically speaking, angle of view and focal length are not directly related. This is because almost any lens will accept light rays from any arc in front of it—an arc approaching 180 degrees. But the practical angle of view is the one that provides rays that can be used to produce an image, and this is influenced by the focal length.

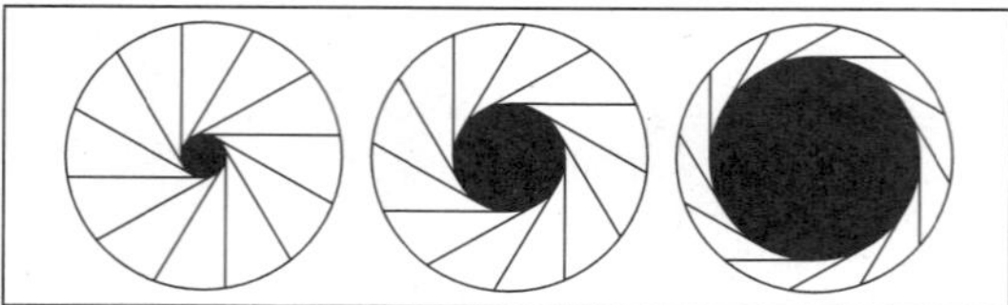

Figure 3.6   *A camera's iris diaphragm opens and closes to control the amount of light entering the lens.*

Knowledge of how f/numbers relate focal length to aperture opening is critical in the film and photography industries because, unlike the video production setting, these professions do not have devices such as waveform monitors to judge light levels. However, even in video production, f-stop information helps to determine depth of field, which in turn helps to control which objects in a scene will be kept in focus. Knowledge of f-stop numbers is therefore a powerful tool in refining shot composition.

AESTHETIC EFFECTS OF LENSES

Picture content is paramount in television and photography. Deciding which objects to include and which to exclude from a scene, or determining which objects to emphasize by focusing on the foreground rather than the middle or the background, can greatly enhance the drama of a scene. Selecting camera angles and deciding how to frame the image will also influence the emotional impact

Does f/number Affect Angle of View?

It may seem that when f/number increases and aperture size decreases, an accompanying decrease in angle of view would occur. But this is not true, for a couple of reasons.

First, *stopping down* the lens—that is, raising the f/number setting to decrease the aperture size and reduce the amount of light permitted to pass through the lens—has no effect on light coming through portions of the lens close to the center. It blocks light only from portions of the lens near the perimeter. Since light from all parts of a scene strikes all parts of the lens, the lens remains able to collect light from all elements of the scene. The net effect of stopping down the lens is to deintensify light from all parts of the scene, not eliminate light from some parts of it. Hence, the angle of view remains constant.

Second, the close proximity of the iris to the lens permits light from the scene to be delivered to the entire focal plane, regardless of how much or how little of the lens is actually used. The location of the iris with respect to the lens is analogous to the location of a window shade with respect to your eyes when you lean on a windowsill. If you pull down the shade to a height just above your eyes, you do not change your view of the landscape outside.

of every shot. In many cases, the impact of lenses on these decisions can make the difference between a barely adequate production and an outstanding one.

Depth of Field

In terms of program content, depth of field is one of the most important characteristics of lenses. Depth of field includes all of the space between the closest and farthest objects that appear to be in focus. Technically speaking, the sharpest plane of focus in the object space is unique; focus immediately deteriorates as one moves away from the focal plane in either direction. But different lens settings yield different rates of falloff of acceptable focus, depending on three variables:

- *Focal length.* When all other conditions are held equal, lenses with shorter focal lengths yield greater depth of field. This is because shorter focal length lenses have more bending power (greater refractive capability). Because of this greater bending power, the image planes for objects at varied distances are closer together, and the image sizes more similar, than for longer focal length lenses.
- *Camera-to-subject distance.* Other things equal, as distance from camera to subject increases, depth of field increases. This is because light rays from distant objects tend to enter the lens in a more nearly parallel fashion, whereas light rays from objects close to the camera create wider angles relative to the axis of the lens and cross at wider angles to focus on the focal plane. For this reason, rays from close objects are still converging when they reach the lens and are not brought to focus as quickly as are parallel rays entering the lens from more distant objects.
- *Aperture size.* When all other factors are held constant, as the aperture gets smaller (that is, as the f/number increases), depth of field increases. This is because rays entering the lens closer to the center converge at the focal point at narrower angles to one another than rays entering from the perimeter.

It helps to remember these principles when planning shots for television. First, controlling depth of field enables you to direct the viewer's attention to preselected parts of the picture. For example, in a product advertisement in which the main shot is the product in the foreground, you may want to make all other objects in the middle and background somewhat blurry. In this case, you might choose a long focal length lens with a large aperture setting and with the product close to the camera.

In contrast, you can increase depth of field to emphasize a grand vista. For example, you can create a prolonged shot showing the central character of a program riding off into a beautiful sunset, with everything in the frame appearing in focus. To do this, you might use a short focal length lens on a small aperture setting, with the main character slowly moving away.

Perspective

Lenses also help to control **perspective,** or the illusion of depth in a two-dimensional picture. This illusion can be created by *linear perspective,* in which

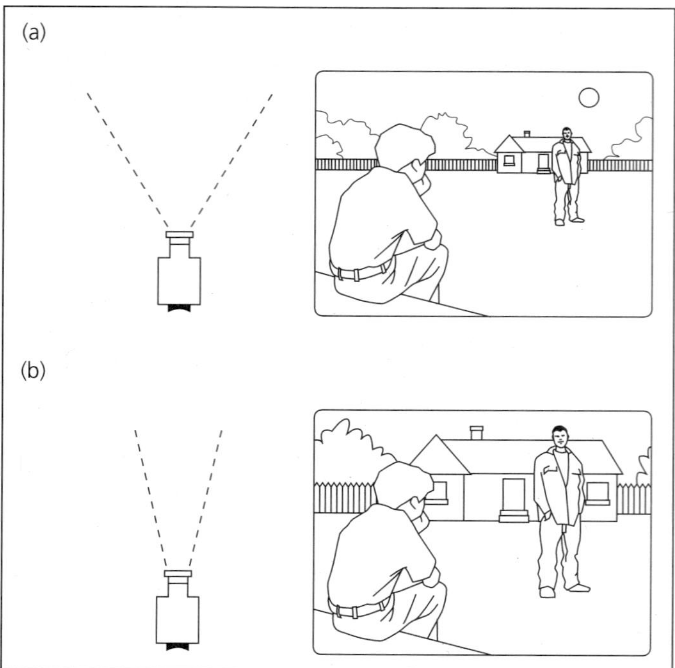

Figure 3.7   *The influence of focal length on perspective. (a) A lens with a short focal length and wide angle of view takes in a good deal of background and makes objects in the background seem relatively small and far apart. (b) A lens with a long focal length and narrow angle of view not only takes in less of the background but also makes the background objects seem larger and closer together.*

apparent distance is conveyed by the relative size and location of objects. When the lines of a roadway appear to coverage at a vanishing point, or the cables of a bridge appear to get smaller as distance increases, we interpret these cues as indicators of depth. Similarly, objects in the bottom of the frame that overlap other objects are usually interpreted as being closer, whereas objects toward the top of the frame that are overlapped by other objects usually appear to be farther away. Other depth cues result from texture and focus elements. For example, more highly defined texture and detail tend to convey closeness, whereas hazy objects appear to be farther away.

Lenses of different focal lengths can produce strikingly different perspective cues. For example, varying focal length can alter the relative sizes of objects, changing the apparent distances between them. As Figure 3.7 shows, shorter focal length lenses expand background space, decreasing the apparent size of background objects while increasing the apparent size of nearby objects. This is because they feature a wider angle of view than the human eye can normally accommodate. Since a wider angle of view takes in more object space, objects in the background appear smaller than they do to the unaided eye. An extreme example of this is an image captured by what is aptly called a *fish-eye* lens.

Conversely, longer focal length lenses tend to compress image space. This is because they have a narrower angle of view than that afforded by the human eye. Since a narrower angle of view takes in less object space, this tends to magnify distant objects and decrease the apparent distance between them. A *telephoto* lens is a lens of this type used to magnify distant objects.

Knowing how to manipulate perspective as a function of focal length is important for controlling image content. It gives the director tremendous flexibility.

ZOOM LENSES

Nowadays virtually all television cameras in both the studio and the field are equipped with *variable focal length lenses,* commonly known as zoom lenses. The **zoom lens** is a complex system of elements arranged to allow the spatial relationship of lens components to change. Zoom lenses therefore offer continuous, smooth variation in focal length between two limits.

The advantage of such a system is that it permits picture composition to change as you change focal length. When you zoom in, objects in the frame become magnified as they appear to move closer to the camera. Conversely, you can decrease focal length by zooming out, thus making the objects in the frame appear to recede into the distance. Throughout these actions, if the image was in focus to begin with, focus will be maintained.

For video, zoom lenses often consist of up to twenty-two elements, eight of which may be made of highly refractive glass. Some lenses may use as many as sixteen different types of glass. When the number of lens elements gets this high, internal reflection of light between elements increases, reducing the total amount of light that gets through the lens to the CCD. When this happens, you may need to increase light intensity to ensure adequate illumination. Further, internal reflections in complex lens systems can cause ghost images and loss of contrast. To reduce such reflections, lens surfaces are given nonreflective coatings.

PROFESSIONAL POINTERS

Steps in Focusing a Zoom Lens

- First, zoom in all the way on the object being pictured.
- Next, use the focus control to focus the lens.
- Then zoom out to compose the desired picture.
- Now the lens will remain in focus throughout the entire range of motion.
- Refocusing is necessary only when the distance changes from the lens to the object being pictured.

There are several advantages to using the zoom lens in television. First, it allows you to "move in" on subjects without physically approaching them. For example, if the subject is a person being interviewed, a slow zoom-in during a key answer can add visual interest and heighten audience involvement without making camera-shy guests feel more self-conscious. Further, the zoom lens permits you to frame content with precision as a program unfolds through time. For instance, in a cooking show, a zoom-in to a tight close-up of a specific ingredient or a detail of the cooking process can help to clarify a demonstration. Similarly, in a drama, a well-timed zoom-out to reveal the perpetrator of a crime can be a crucial part of the action. Zoom lenses are also especially good for large public events, such as parades and sporting events, where the work space may be confined but the coverage area is large.

Among the aesthetic disadvantages of the zoom lens (in addition to the technical problems of lower light levels) are the abnormal shifts in perspective that occur while zooming. As you zoom in, focal length increases; this means depth of field decreases, distant space compresses, and the relative sizes of distant objects increase. Conversely, as you zoom out, all of these effects are reversed. Although these factors may not be noticeable during small increments of change, they can influence the quality of the overall production. The convenience of the zoom lens should not tempt you to ignore careful placement of the cameras, a subject we will discuss in detail in later chapters.

CONTRAST AND COLOR

Besides knowing about lenses, videographers need to understand the camera's ability to render brightness and color information accurately when different light levels are presented. One basic characteristic of a camera is the minimum amount of light it requires to create a picture. We call this the **operating light level** of the camera, and it varies quite a bit from one camera to another.

Cameras also differ in **contrast ratio,** the difference in brightness between the lightest and darkest possible images the camera can produce. In color cameras, the optimal contrast range is roughly 30:1 or 40:1 (or about 4½ to 5½ f-stops). This means the brightest possible image the camera is capable of producing is no more than thirty or forty times brighter than the darkest, a narrower range than that of the human eye.

Knowing this helps you set limits on what can be differentiated for the viewer. For example, if you are displaying two different sets of silverware on crushed velvet, with one set on black velvet and the other on navy blue, the limited contrast range of the camera may make them indistinguishable.

White Balance

Because the color camera breaks light into three separate signals (red, green, and blue), it is necessary to balance the strengths of these signals correctly to maximize accuracy in color reproduction. Accurate color is especially important in rendering flesh tones, since large discrepancies in such cases look unnatural.

Adjusting the levels of signal strength to enable cameras to render colors most accurately is called setting **white balance**. In a studio, white balance is set by focusing all cameras on a white card under the same lighting conditions that will be used for the production and adjusting their color at the camera control unit until an image is obtained that most closely matches the color of the original card.

Using white as a color reference for the camera makes sense because, as we already know, in light-based systems such as video, white is produced by combining all three additive primary colors—red, blue, and green—in just the right amounts. Hence, if levels are set so that the video image of a white card closely matches the white of the actual card, we get accurate colors across the rest of the spectrum. Further, in studio productions using more than one camera, setting white balance for each camera enables subjects and scenery to look the same across all cameras. This is critical for avoiding jarring differences in color rendition of a subject when cutting from one camera to another.

In field settings, in cases where no camera control unit is available, setting white balance is still necessary to capture colors accurately. To do this, portable cameras have white balance controls. To set white balance in the field, focus the camera on a white card or other convenient white surface under the lighting conditions you intend to use for the shoot, and press the white balance control button. The camera will set the balance and let you know when it is ready.

In addition to setting a white balance, the camera sets a black balance. In some cameras black balance must be set manually, but in most cases the camera does this automatically when it is capped.

Color Temperature

In normal daylight, objects are lit with radiation from all parts of the visible spectrum. It is under these conditions that we tend to fix the colors of objects. However, the colors of objects may change under different lighting. Imagine trying to identify the colors of different crayons without their wrappings, lit only by a strong, blue, artificial light.

One reason colors of objects change under different lighting is that, unlike sunlight, artificial light rarely includes wavelengths from all parts of the visible spectrum. In fact, even artificial light meant to simulate the natural white light of the sun rarely matches all of the sun's spectral qualities. Therefore, when we use artificial light to illuminate a scene for television, we must compensate somehow for the missing parts of the spectrum if we want to render colors accurately.

To help achieve this, the dominant hues of various light sources have been classified by their **color temperature,** which is described in terms of a standard scale of values expressed in degrees Kelvin (K). By this standard, ordinary daylight is located in the range of values between 4,500 and 6,200 K (5,600 K average), with the light of a clear blue sky located as high as 10,000 to 12,000 K. Incandescent studio lights are generally located at around 2,800 to 3,200 K, and fluorescent lights range from about 4,000 to 7,000 K (see Figure C.2 in the section of color plates). Notice that on the Kelvin scale, bluish light has a higher color temperature than reddish light. This is the *reverse* of the way we normally use the terms *warm* and *cool* in reference to color.

f you are shooting in a bright interior space where the light quality looks very much like the sunlight outdoors, can you simply move outdoors without color-balancing the camera again? Can you trust what your eyes tell you about the similarity of the light? The answer is no.

Compared to sunlight, incandescent light contains more orange-yellow color. This fact is not always evident to our eyes, but to a television camera the differences are quite apparent. If you color-balance with incandescent light and then go outdoors without color-balancing again, the material you shoot will take on a bluish cast. Conversely, if you color-balance for daylight and then come indoors without rebalancing, the indoor footage will appear more orange than it should.

What about mixing natural and artificial light? Since natural light differs so significantly from artificial light, it is best to avoid illuminating a subject for video with different types of light at the same time. Doing so confuses the camera.

Video cameras are normally set for standard tungsten studio light, around the 3,200 K region. Field cameras are equipped with a color filter wheel that enables you to shoot outdoors in natural light by rotating a daylight filter into the path of the lens. However, because different light sources vary so much in color temperature, it is important to balance the camera in the same light you will use for the shoot.

Even within the category of artificial lights, color temperatures may vary widely. Whereas quartz studio lights operate in the 3,200 K range, high-intensity arc lights operate at about 6,600 K. Obviously, if you make a transition from one light source to another, you may need to rebalance or change filters.

Filters

As the name implies, a **filter** is a device for eliminating unwanted portions of things that pass through it. Photographic filters are pieces of colored transparent material (usually glass or plastic) placed over lenses to absorb some wavelengths of light while permitting others to pass. Among the functions filters serve are to increase contrast between similar colors, reduce haze, reduce brightness in a scene, and create special effects. Other filters are constructed to pass only wavelengths from a light source that vibrate in a particular plane or reflect at a particular angle, thus eliminating glare and certain reflections. These are called **polarizing filters**.

In addition to using filters on the cameras, video often employs a type of filter called a **gel,** a material (formerly gelatin but now usually glass or plastic) placed over a light to color it. Gels on lighting instruments can be used to set a mood or create special effects.

KEY TERMS

law of reflection *(42)*   zoom lens *(49)*
inverse square law *(43)*   operating light level *(50)*
aperture *(44)*   contrast ratio *(50)*
focus *(44)*   white balance *(51)*
focal length *(44)*   color temperature *(51)*
angle of view *(44)*   filter *(52)*
depth of field *(45)*   polarizing filter *(52)*
f/number or f-stop number *(45)*   gel *(52)*
perspective *(47)*

QUESTIONS FOR REVIEW

1. In television production, why is it helpful to know the law of reflection and the inverse square law? Give some examples of how these principles would help the producer-director in planning productions.

2. What does changing the f/number setting on a lens do to the aperture size? To the angle of view?

3. How do focal length, camera-to-subject distance, and aperture size influence depth of field? What combination of factors maximizes depth of field? What combination limits it?

4. How does a zoom lens alter perspective cues as you move in on a scene? How do perspective cues change as you zoom out?

5. In what way does a camera need to be "balanced" for color? Why?

4 Lighting Equipment and Design

In recent years, advances in video technology have made it possible to produce cameras with greater sensitivity to light than were available in the past. As a result, the basic light levels cameras need to "see" have decreased. But this change does not mean we can be less careful in planning lighting designs. What principles can guide us in designing a sensible **light plot,** that is, a precise layout or diagram showing the types, sizes, positions, and directions of all the lights we will use? How can we use lighting to enhance the purpose and meaning of a television program?

In addition to providing visibility, lighting enhances the illusion of depth of objects and performers. Lighting also adds meaning, visual beauty, mood, and dramatic dimension to every scene. Lighting provides composition by incorporating shadow and highlights as well as brightness and darkness, thus avoiding giving every object the camera sees equal status. A picture composed with light and shadow can direct the viewer's attention to key parts of the image, selectively adding or reducing emphasis.

The process of creating a lighting design begins by reading through the script. The script should indicate set elements, talent requirements, subject matter, content, sequence of events, tone, and purpose of the program, all of which help determine lighting needs. The script might also indicate time of day, locale, and the presence of specific light sources (such as sunlight or lamplight) featured in a scene. It may give important information about talent movement, which can also influence lighting decisions. The light plot is further determined by the physical limitations of cameras and the physical constraints of the set or the field space (for instance, ceiling height, size of the studio, or location and direction of the sun).

It is essential to develop light plots holistically, that is, to recognize that lighting decisions affect, and are affected by, all the physical and symbolic elements of every production. Therefore, lighting design should be viewed as an integral part of the overall production process. After surveying the common types of lighting equipment used in video production, this chapter presents a rationale and techniques for creating lighting designs that are pleasing, efficient, and consistent with program content and objectives. The topics include:

LIGHTING EQUIPMENT

lighting grid • dimmer board • lighting instruments • additional lighting equipment

FUNDAMENTALS OF LIGHTING DESIGN

naturalistic lighting • a sample naturalistic light plot

DEPARTURES FROM NATURALISM

flat lighting

MOTIVATING LIGHT SOURCES

LIGHTING IN MORE COMPLICATED CASES

FIELD LIGHTING

procedures • equipment

In this section, we look at lighting equipment used in studio production. (We discuss field lighting later in the chapter.) Among the items we will examine are the lighting grid, the dimmer board, common types of lighting instruments and their component parts, and peripheral equipment and hardware.

Lighting Grid

The studio environment contains a **lighting grid** (Figure 4.1) that consists of a series of parallel or cross-hatched, sturdy metal bars hanging near the ceiling or affixed to the studio walls close to the ceiling. The grid provides numerous locations for hanging lighting instruments above sets and talent on the studio floor. In addition to the grid itself, electrical outlets are spaced near each bar at regular intervals to provide electrical power. Lighting grids make it easy to light sets and talent from above. Lighting grids also get lights and cables off the floor so that cameras can move more freely around the studio floor.

Dimmer Board

Electrical outlets are wired to a **dimmer board** (Figure 4.2) equipped with switches and faders that permit single instruments or groups of instruments to be turned on and off instantly or faded up and down gradually. The dimmer board may be located in the studio, the control room, or some remote location. Many dimmer boards permit the grouping or *ganging* together of selected instruments, which can then be manipulated with a single fader or controller. Some dimmer boards use rheostats as dimmers. Others use computer

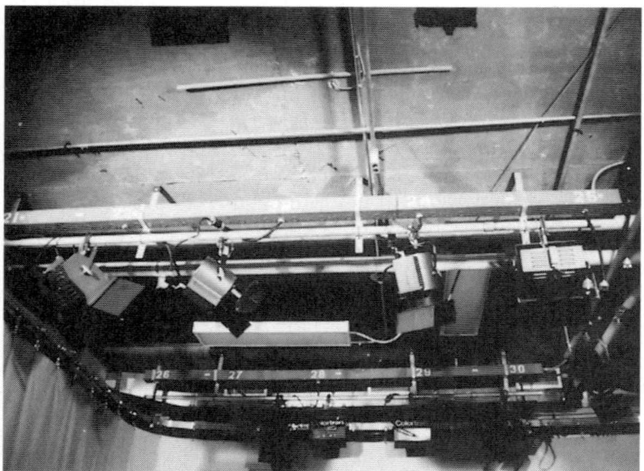

Figure 4.1 *A lighting grid on the studio ceiling, with lights hanging from it.*

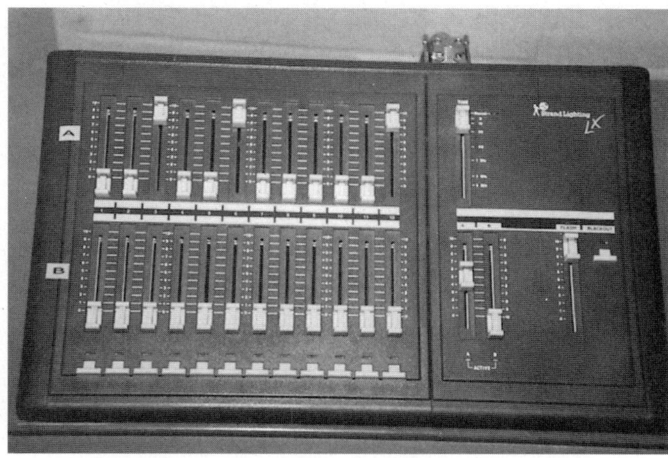

Figure 4.2   *A dimmer board.*

software, making them capable of executing complex instructions. Many boards provide a master switch for controlling electrical power to all instruments connected to the system.

Lighting Instruments

Among the most commonly used lighting instruments are those that provide directional sources of light, called *spotlights,* and those that provide more diffuse sources of light, called *flood lights.* The most common spotlight used in television production is the **Fresnel spotlight** (pronounced "fra-nell"), named after the French physicist Augustin Fresnel, who invented its special lens (Figure 4.3). The major components of the Fresnel spotlight that enable it to serve as a directional light are its lamp, reflector, lens, and focusing apparatus.

The Fresnel lens has a plano-convex shape with stepped concentric rings, enabling it to throw a beam of directional light onto the subject. A reflector located behind the lamp directs light out through the lens. Turning a focusing knob changes the spread of the beam by moving the lamp and reflector either toward or away from the lens. Focusing is made possible by mounting the lamp socket on an adjustable worm gear connected to the focusing knob at the back of the instrument housing. Some instruments use a sliding mechanism to move the lamp. Either way, as the lamp is moved closer to the lens, the beam spreads out and becomes less intense as it covers more area. As the lamp moves away from the lens, the beam narrows and becomes more intense as it *spots down* to cover less area. Hence, intensity and coverage area can be controlled.

In addition to the components just mentioned, the Fresnel spotlight is often fitted with external, adjustable, black metal flaps, called *barn doors,* which provide a means of controlling the spill of light from the instrument. It is also fitted with a *gel frame holder* (the white area inside the barn doors in Figure 4.3b), which permits the attachment of color filters, as well as *diffusers* or *scrims,* neutral-color filters for softening or reducing light intensity. Fresnel lights commonly come in 500-, 750-, 1,000-, and 2,000-watt power ratings, with 6-,

---

PROFESSIONAL POINTERS

Working with Lighting Equipment

- To avoid burns, wear protective gloves when working with hot lights.
- Unplug instruments before changing bulbs (even ones that are not hot).
- Don't touch even cold bulbs with your bare hands. When installing them, hold them with a soft cloth, paper towel, or even the box they came in. This will keep oil on your hands from getting on the glass, which may cause the bulbs to shatter when they heat up.
- To avoid shocks, turn off the power at the dimmer board before connecting or disconnecting lights.
- If you climb a ladder to move a light, have a helper standing by who can "foot" the ladder (stabilize it with his or her foot) and take instruments out of your hands before you climb down.
- Make sure the ladder you use is in good condition and is approved for electrical work.
- When reconnecting lights to the grid, always reattach the safety chain to keep the lights from falling on people below.
- Before turning on lights, be sure you have not exceeded the safe power limits of the system.
- Periodically check the dimmer board and circuit breakers for proper functioning to avoid risk of fire if you accidentally overload the system.
- If you use light stands, secure them with sandbags or tie-downs, especially in high-traffic areas, and put up proper warnings so that stands do not get knocked down.
- Electrical cables running across the floor can trip people. Secure your lighting wires, cables, and extension cords with gaffer's tape or tunnel tape to minimize hazards.
- Protect cables from damage with cable troughs, and hide them from view with rubber mats.
- To keep cables from overheating, don't bunch them together.

8-, and 10-inch lens diameters. Other ratings and sizes are also available. Selecting the appropriate instrument depends on studio size, limitations of the lighting board and power source being used, and the nature of the production task at hand.

Compared to spotlights, flood lights provide a more diffuse, nondirectional base light to large areas. They are often used to light large areas of scenery and backdrops (Figure 4.4). Flood lights are often called *fill* lights or *soft* lights, since they are used in front of talent and opposite spotlights to fill in and reduce the harsh shadows created by the spotlights.

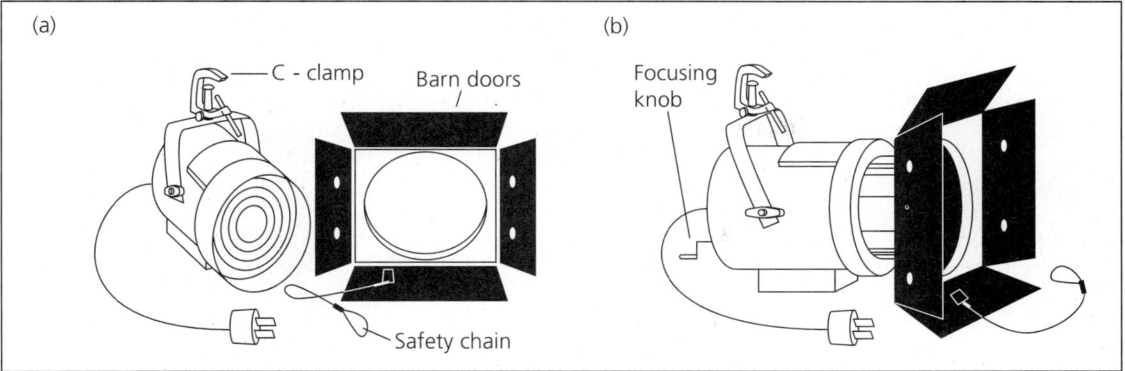

Figure 4.3   *Fresnel spotlights. (a) With the barn-door attachment open, the concentric-ring lens can be seen. At the top is a C-clamp, often used for attaching lights to the lighting grid. Also note the safety chain, shown resting in front of the light. (b) Here the barn doors are in position, ready for use. A gel frame holder is visible between the barn doors and the lens.*

Fill lights are comparatively simple devices, consisting of a lamp (available in wattages comparable to those for spotlights) and a reflective surface inside the instrument housing. Because of the shape of the housing, some fill lights are called *scoops* (Figure 4.4a). Those with a more boxlike shape are called *broads* (Figure 4.4b). Fill lights offer few options for controlling spill and intensity. Of course, if the light is attached to a dimmer, you can control light intensity to some extent by dimming; but the more you dim a light, the more you change its color temperature, and this can create unwanted effects in the way the camera reproduces colors. A better way to control light intensity is to either change the distance from the light to the subject or use scrims or diffusers.

Another, more specialized type of lighting instrument, with a sharp-edged and highly controllable beam of light, is the *ellipsoidal* spotlight (Figure 4.5), so

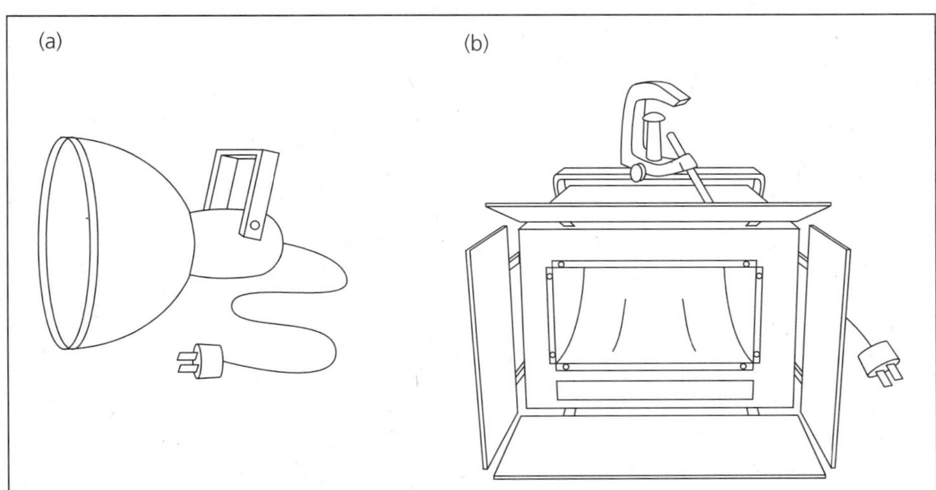

Figure 4.4   *Types of flood lights. (a) A scoop. (b) A broad.*

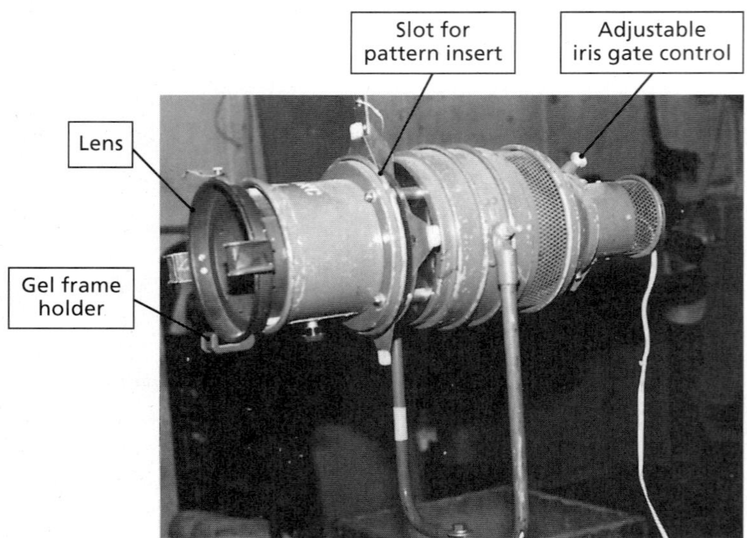

Figure 4.5   *An ellipsoidal spotlight.*

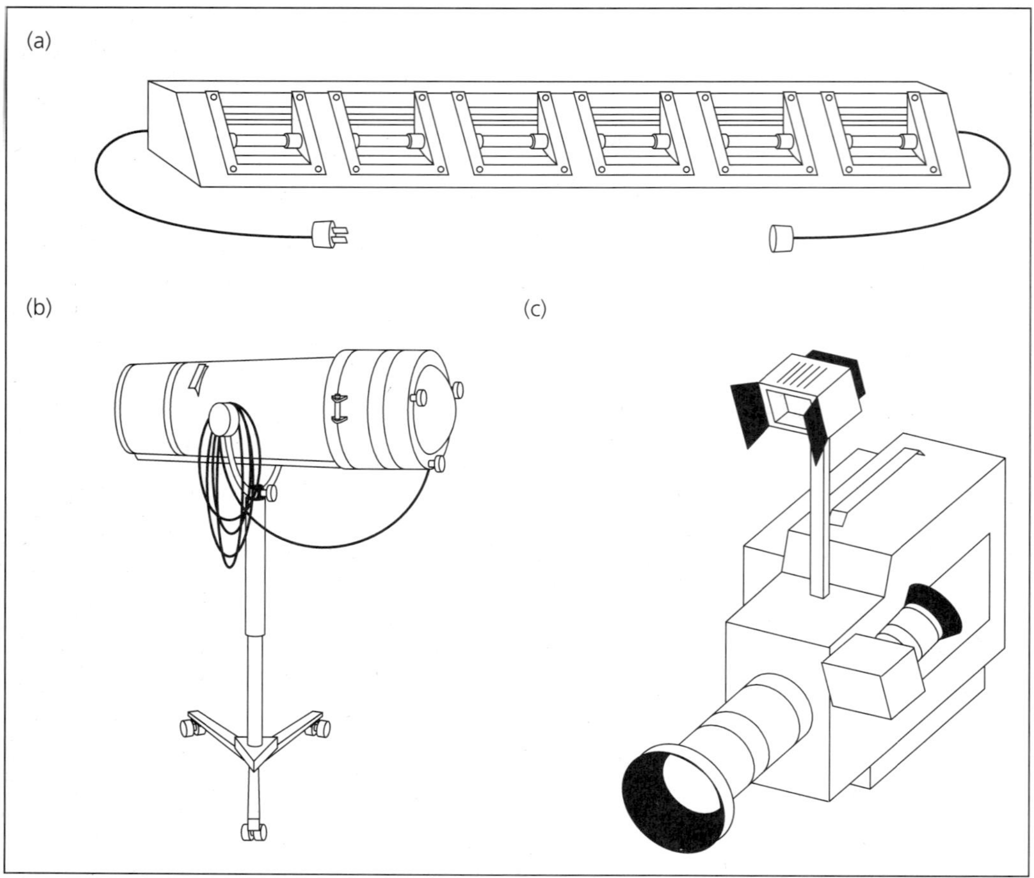

Figure 4.6   *Other common types of lights. (a) A strip light. (b) A follow spot. (c) An eye light (the small light mounted on top of the camera).*

named for the elliptical shape of the reflective surface inside the instrument housing. The ellipsoidal spotlight sends an intense beam of light through an adjustable iris that controls the diameter of the light beam. After passing through the iris gate, the light beam may be further shaped by a customized pattern insert, a metal disk with shapes cut out of it; this insert is called a *pattern* or *template*. The light is then focused by a large, compound lens. This arrangement makes the ellipsoidal spotlight useful for projecting patterns onto backdrops and other surfaces.

Several other lighting instruments deserve mention. These include strip lights, follow spots, and eye lights (Figure 4.6). A *strip light* is an oblong box containing a series of small lights that can be fitted with colored gels and used to light sets and backdrops. These small lights may be either hung from the lighting grid or set up on floor stands or on the ground close to the surface they are lighting, hidden by scenery and out of camera view. Strip lights are also useful for lighting sets in highly stylized productions, where they can add a festive and decorative accent; for instance, they can be used on runways for fashion shows or to create traditional footlight effects for performers.

Follow spots are powerful, stand-mounted spotlights that require an operator. They are useful for following action on the set, especially in musicals and variety shows. Follow spots throw a powerful beam of light with an adjustable sharp or soft edge.

Finally, *eye lights* are small, low-power lights mounted on top of cameras. They are commonly used to provide additional sparkle to subjects' faces and eyes.

Additional Lighting Equipment

In addition to the lights themselves, other types of equipment are critical for rendering high-quality lighting designs. Among the more important is the **light meter** (Figure 4.7), which measures light in terms of *lux* or *foot-candles,* standard measures of light intensity falling on a given area. A foot-candle refers to the light intensity produced by a standard candle (specified precisely in terms of size and material) at a distance of one foot. One lux equals about one-tenth of a foot-candle.

Light meters are of two types: incident and reflected. An *incident* light meter measures the amount of light falling on a subject. To use one correctly, aim the meter's sensor at the principal camera from the place the subject will be. A *reflected* light meter measures the amount of light reflecting from a subject. To use one properly, aim the meter's sensor at the subject to get a correct reading. Light meters are useful for locating hot spots and dark spots in a production area.

Remember, however, that surrounding light conditions can change from moment to moment and from camera shot to camera shot, and that extremely dark or bright backgrounds can affect the accuracy of meter readings. Remember also that various cameras may have different performance characteristics, such as different operating light levels and contrast ratios, and may therefore perform more or less adequately in the same light. For these reasons, light meter readings alone cannot guarantee desired results.

Therefore, it is essential to check the final court of appeals, the **waveform monitor** (Figure 4.8a) on your camera control unit. The waveform monitor

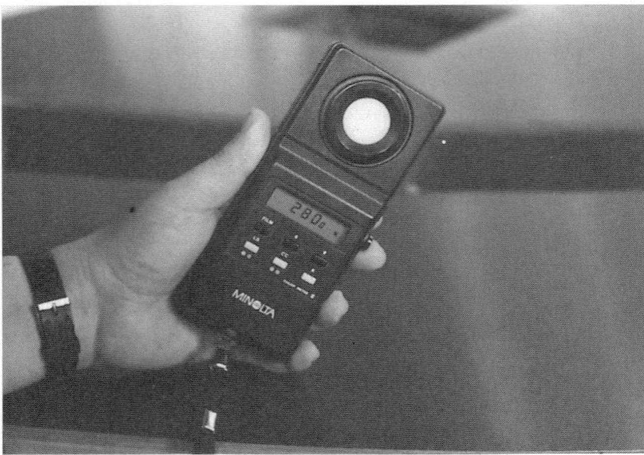

Figure 4.7   *A light meter.*

graphically displays the white and black levels in the video signal, making it possible to control brightness, which must be kept within the proper range to maintain video quality. Two controls on the camera control unit enable you to maintain brightness levels: the *pedestal* control and the *iris* control, both of which should be adjusted while referring to the waveform display. Adjusting the pedestal and iris controls is called *shading* the camera. In addition to providing optimal contrast range, proper shading reduces or eliminates jarring brightness differences when cutting between different cameras during a show.

To set brightness levels most efficiently, it is best to use a *chip chart* (Figure 4.8b), which displays shades of gray from black to white. The darkest area on the chart is called *TV black,* and the lightest is called *TV white.* To use the chart,

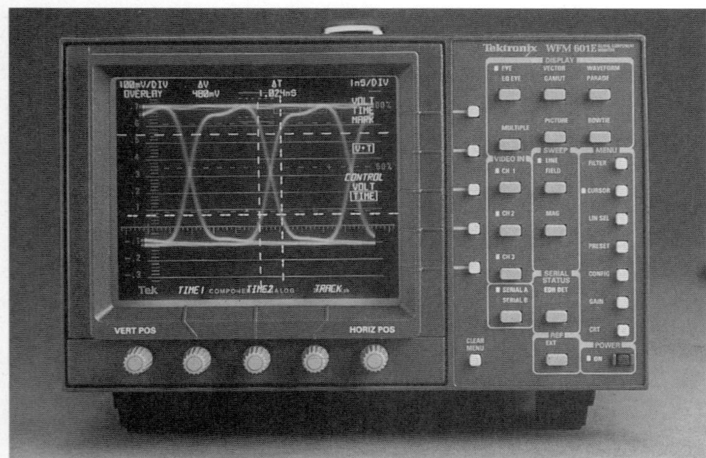

Figure 4.8   *Devices for setting white and black levels. (left) Waveform monitor display produced by training the camera on a chip chart such as the one in (right), with a nine-step gray scale.*

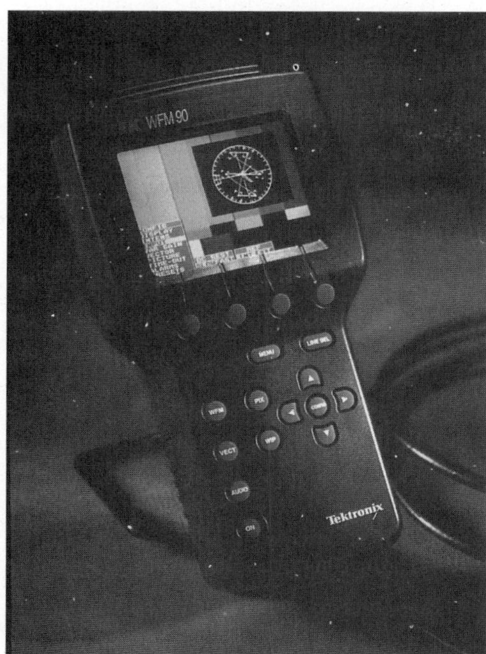

Figure 4.9   *A vectorscope.*

place it where talent will appear with the lights turned on as they would be for the actual program. Then focus the camera on a full shot of just the chart, and adjust the pedestal and iris so that the brightest portion of the picture registers 100 on the waveform monitor and the darkest portion registers 7.5. This procedure ensures that you will get the proper contrast range for most situations.

Just as a waveform monitor and chip chart are used to set brightness levels to ensure the best possible contrast range, a **vectorscope** (Figure 4.9) and an electronically generated set of *color bars* (see Figure C.3 in the section of color plates) are used to balance color reproduction. The vectorscope displays six small squares inside a circle, marking the areas where the three primary and three secondary colors should appear when colors are rendered accurately. Adjustments are made to align the color bars with their designated locations. When all six colors appear in their proper locations, colors seen by the camera should be accurately reproduced.

Several other types of peripheral equipment are also important:

- Reflector boards (Figure 4.10), boards or sturdy cards with bright white, silver, or even mirrored surfaces, are used to bounce light onto a subject. They are frequently used in field settings to alter the effects of sunlight.
- Rods and pantographs (Figure 4.11) are expandable devices used to lower lighting instruments from the lighting grid to a desired height closer to the studio floor. Pantographs may also be used to adjust the positions of mounted television monitors.
- C-clamps (see Figure 4.3a) are screw-down devices used to affix lighting instruments to lighting grids and stands.

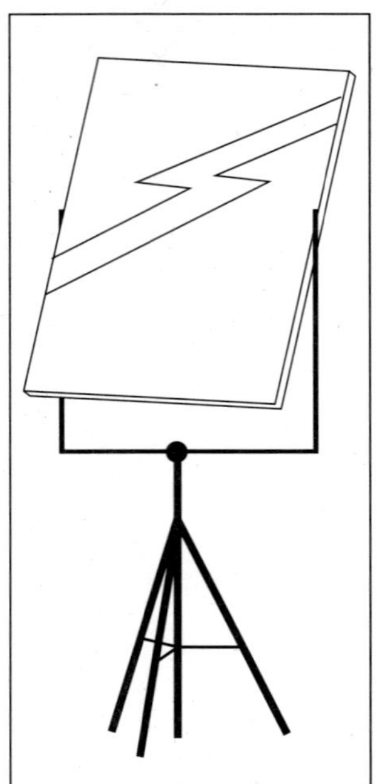

Figure 4.10   *A reflector board.*

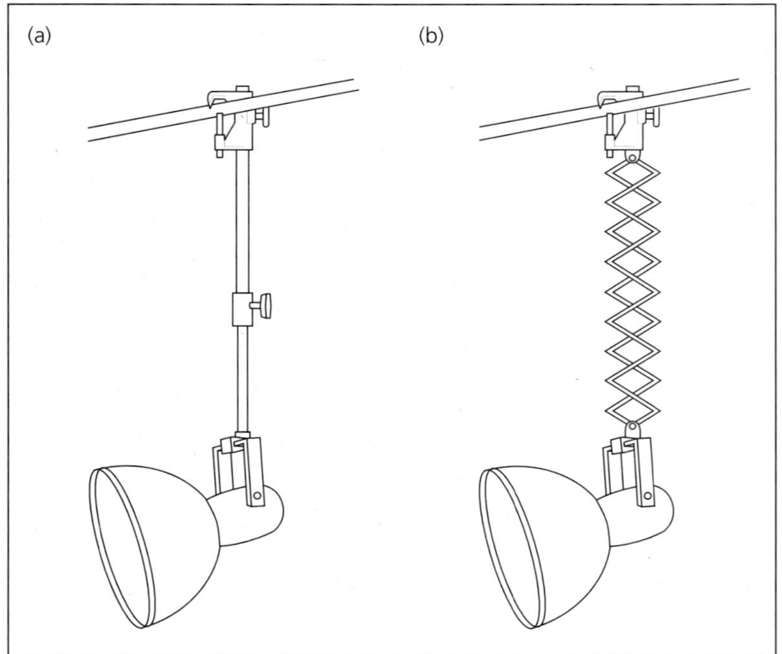

Figure 4.11   *Devices for hanging lights. (a) Rod. (b) Pantograph.*

- Safety chains (see Figure 4.3a) keep instruments from falling if their C-clamps come loose.

FUNDAMENTALS OF LIGHTING DESIGN

At the simplest level, lighting design principles are usually grounded in **naturalism**. The naturalistic approach recognizes that over millions of years we have been conditioned to see objects illuminated from above by a single main source of light. This is because we inhabit a solar system featuring only one sun, which provides our main source of light. Normal lighting, to us human beings, must be consistent with this arrangement.

Theater people recognized this principle long before the advent of television. Traditionally, stages had been lit from below, with footlights, but by the 1820s, when more powerful gas lights replaced candlelight, theater critics began to note the advantage of the chandelier. Chandeliers threw light on the faces of the actors from above, which seemed natural. Strong footlights, on the other hand, inverted the natural shadows of the face and distorted the actors' expressions.

As another example, think of children playing with a flashlight, scaring one another by aiming the light up at their faces from below their chins. Such lighting appears unnatural because it reverses the normal pattern of light and shadow on the face, subverting our normal expectations and at times creating monstrous effects. Figure 4.12 illustrates this phenomenon.

In addition to being generally located above us, the sun is also millions of miles away, and its light passes through all of the earth's atmosphere before

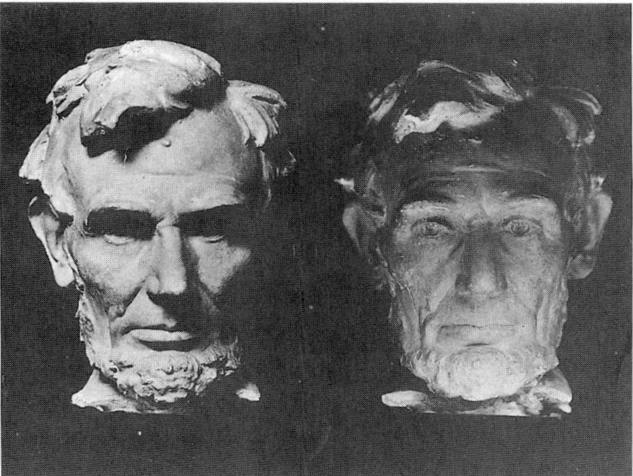

Figure 4.12   *Two photos of the Lincoln Memorial in Washington, D.C., illustrate (left) the natural effect of lighting from above and (right) the unnatural look caused by lighting from below. Note how Lincoln's expression appears to change.*

reaching us. As a result, some light is diffused, having been deflected repeatedly at odd angles. It therefore strikes objects in our vicinity on all sides, reducing harsh shadows. Even when the sky is extremely clear, objects rarely appear to be illuminated on only one side and utterly dark on the other. Rather, they appear to be illuminated on all sides, but with the sunward side more brightly lit than the others. On overcast days, the directionality of the sun's light can be completely eliminated, making all the light we see appear diffuse and directionless. Under such conditions, objects may appear to be illuminated evenly on all sides.

In summary, sunlight provides both the main source of "hard," directional light from above—the **key light**—and "softer," more diffuse (less directional) **fill light** that reduces harsh shadows throughout other parts of a scene. This arrangement constitutes a naturalistic lighting scheme. You can simulate naturalism by following this simple rule: *in the absence of compelling reasons not to do so, use lighting designs that imitate the way objects are lit in nature.*

Following this rule is not as easy as it sounds, however. First, in designing light plots for television, we cannot simulate natural light using a single instrument. We must use more than one. When a single light shines on a subject from nearby, it creates unnaturally sharp shadows, since the light does not have enough distance to diffuse adequately through the atmosphere before reaching the subject. That is why designing lighting schemes is *always* a departure from "reality," even when the program is nonfiction and the artistic goal is realism.

Second, when natural shadows extend from objects on a bright day, we see *only one shadow* for each object. Duplicating this effect is difficult when lighting for video, because we often use more than one light per object, and *each time we aim a light at something, we create a shadow somewhere.*

Third, in nature, the *direction* of shadows cast by objects in our field of view is consistent; that is, shadows are always cast on the side of objects opposite the sun. If a shadow from one object is cast from left to right, shadows from all objects in the vicinity should also be from left to right. In addition, the shadows will be parallel with one another, since light rays from distant light sources such as the sun are parallel. Therefore, when we use multiple instruments to light a set, we must ensure that the shadows seen on air are cast consistently or we can quickly break the illusion.

For all these reasons, controlling shadows is as critical a part of lighting design as controlling light. To do this, you must know not only where the set pieces and the talent are with relation to the lights but also where the cameras are. You must also know how much of the scene the widest shot includes. Knowing the extent of the widest or *master* shot enables you to plan lighting schemes that direct unwanted shadows either off the set or beyond the reach of the cameras. Conversely, using the same principles, you can also direct shadows onto parts of the set where they are needed. Let the program purpose and content dictate these decisions.

Naturalistic lighting provides an attractive, visually satisfying, sculptural appearance of the talent and objects seen on camera, enhancing the sense of depth for the viewer. It is especially useful for stationary talent in limited performance areas. Programs that generally use such lighting include news, sports, public affairs, information, interview, talk, and audience participation (quiz and game) shows.

A Sample Naturalistic Light Plot

Let us apply the principles and use the equipment mentioned so far to light an actual program: a simple, two-camera interview program for two stationary on-air talents, a host and a guest author. The approach presented here can be adapted for any program featuring stationary talent, including panel discussions, audience participation programs, and game shows.

Figure 4.13 shows a rough floor plan of the program. Notice that the subjects are in the same relationship to each other as Jay Leno or David Letterman would be with an interview guest. The host is seated behind a desk with a microphone positioned in front. The floor plan also shows the location of two cameras, a backdrop, and a stand for a program graphic—in this case, probably a book jacket, since we are interviewing an author.

The backdrop in this case is a stretch of black fabric extending around the perimeter of the studio, from the floor to the top of the set space. Alternatively, we could use a **cyclorama** (or *cyc* for short, rhyming with "bike"), which consists of a nonblack, continuous, seamless, opaque fabric similarly stretched around the perimeter of the studio, creating the illusion of endless depth. Cycs are often gray, and they may be lit with colored strip lights to create attractive backgrounds. In some studios, instead of fabric, the cyc is a smooth plaster wall, sometimes called a *hard cyc*. Cycloramas provide a smooth, unobtrusive background for the talent. When a black backdrop is used, only the talent, chairs, and desk are seen, since black is invisible on television.

Our goals in naturalistic lighting will be to flatter the talent and eliminate spurious shadows that might distract the viewer from the talk. First, we determine which camera will capture shots of just the guest in close-up (the guest's

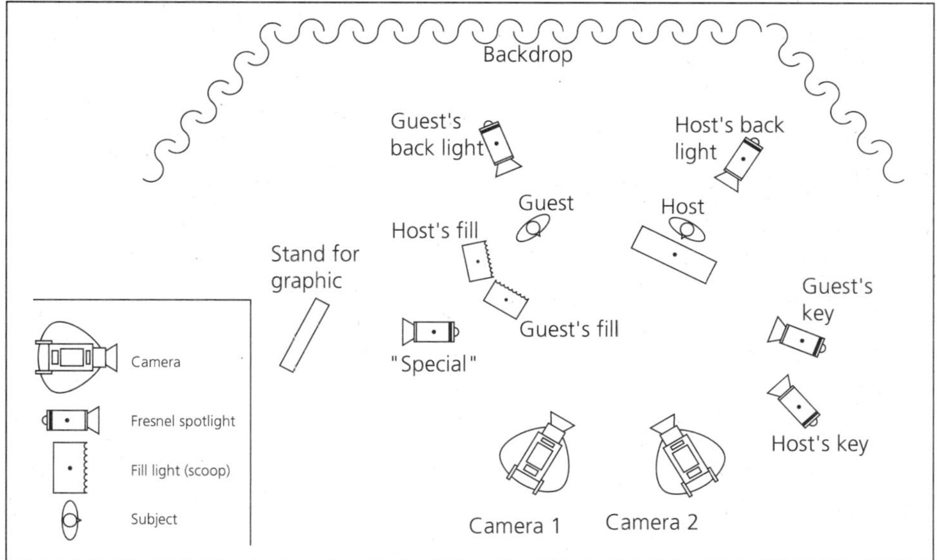

Figure 4.13 *A floor plan and lighting plot for a simple, two-person, stationary-talent interview program.*

one-shot) and which camera will carry the host. After seating the guest on the left, we decide the guest's one-shot will be carried by camera 2, as shown in Figure 4.13. We select camera 2 as the guest's principal camera because camera 2 is positioned to get a fuller front-face shot of the guest than camera 1, which can get only profile or three-quarter guest shots. For the same reason, we use camera 1 for the host's one-shot.

This arrangement is called *cross-shooting* because the imaginary lines from the cameras to their assigned talents cross each other. If we had used camera 2 for the host and camera 1 for the guest, we would have been *parallel-shooting*. Parallel-shooting, which provides profile and three-quarter images of each person, would have given us less intimate images. For an interview program, cross-shooting is unarguably the better choice.

Now that we know where the cameras are with respect to the talent, we can set the lights. The key lights, those with the most directional and concentrated beams, are the lights that are hardest on the talent. The guest's key light is positioned above and in front of the guest. It angles downward at roughly a 45-degree angle. It is also set at about a 45-degree horizontal angle from the imaginary line between the guest's head and camera 2. The same is done for the host's key light.

The 45-degree angles are a rough rule of thumb, a starting point. The vertical angle simulates sunlight, but, after all, the sun changes its angle as it passes through the sky. Therefore, we can feel free to position the lights to maximize the flattering effect we are trying to achieve. For example, if the guest is an attractive actress, it makes sense to go for the beauty shot. We may start with the 45-degree vertical angle and see what improvements we can make by departing from that angle. Bone structures differ from person to person, as do program purposes, so it's a good idea to experiment and check the effects by looking at the program monitor.

As the brightest source of light, the key light casts the sharpest, darkest shadow on the opposite side of the guest's face (especially below the nose and chin and onto the neck). To soften this shadow, we position a fill light on the opposite side of the talent, also at horizontal and vertical angles of about 45 degrees. The fill light is more diffuse and less intense than the key light. Therefore, we may need to place it closer to the talent to achieve the right intensity. The purpose of the fill light is to reduce unnaturally sharp shadows made by the key and to illuminate the rest of the guest's face. Together the two lights should provide a nicely sculpted image for the camera.

Finally, to create the illusion of depth and to separate the image of the talent from the background, we use a **back light** to illuminate each talent's back, especially the head and shoulders. We do this by positioning a light directly behind each talent, aiming toward the camera but tilted down at about 45 to 60 degrees from the horizontal plane. It is common to tilt the back lights down toward the floor at an angle as great as 60 degrees to reduce the chance of spilling light into the camera lens. Since the back light is aiming in the direction of the camera, the extra angle is justified. To further guard against sending incidental light from the back lights into the cameras, we will use spotlights with barn doors. (If you have a choice, *never use a scoop light or broad as a back light.*)

This arrangement of key, fill, and back light on each talent constitutes a **three-point lighting** scheme, and it completes the basic lighting job. Three-point lighting is the bread-and-butter arrangement for stationary talent in video production. It provides a natural, aesthetically pleasing presentation of the subject. Notice the triangular relationship of the key, fill, and back light. Notice also that each talent's principal camera (the one that gets the subject's one-shot) is centered between his or her key and fill lights.

As Figure 14.3 shows, we have used a three-point lighting scheme to light each talent separately; that is, we have six lighting instruments. It is possible that with slight adjustment, one set of lights may be enough for both talents, since the guest and host are stationary and close together. The advantages of using only three instruments would be that we would use less electricity, set up in less time, and provide consistency in light locations (since both talents would receive their key light from the same place). However, disadvantages might include a lack of coverage if the guests shift around and an increased risk that the host will cast a distracting shadow on the guest if the host moves. On balance, these disadvantages generally outweigh the advantages, so the professional approach is to use separate key, fill, and back lights for each talent whenever possible.

To determine whether the lighting is satisfactory, examine the line or program monitor, which should be checked for each camera before show time. Under no circumstances should you trust the way things look to your eyes. When you check camera shots on the monitor, make sure you have acceptable coverage for the talent and the set. Look for *specular highlights* (areas of extreme brightness) or unwanted shadows.

We are not finished with our lighting job, however, until we have lit the book jacket for an insert shot during the interview. The light for the book jacket is known as a *special,* since it performs such a singular function. Figure 4.13 shows the location of the special to the right of the stand, following the same rule of 45 degrees described for the talent. We could have placed the light for the graphic on the other side of the stand, but that might bounce re-

Eliminating Unwanted Shadows

Let's assume that after setting up the three-point lighting scheme shown in Figure 4.13, we find unwanted shadows on the backdrop. How can we eliminate them? There are various solutions:

- We can wash out the shadows by shining additional lights on them. This solution is not usually recommended, though, because it uses additional lighting instruments and more power.
- We can reposition the lights, placing the key lights higher than 45 degrees, causing the shadows to fall on the floor outside the range of the camera. However, this is often murder on the talent, since it lights them from a higher angle, elongating shadows on their faces, especially below the nose and eyebrows.
- The best solution in most cases is to move the talent farther from the background and then reset the lights.

Usually unwanted shadows can be avoided by placing talent no closer than six feet from the backdrop or walls of the set. Since most people are less than six feet tall, this is a good rule of thumb, even for standing talent, especially if you keep the angle of attack of the lights at around 45 degrees.

flected light from the graphic onto the set, causing unnecessary glare or hot spots. We might also have placed the special light directly over the camera, but that would have increased the risk of bouncing unwanted light from the book jacket into the camera lens. Instead, our placement of the special light sends unwanted reflected light off the set and away from the camera, thus lighting the book jacket without causing unnecessary problems.

Finally, if we used a cyclorama as background, we might wish to light it with some scoops or colored strip lights. (Figure 4.13 does not include this option.) Whatever lights we might use on the backdrop and other peripheral set areas, they should be separate from the lights used on the talent areas. The reason is that sets and scenery are usually more reflective than talent and costumes, so the intensity of light cast on them must be kept lower. Further, since talent action is where the main interest most often lies, the talent generally should receive the most light. Lighting set and talent areas separately becomes even more important when performers move around, since it becomes more difficult to keep annoying shadows out of camera range.

Beyond the basic lighting plan just described, additional lights, called *sidelights* and *kickers*, may be used at certain times. **Sidelights** are directional lights coming from the side of the subject, providing additional fill and shape to the talent or object being pictured. **Kicker** lights, which are positioned a bit lower and more toward the back, serve to emphasize the edges of the subject. One common application of such lights is to highlight a model's hair in a shampoo commercial. Such lighting may also be used in a dramatic presentation to show the outlines of a villain committing evil acts; in these instances, the key and fill lights may be dispensed with entirely.

DEPARTURES FROM NATURALISM

Naturalistic lighting, as we have described it, provides the **base light,** that is, the light needed for basic visibility. It also supplies a sculptured, attractive look to subjects and a sense of depth for the talent and sets. However, at certain times departures from or additions to naturalism are justified. In this section, we consider some of the more compelling reasons to break from naturalism.

First, video, more than any other visual entertainment medium, relies on the close-up for emphasis. The close-up, as one network lighting designer put it, is the "money shot." Even sports programming now employs longer lenses and image stabilization systems to get more close-ups. In fact, some professionals believe that even if the advent of larger screens and higher resolutions causes a shift to a more cinematic style, using more panoramic and wide-angle long shots, the close-up will continue to be the most important shot in video. Hence, it is the appearance of the talent in close-up that is often the primary consideration. Often, if the talent's appearance is improved by deviating from naturalism (for instance, by departing radically from the standard 45-degree angles), the director may want to do so.

Second, the standard naturalistic angles can be hard on talent. While there are twenty-four hours in a day, only a few of them are the "magic hours" that photographers and video producers talk about—those hours when natural light comes from low-angle directions. It is sometimes desirable, therefore, to use lower lighting angles to get more pleasing effects. Although lower angles do cause more shadow problems, these can usually be solved by careful placement of lights and enlisting the help of the set designer and director. For example, if a low key light used to improve the talent's appearance causes an unwanted set shadow, the shadow may be hidden from view by the talent's own body.

Third, people don't always want to see only what is "normal" when they watch video. Normal is what is out the window, but special is on television. Therefore, in many cases, people on video are made to look better than they might look in person, just as skies may be made to appear bluer. In short, everything may be made more attractive (or even uglier) than normal to drive home an emotional point.

Additional departures from naturalism may be dictated by the program genre. A horror story may use multiple shadows, lighting from below, and bizarre colors, intensities, and other special effects. An outer space saga, not surprisingly, may call for "other-worldly" lighting. Light and shadow can be used for dramatic emphasis, such as when a climactic point is reached in a video drama and the lights are killed everywhere except in one spot. Some videographers may even resort to unnatural lighting schemes just to give a distinctive look to their talent or show. All such lighting decisions should be based on the meaning of the work.

Finally, from a purely tactical perspective, some production formats dictate against using naturalistic lighting schemes, especially those featuring moving talent in multicamera settings. Programs in this category commonly include situation comedies, sketch variety series, dramas, and soap operas. These programs intercut shots from several cameras in different locations, perhaps all shooting the same performer in the same performance area from different angles. Seeing cuts of the same set area or performer from different camera angles may be too jarring if highly discrepant light intensities are used. For this reason, instead of naturalistic light, a "flatter" lighting scheme is warranted.

Flat Lighting

The renowned cinematographer Karl Freund introduced a **flat lighting** technique for the "I Love Lucy" series in the 1950s. The program's producer, Al Simon, approached Freund and asked if he could "film with three or four 35 mm cameras, in front of an audience, like a stage play, without stopping, resorting to retakes only under the most dire circumstances." Freund responded, "You can't do that. Every shot requires different lighting. You couldn't photograph three or four angles at the same time and come up with a decent piece of finished film" (Andrews, p. 49).

Despite his objections, Freund developed flat lighting to deal with the problem. To cover talent action over almost the entire set, Freund made the light intensity uniform over the total area at all times. Set illumination was mostly from overhead. The only light from a lower level came from a portable fill light mounted just over each camera. With overhead lighting, cameras could dolly over the studio floor unobstructed. Cuts could be made from one camera to another without creating jarring differences in light.

Freund's system of flat lighting is still in use today, as evidenced by its use for studio scenes in programs such as "Friends" and "Coach." In multicamera productions, the most compelling reason to use flat lighting is to minimize light level differences when cutting between cameras shooting the same scene from very different positions. However, this does not mean fill lights are the only acceptable tool. Spotlights may still be used for acting areas and smaller set spaces.

MOTIVATING LIGHT SOURCES

Some scripts refer to special light sources indicating time or place, called **motivating light** sources. Motivating sources include different types of natural light (early sunrise, midday, sunset, cloudy daylight, moonlight) or artificial light (lamplight, candlelight). More exotic effects, such as lightning, are also motivating sources. The illumination provided in simulating such sources should be convincing, but should not draw attention away from the main action. The following sections offer techniques for creating some of the more common motivating light effects.

Sunlight and Sunset   Aim a highly directional beam—uncolored light from an incandescent lamp—onto an acting area or an essential piece of scenery. It should be the brightest light on the set. Use amber gels of varying grades to indicate sunset, and vary the angle of attack to match the angle of the setting sun. If you use more than one instrument for this effect, they should be parallel. The best instrument is a powerful concentrating instrument mounted far from the set. A follow spot or a high-powered spotlight is a good choice. If you use a projector, you can control the spread of the beam with a funnel, which is a cylinder fitted over the light to control unwanted spill. Use dimmers to control changes in brightness.

Moonlight   Moonlight may be rendered as sunrise and sunset are, except for adjustments in intensity and color. Adding a steel blue gel with a color temperature comparable to moonlight will normally suffice. Aim the motivating spotlight at the acting area. On the other side, use spotlights in the flood-beam setting for fill, outfitted with a lighter blue gel at lower light intensity.

General Daylight   Regular daylight requires a virtually shadowless, broad, general source. It should appear as light coming in from all directions. Avoid directional light. Intensity should be uniform. Color regular incandescent fill lights with light steel blue gels. A large reflecting screen can help create the look of light entering through a window, as can several small floods equipped with gels of the same color. You can imitate changes in daylight from dawn to sunset by progressively adding light blue- to violet- to steel-colored gels. To achieve the look of an extremely hot day, simulate general daylight with light amber.

Fixture Lighting   In most cases, lighting fixtures that appear on camera, such as table lamps in an interior scene, are used strictly as motivating lights. Unless they contribute a dramatic effect, they are not used as major lighting devices for the acting area. To reduce hot spots, keep the wattages in the fixture lights to a minimum: 25 to 40 watts are usually sufficient to add a necessary accent to a scene. Light the acting areas from hidden sources.

On the other hand, fixture light can occasionally add useful illumination to a scene. Such lights are called *practicals*. An example of a practical light would be a lamp over a music stand to help the performer see the music. Frequently the location and reflective power of the scenery will determine the use of such fixtures. To control intensity in these cases, use a dimmer. Another way to control fixture light is to cut away the back of a lampshade (on the side away from the cameras) to increase the intensity of light falling on a piece of scenery without letting the light spill directly into the camera's view.

LIGHTING IN MORE COMPLICATED CASES

How can you apply the basic principles of lighting to more complicated productions, such as a sitcom or soap opera? Lighting for drama or comedy is a challenge, because such programs feature talent moving through and interacting with the set, and they often require the use of motivating lights as well as lighting changes to indicate different times of day. Here are some guidelines.

First, meet with the director and the rest of the creative team, including the producer and the scenic and costume designers, to study the script, floor plan, and talent needs. Determine the subject matter, treatment, and tone of the production. Take inventory of the available lighting hardware, and become familiar with the space you will light.

From the script and floor plan, determine where the cameras and talent will be, and begin forming a rationale for lighting. Consider how you will provide base light. Then work on providing additional special lighting needs, including motivating lights and any special instruments. For example, the locations of doors, windows, and set pieces will help you decide where to place instruments to provide sunlight, moonlight, and fixture lights. Let the tone of the program dictate the mood you will try to convey, such as whether the lighting should be festive or somber.

Arrange for unobstructed access of the cameras to all of the locations where they need to go to maximize shot variety. Note all camera angles, and light accordingly to avoid spilling light into the camera lenses. Also note where key speeches or action take place, and make those areas a central focus of your lighting design. In areas where talent comes close to the set, such as doorways, fit lights with funnels or barn doors to shape and limit their throw and to reduce or eliminate unwanted shadows. Light the set areas separately from the talent positions to control shadow problems. Remember, it is generally best to keep light intensities on sets lower than those used for talent, because sets tend to be more reflective than talent and the talent is where you want to fix the audience's attention. If appropriate, consider using flat lighting for the acting areas to keep intensities uniform over that space, thus reducing jarring differences when cutting from one camera to another.

When the light plot is complete, use the rehearsal to observe the peformers as they move about the set. As they move, judge how they look on the cameras intended to cover them, both in the master shot and in close-ups. Fix any dead areas where lighting is dark and uneven. *Spike* (mark) spots where performers deliver speeches with bits of masking tape. Observe the talent on these spots, and improve the lighting if possible. As the action is covered, keep watching for spots that may be too bright or too dull. Use both a light

I N D U S T R Y
voices

Lonnie Juli
Lighting Director, "The CBS Evening News with Dan Rather" and "48 Hours"

Q: What's your background? How did you get here?

A: I'm a native New Yorker with a bachelor of fine arts degree in theatrical production from Hofstra University. I specialized in theatrical lighting. I had worked in and around the theater and TV industry. I was one of those theater designers who swore I'd never wind up in television, and here I am.

If there was anything that helped my success, it's that I had a slightly different take from those trained technically. We were trained artistically. We didn't worry about f-stops or color temperature until later on, and then it was a matter of adapting that sensibility to the limitations of the video system. One of my teachers at Hofstra always felt that if you weren't going to learn to do almost everything in this business, you might as well get ready to sell shoes. The openings aren't always there when you're ready for them, so you have to be ready to do different things. Aesthetically, I was trained well, and it paid off.

Q: Does that mean you take a holistic approach and that lighting design is not an isolated component in your work?

A: Lighting is not an isolated part of production. Everything we do in this craft is collaborative. Even in a one-person show, there is still a writer and someone to pick out the stool the talent sits on. Everything we do affects everything else, and we try to coordinate that.

Q: How would you characterize the work you do as a lighting designer?

A: My title is *lighting designer* or *lighting director* on the evening news, but the business we're really in is photography. We're taking a three-dimensional scene or subject and rendering it photographically in a two-dimensional medium called television.

One of my retired colleagues always used to say, "It's just photography," and he was right. It really just is, and if you understand the limitations and capabilities of the system, you have more success.

Q: How do you get brought into a project? Describe the process.

A: Like the Bible, it all starts with the word. First, there's a script or a concept. And with the direction of the director, you interpret the script in terms of establishing mood and time. Then you analyze the script in terms of individual scenes if it's arranged that way. Then you decide, improve, modify, and update your vision, always subject to the director's vision about how each scene will look and how it will feel. I say "feel" because, while it's a visual medium, you try to create a tactile sensation.

Then, depending on what equipment you have, or can get, or invent, or are waiting to have invented, or can fake by yourself, you create *not* a completely naturalistic rendition of the scene you want but the stimuli that convince viewers that they're experiencing what you want them to experience. If you break it down and analyze it, a lot of our stuff is scientifically inaccurate, and much of the time it's *purposely* inaccurate, because if it were accurate, people would not notice the things we want them to notice.

If you've ever read Shakespeare, you realize that his audience was a very informed audience. There was not a whole lot of stage direction. The audience knew what to expect, and they gave themselves freely to it. But modern audiences are more resistant, so you sometimes have to trick them into arriving where you want them to be.

Q: Is that a cultural difference, or is it that today's audiences are so inundated with other media?

A: I think we're inundated and overstimulated by media. Also, our audiences today are more visual, which is good for people in my business. What I do is try to present the audience with stimuli they would expect if they were in the setting we are trying to create. That's why directors of photography and special-effects directors on film can get people to believe they're seeing characters floating in space or walking on Mars—places where we don't even have a good idea of what they look like. But we hope we know what people would *expect* them to be like.

Q: Isn't it a bit presumptuous to say you know what the audience expects to see?

A: It goes back to what we were taught about stage direction. You direct from the audience's seat, from the audience members' perspective. So you ask yourself, "If I were sitting in the audience, and I was going to Mars, what would I expect to see?"

If you're shooting a winter scene in the summer on styrofoam snow, you have to ask yourself, "OK, what makes me feel like winter?" Then you go with low-angle light, long shadows, certain cooler color temperatures to make the shadows look cooler. Are winter shadows really that way photometrically? Maybe not a whole lot. But do we expect them to be? Yes! So we cheat. We cheat all the time.

Q: What changes in the industry have you noticed over the last ten years?

A: Technology has changed a lot. Years ago, the only way we were able to get daylight artificially was with what we called long-arc lights, carbon arc lamps that are used in searchlights. It wasn't that accurate, and it had to be filtered somewhat. Since then, several other technologies have come out, most notably HMI lights. HMIs vary a little bit in critical applications—you still have to make adjustments, using color filters—but they work much better. They let you create hard shadows similar to sunlight.

Other new technologies include robotic light fixtures that can pan, tilt, iris, change color, and focus from a remote location. Whereas in the past we had to spend a lot of time on ladders, now we can program hundreds of individual cues into a light. Varilite is a popular product name for such instruments. We also have more projection options available for backgrounds than before. But projection technology is not as important as it used to be because of matte photography and newer chroma key technologies like Ulti-matte.

Q: Are the matte techniques all done in postproduction these days?

A: Some, but not all. For the recent election coverage, for example, CBS used a virtual reality set where all the scenery was generated in real time.

Q: Are you as a lighting designer upset with that change?

A: No, I think we did well with it for an early point in such technology. I got to offer the folks who worked on it some key information about how light would react in those situations. I worked with the virtual set designers so that the pieces they were generating looked as if they existed in the real world, not like a Nintendo game.

Q: I also understand that the new chip cameras have digital circuitry that enables you to do cosmetic things with faces.

A: Sure. Those cameras have analog output, but they have two digital detail circuits, one that enhances detail from light to dark and one that samples skin tones. This allows you to limit detail, helping eliminate unwanted lines from the face. In film and cinematography, we were using silk stockings over the lenses, stockings the same color as the face to soften those tones. So the effect is not new to me as a lighting director. While it's a new technology, it is not a new effect.

Q: What are the major differences between lighting in the studio and in the field?

A: The biggest difference is that in a studio there's more control. In remote lighting, you have scrims, flags, and stands, and you have to deal with power supply considerations, weather, sun direction, rain, cloudiness. You must cope with adversity. By contrast, in a studio, everything you need is there, hanging, waiting to be used.

I love working remotes because you live or die by your wits. Sometimes you say, "There's absolutely no way to hang a thirty-pound light over there," and a half-hour later, there's a thirty-pound light hanging there. What I pride myself on is that my standards are just as high in the field as in the studio. We have a sign in the newsroom that says, "Adequate is inadequate," and that's absolutely true. If you're not going to do your absolute best, if you're not going to make people look better than they look in real life, if you're not going to enhance the experience of the viewer and assist in telling the story, then don't bother.

meter and the appearance of each camera shot on the program monitor to make adjustments. To make adjustments, you can move lights, tune them, or cover them with gels or scrim to color or soften them. Use all the tools available to you to make the lighting scheme convincing and appropriate to the subject matter, tone, and goal of the production.

When you are satisfied that you have a light plot that does what it is supposed to do, log the location of each instrument on a lighting grid. Note dimmer levels. If any lights are grouped together, note them. This logging of the light plot will be helpful if lights burn out and need to be replaced quickly. The light plot should be complete enough so that if the lighting designer is unable to report to work, the crew member taking his or her place can set lights for the production by referring to the floor plan, light plot, and log.

FIELD LIGHTING

Lighting in the field presents unique challenges. You may decide to use the available natural light if the shoot is outdoors. You may decide to use the available artificial light indoors. Or you may choose to supplement the location's light with lights that you bring to the shoot. If you choose to augment the available light, you will need access to a power source. This may mean simply bringing a light kit and extension cords and using wall current at the location. But if wall current is not available, you may need to bring a generator. If you use a light kit, bring extra lamps in case some burn out.

Procedures

Scout the location in advance, at the same time of day as the shoot. Test outlets you may wish to use to make sure they work, and confirm whether they can handle your power needs. The purpose of scouting is to eliminate surprises.

If you decide to use available natural light outdoors, note the direction of the sun, and keep in mind that it changes as the day progresses. Weather conditions can also change, and this can cause continuity problems if you shoot scenes on different days. If you use artificial light to offset differences, remember to use the proper light for a color temperature match. If possible, shoot at the same time any scenes that will be viewed together so that the sun will be at roughly the same place. If you shoot scenes meant to be viewed together, you may also need to wait for similar weather conditions to get what you want.

If you cannot get the cooperation of nature, you may wish to consider changes in the script to accommodate shooting conditions. Obviously, flexibility is a virtue in location shooting. Either way, leave enough time in the production schedule to deal with such contingencies. If you are waiting for a rainy day to do a crucial scene, plan the rest of the shooting schedule so that you can get other things done while you watch the weather report.

If you shoot with indoor lights near a window with sunlight streaming in, shield the window to keep it from throwing off the color temperature, or cover the window with the proper filter to correct for color temperature differences. Similarly, when shooting outdoors, be aware that extremely bright sun can cause contrast ratio problems, making your talent look washed out on the sunward side and underexposed or dark on the opposite side. To correct such problems, you can shield the talent by blocking off direct sunlight with large, sturdy, opaque cards called *flags* or by moving the shoot to the shady side of a building. Or you can use reflectors to fill in and reduce harsh shadows on the performers' faces opposite the sun. Seek a solution that brings the picture into a tolerable contrast range.

Remember that the canons of lighting do not change from studio to field. Control of light and shadow is still the goal. If you are striving for naturalism, light from above, and control intensities and shadows. Finally, be sure to set white balance for the camera(s) you are using each time you change location or lighting so that the color rendition of your video image is acceptable and consistent from setup to setup.

Equipment

Field lighting requires the use of some unique equipment. Among the tools used in the field are portable lights; light stands; booms, clips, and braces; and flags and reflectors.

Portable lights offer several advantages over studio lights when doing remote video shoots. First, portable lights are lighter in weight than studio lights. Second, they use normal wall current and do not require any special adapters to be plugged into normal wall outlets. Third, they come in kits that provide custom-designed stands for the lights to make them easy to use. In fact, much of the equipment designed for field use is also used in studios because it is convenient. Figure 4.14 displays some of the basic elements of a field lighting kit.

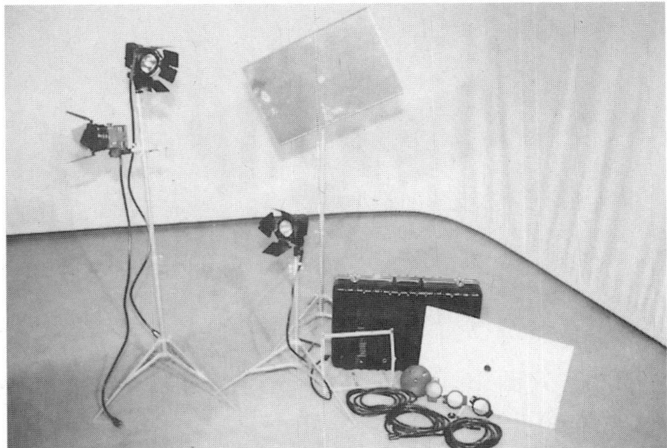

Figure 4.14 *Some of the typical contents in a field lighting kit: portable lights, light stands, reflector boards, extension cords, carrying case. Also visible are a frame for cards (the square item leaning against the carrying case), some filters (in front of the white reflector board at right), and a clip (attaching the light at the far left to its stand).*

Light stands, depending on the brand, enable you to raise lights to heights of up to fifteen feet or lower them to about two feet. The stands are very lightweight and retractable for easy storage. For especially tight spots, clips and braces are available for mounting lights in hard-to-reach places. *Booms*—long, movable arms—can be rigged to make lights available in set areas where you don't want the light stand to be visible in a shot. Flags, as mentioned earlier, are clip-on black cards that attach to stands that you can place in an area to block light from entering the set or camera. Conversely, you can use reflectors to increase the effects of sunlight on a subject by creating fill light for the side of the subject opposite the sun. Some reflectors are shaped like umbrellas (those made for artificial lights); others are simply flat cards with a foil or highly reflective white surface.

A special outdoor light made with metal halide material is worth noting. Called an **HMI lamp,** this type of instrument provides light in the color temperature range of about 5,000 to 5,600 K, making it suitable for simulating daylight, which spans from about 4,500 to 6,200 K. In addition, HMI lamps do not get as hot as incandescent lights do, chiefly because they work in concert with a *ballast* that serves to limit current through the lamp. HMIs are most often used in high-end film productions.

KEY TERMS

light plot *(54)*	cyclorama *(67)*
lighting grid *(56)*	back light *(68)*
dimmer board *(56)*	three-point lighting *(68)*
Fresnel spotlight *(57)*	sidelight *(70)*
light meter *(61)*	kicker *(70)*
waveform monitor *(61)*	base light *(70)*
vectorscope *(63)*	flat lighting *(71)*
naturalism *(65)*	motivating light *(72)*
key light *(66)*	HMI lamp *(78)*
fill light *(66)*	

QUESTIONS FOR REVIEW

1. List the standard items of lighting equipment found in a television studio.

2. In addition to providing visibility, what can lighting do to enhance a video presentation?

3. How can an understanding of naturalism guide the lighting designer in planning lighting designs for television?

4. What aesthetic considerations and physical constraints justify departures from naturalism in planning lighting designs?

5. Sketch out some ideas for lighting an indoor basketball game. What could you do, for instance, to reduce problems caused by the highly reflective floor? How would you deal with the color temperatures of the lights provided at the arena?

5 Using the Camera

hen in use, the video camera is more than an imaging device. Besides the lens and electronics that enable the camera to register an image, the camera has additional mechanical and electronic equipment that allows you to control its position and movement and adjust the image composition. These features range from mounting devices such as pedestals and body braces to viewfinders, headphone jacks, and even prompting devices for cueing talent.

In addition to understanding the camera components themselves, a video production person needs to understand the specialized language used to talk about them. "Truck left," you will hear. "Get an establishing shot and then zoom in for a close-up." In keeping with our general approach, though, this chapter will cover more than the devices and the language with which you need to be familiar. We will also discuss the design elements and aesthetic considerations that go into picture composition. The topics in this chapter include:

CAMERA MOUNTS AND CAMERA MOVEMENT

tripods, wheels, and dollies • studio pedestals • mounting heads • cranes and jibs • body and vehicle mounts • the language of camera movement

ELECTRONIC COMPONENTS OF CAMERAS

viewfinders • zoom and focus controls • tally lights • intercoms and headsets • cue cards • TelePrompters

BASIC DESIGN ELEMENTS OF PICTURE COMPOSITION

line and shape • texture • pattern • color and contrast • depth and perspective • placement of key elements • balance • unity and variety

PICTURE COMPOSITION IN VIDEO

framing • using the z-axis • focus • camera angle • types of camera shots

A FINAL NOTE ABOUT PICTURE COMPOSITION

CAMERA MOUNTS AND CAMERA MOVEMENT

The equipment that physically supports the camera provides stability, ease of operation, access to different locations, and smooth, controlled motion in the video image. Using appropriate mounting equipment makes camera operation safe and easy and helps you achieve the program you want. The following sections cover the basic types of mounts, the movements they facilitate, and the language used to describe camera movement.

Tripods, Wheels, and Dollies

The camera may be mounted on a **tripod,** a three-legged support with adjustable extensions, usually tipped with spikes or rubber ends for stability (Figure 5.1). Each pod can be extended to a different length to accommodate uneven terrain. Tripods can be placed on a *spreader,* a device that limits the spread of the tripod legs so they remain stable under the weight of the camera.

For leveling the camera on uneven ground, most field tripods come equipped with a bubble level similar to that used by carpenters. If you are using a tripod without a level indicator, you can still level the camera by aligning the top or bottom edge of the image in your viewfinder with a true horizontal in your set (such as the horizon) or by dropping a plumb line into your shot and aligning it with the left or right vertical edge of the image.

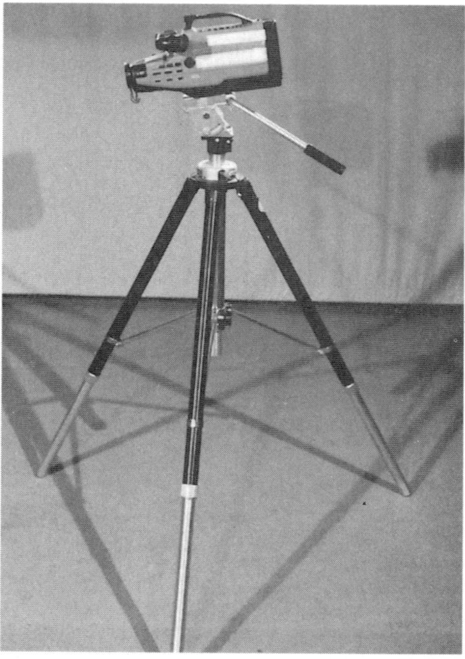

Figure 5.1 *A camera on a tripod.*

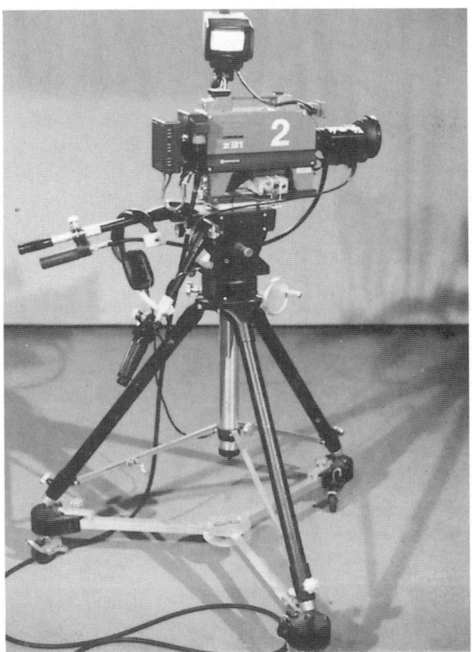

Figure 5.2 *A camera tripod placed on a dolly for easy movement in the studio.*

For easier movement, especially in the studio, tripods can be placed on wheels, using a *dolly* (Figure 5.2). Studio tripods generally have brakes on each wheel, and they often come with wheel covers that help to protect the wheels from obstructions. The need to avoid obstructions is a major consideration, even on a studio floor that appears flat and clear. Because television is a low-resolution medium, much of what the camera sees is shot in close-up, where even small aberrations in camera movement may be magnified many times in the picture. Wheel covers are helpful because they tend to push away obstructions such as cables rather than letting the camera roll over them. Still, it is best to have a crew member act as a cable puller.

Studio Pedestals

A more sophisticated class of camera mount is the **studio pedestal**. Studio pedestals are more versatile than tripods in that they permit you to raise and lower the camera while on air. The *counterweight* pedestal uses weights to maintain a safe center of gravity as the camera moves smoothly up and down. The *pneumatic* pedestal uses air pressure to raise and lower the camera (Figure 5.3).

Studio pedestals also have mechanisms that set the wheels parallel to one another. The direction of the wheels (or *casters*) is controlled by a steering ring indicating the position of the wheels (see Figure 5.3), making it easy to push the camera along a predetermined course. Casters also have brakes, as well as a cable guard that functions in the same way wheel covers do to keep obsta-

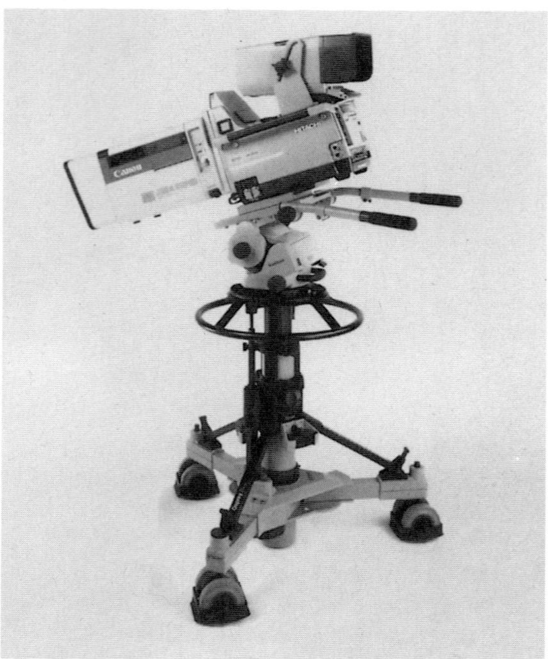

Figure 5.3　*A studio pedestal.*

cles out of the way. Some automated or robotic pedestals are controlled by computers.

Studio pedestals offer many advantages, but they also have some drawbacks. For example, counterweight pedestals are big and heavy, especially when loaded down with equipment. When fully loaded, they may be too heavy for a single operator to handle easily. They may also function poorly if used as mounts for lighter field cameras, for which they were not designed. Pneumatic pedestals are lighter and easier to move than their counterweight cousins, but they too have drawbacks, such as their tendency to leak air from their pressure units. This means the air must be replaced from time to time, which requires a compressor.

Mounting Heads

A **mounting head** (Figure 5.4) is used to fasten the camera to the tripod, pedestal, or other support device, enabling the camera to move in a controlled manner. Several major types of mounting heads deserve mention:

- *Cam friction head:* Often used with studio cameras, these heads use cylinders called *cams* to allow camera movement. Drag controls provide the friction necessary to keep the motion smooth and maintain the camera's shifting center of gravity during tilting.
- *Fluid head:* Suitable for field production because of their light weight, these heads use springs in heavy oil to provide drag control.

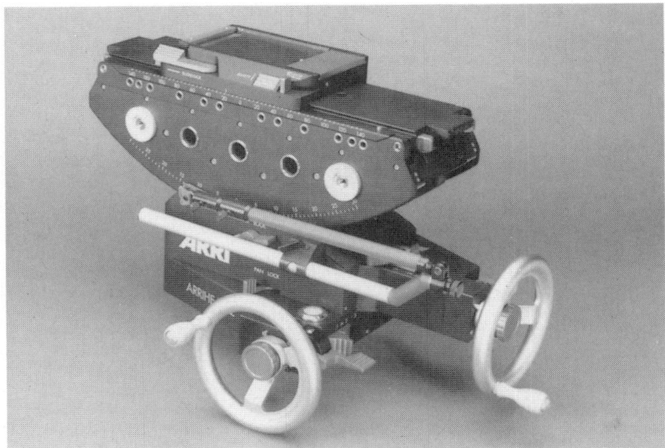

Figure 5.4 *Mounting heads are used to fasten a camera to a tripod or pedestal.*

- *Friction head:* Often used in field production, mainly on tripods, friction heads are equipped with a handle for moving the camera into a selected position. They are best suited for jobs requiring only limited movement.
- *Geared head:* Two geared wheels deliver extremely smooth movement, but this type of head is not suitable for fast adjustments.

Mounting heads can also be programmed to perform automated movements.

Cranes and Jibs

As Figure 5.5 illustrates, the studio **crane** features a counterweighted arm mounted on a heavy, wheeled dolly. The crane permits you to move the cam-

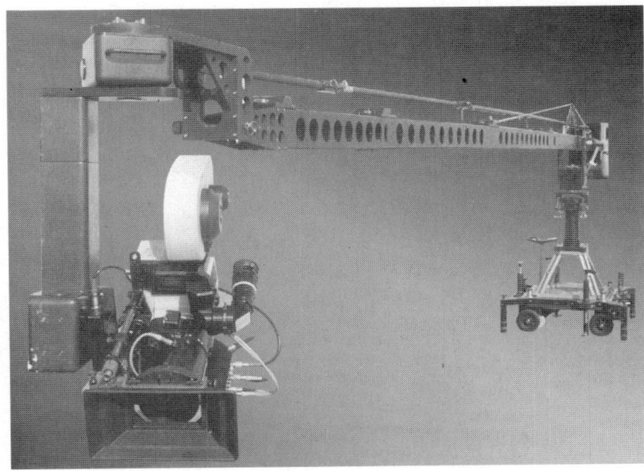

Figure 5.5 *A studio crane.*

era from roughly one to twenty-five feet above the ground or studio floor, depending on the model. With a crane, you can rotate the camera through a complete circle while raising or lowering it. This makes quite dramatic camera work possible. The crane allows you to cover large vistas of ground action, and it is very useful for *follow shots* (shots in which the camera follows a moving subject, such as a football play that moves downfield).

However, cranes have some disadvantages. First, they require a large space for use and storage. Second, at least two operators are needed, one to control the camera and another to move the crane. Third, the operators may need extensive practice together to master the shots you have planned. Finally, achieving the shots you want may require a separate monitor for each operator.

Since the advent of lighter cameras, the **jib arm** camera mount (Figure 5.6) has arrived to provide the same advantages the crane does, but with a single operator. With a jib, you operate from a ground location in front of a monitor, moving the camera by remote control.

Body and Vehicle Mounts

The move toward lighter cameras has also made it possible to mount cameras directly on the operator's body. **Body mounts** (Figure 5.7) can vary from simple, inexpensive models to sophisticated, expensive ones, but they generally fall into two categories: (1) shoulder mounts and (2) body braces that use a

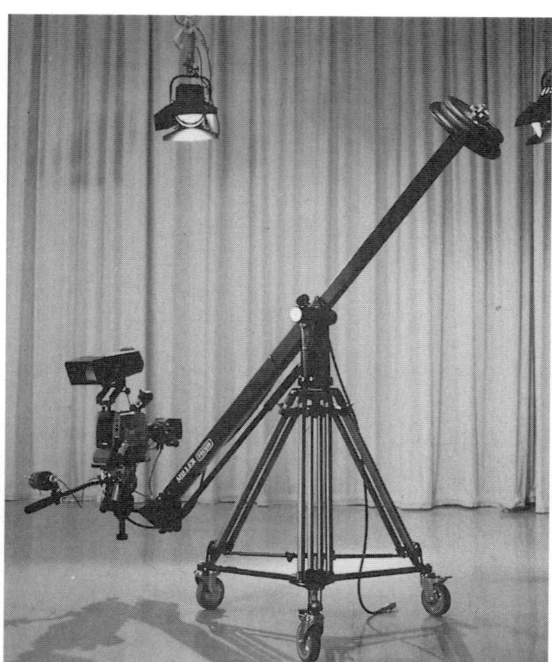

Figure 5.6   *A jib arm camera mount.*

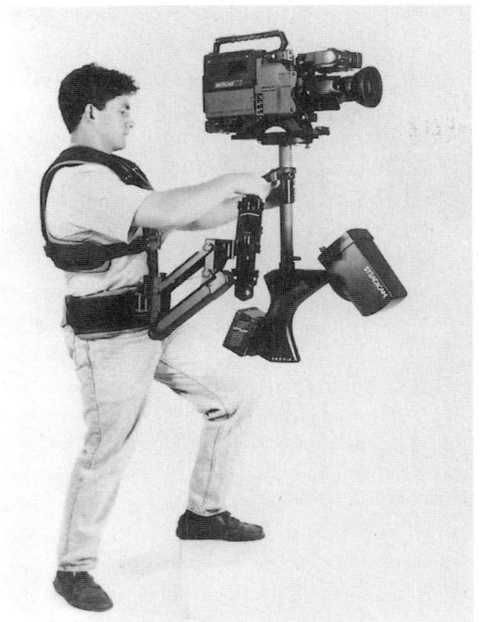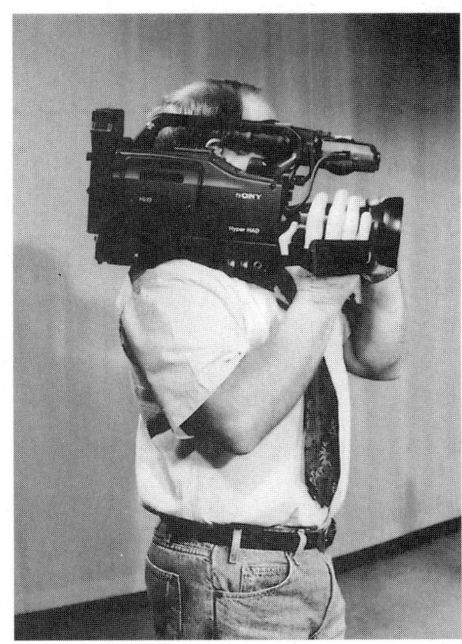

Figure 5.7   *Body mounts.*

harness or belt. Shoulder mounts place most of the camera's weight on the shoulder, freeing the hands for framing and focusing work. Body braces provide greater stability than shoulder mounts, but may restrict freedom of movement and may prove uncomfortable.

Among the most sophisticated body mounts are the *Steadicam* and the *Panaglide*. These trade names refer to body mounts that permit you to cancel out, within limits, the jarring motions of walking or running while shooting video. They use a spring arm and stabilizers to produce smooth motion. However, they are expensive, are quite heavy, and require a great deal of effort to use. Nevertheless, they allow you to get high-quality action and follow shots. If the camera needs to be attached to a moving object such as a car or a helicopter, special camera clamps and beanbags are available.

The Language of Camera Movement

The various camera mounts all serve to keep the camera steady while allowing it to move in certain ways. In fact, the essence of television, in addition to sound, pictures, and color, is movement. Unlike paintings or still photographs, video images change from moment to moment. Further, whereas a painter can paint alone, television is still a largely collaborative enterprise. For these reasons, it is essential that both directors and camera operators understand the jargon used to describe camera movement.

Table 5.1 describes and illustrates the standard types of camera movement. These are terms you need to know so well that they become automatic.

TABLE 5.1

Terminology for Camera Movement

Pan	Turn the camera from side to side on the mounting head while keeping the tripod or pedestal stationary. You *pan right* by moving the lens to the right as you look at the camera from the operator's point of view. Conversely, you *pan left* by moving the lens to the left from the operator's point of view.	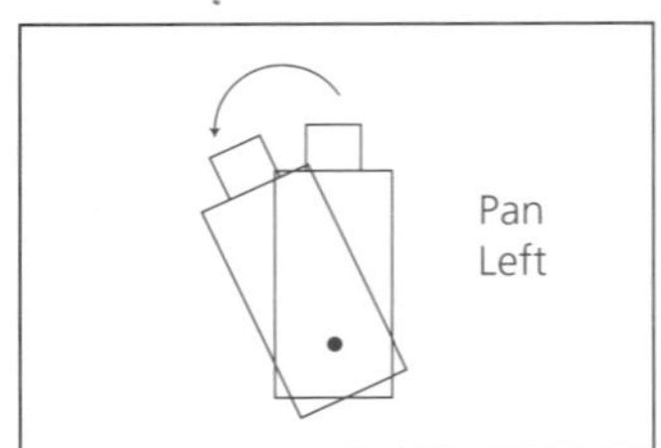
Tilt	Tip the camera up or down on the mounting head while keeping the mount stationary. You *tilt up* by moving the lens upward. You *tilt down* by moving the lens downward.	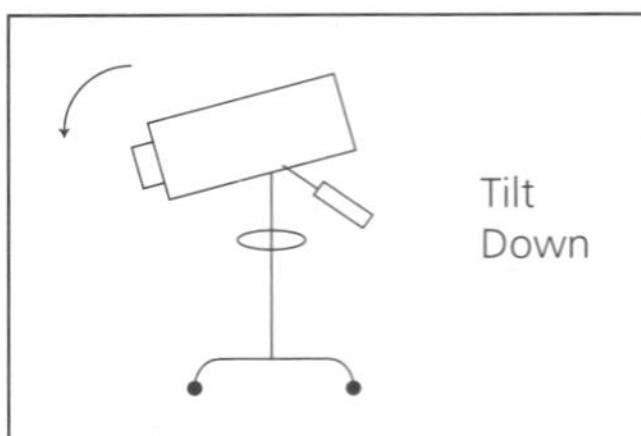
Cant	Take the camera off the horizontal so that the subject appears to be tilted on screen.	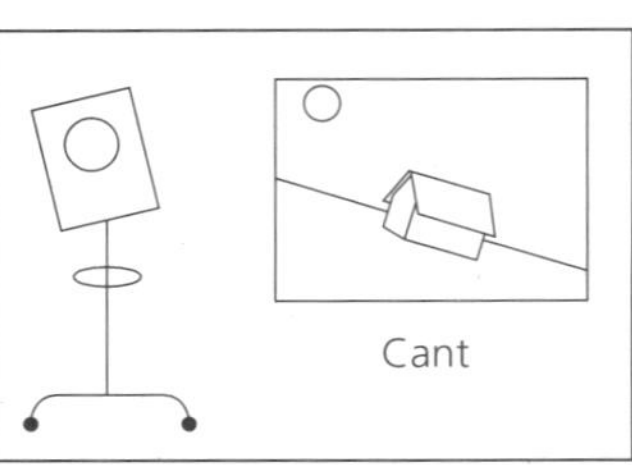
Zoom	Change the focal length of a variable focal length lens while keeping the camera stationary. *Zooming in* means moving the lens elements to a narrow angle of view, making the scene appear to move closer to the viewer. *Zooming out* means moving the lens elements to a wide angle of view, making the scene appear to move farther away.	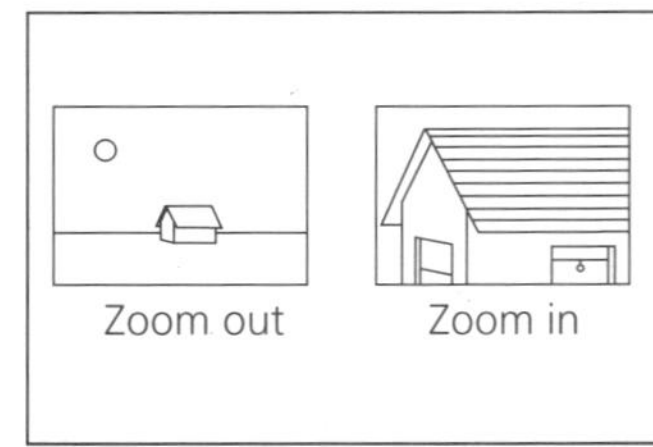
Dolly	Move the entire camera and camera mount toward or away from the subject being pictured. You *dolly in* by moving the camera and mount closer to the subject. You *dolly out* by moving the camera and mount away from the subject.	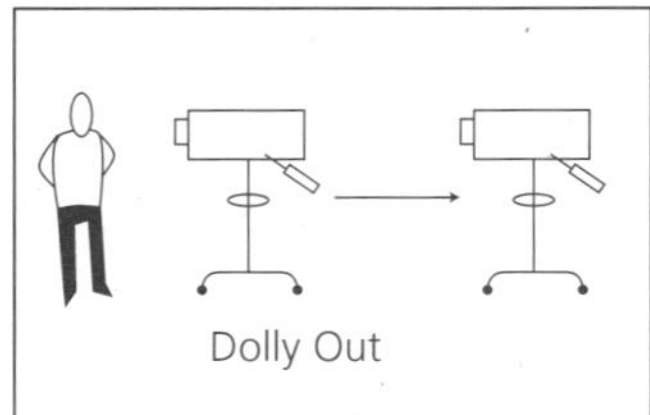

Terminology for Camera Movement (continued)

Truck
Move the entire camera and camera mount laterally with respect to the subject being pictured. You *truck right* by moving the camera and mount to the right, from the point of view of the camera operator. You *truck left* by moving the camera and mount to the left, again from the point of view of the camera operator.

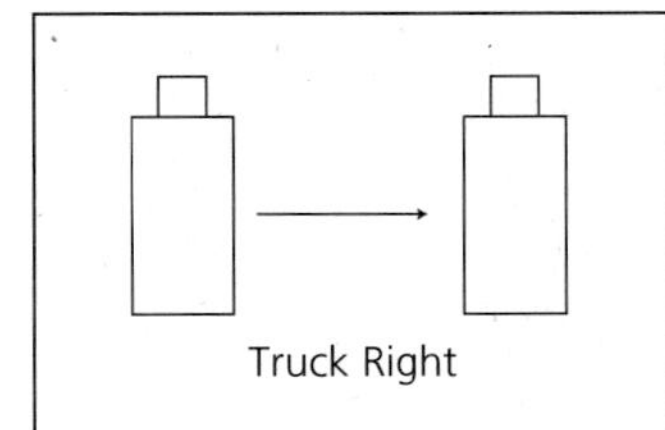

Arc
Combine the features of the pan and truck to produce a curving movement. The curve can be specified to be tight or wide, toward or away, or to the right or left.

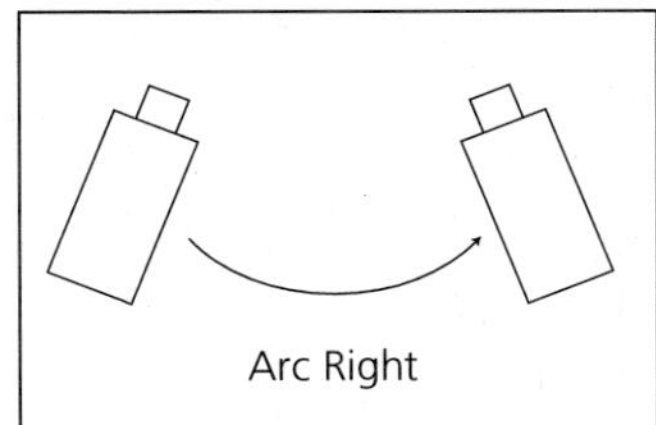

Ped
Raise (*ped up*) or lower (*ped down*) the camera on its pedestal.

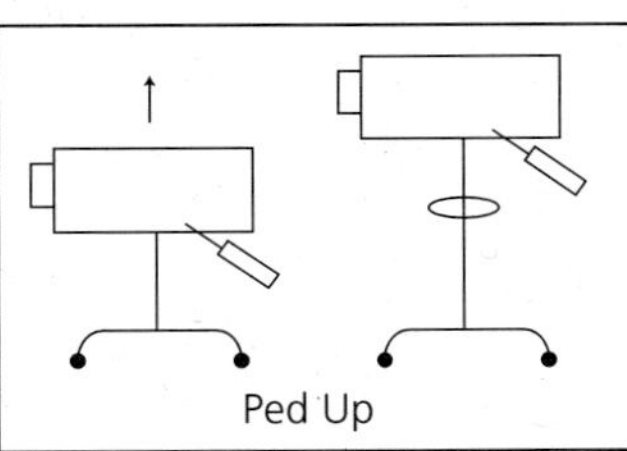

Crane or boom
Raise or lower the entire camera on a camera crane. You can *crane* or *boom up,* and you can *crane* or *boom down.*

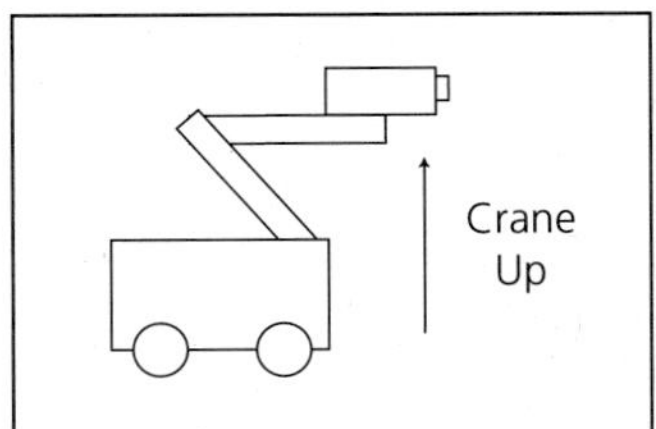

Tongue
Move the entire camera from side to side, like trucking but with the use of a crane. You can *tongue left* or *tongue right.*

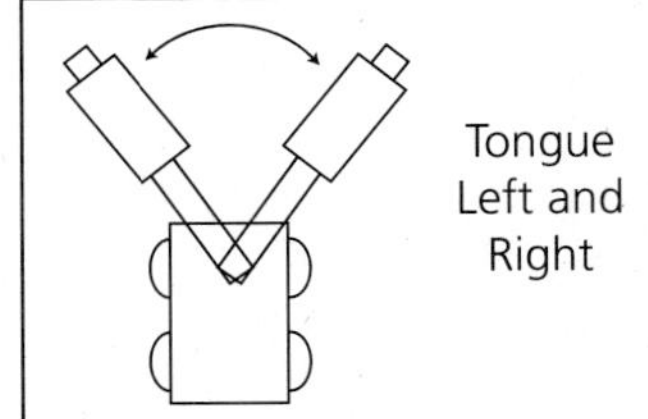

ELECTRONIC COMPONENTS OF CAMERAS

Viewfinders

The studio camera is equipped with a **viewfinder** (Figure 5.8), mounted on the operator side of the camera, to display what the camera is seeing. The viewfinder is a television monitor, usually black and white. Measuring about seven inches diagonally in studio models, it makes it easy to view with both eyes from a distance of about two feet. In addition to showing the operator what the camera is seeing at every moment, the viewfinder can switch or combine its image with those from other cameras, making it possible to see how composite images will appear on screen.

Viewfinders must be adjusted for contrast and brightness to be able to display optimally clear images. They should also be periodically adjusted to ensure that they display the same field of view the *line monitor* (the actual program monitor) does. Significant discrepancies can result in poor picture composition.

In field cameras, the viewfinder is usually a 1.5-inch monochrome screen with a magnifying eyepiece designed to touch the face (Figure 5.9). When used with the eyepiece, the screen is seen with one eye. In many cases, however, the eyepiece can be swung away to permit the operator to look at the screen with both eyes from a slightly greater distance. Field camera viewfinders also include status displays that indicate battery power, adequacy of the light level, and audio level, and offer reminders and warnings about white balance and tape mode.

Zoom and Focus Controls

Studio cameras usually provide zoom and focus controls on the panning handles (Figure 5.10). These controls permit you to adjust the framing and focus of the camera to compose the desired picture.

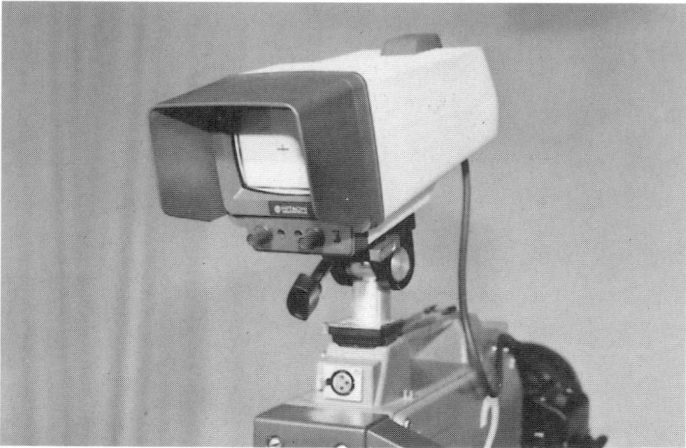

Figure 5.8   *A viewfinder of a studio camera.*

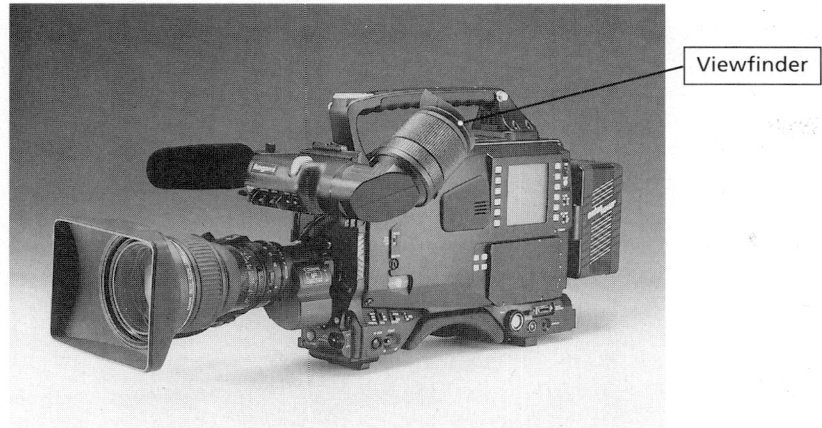

Figure 5.9 *A typical field camera's viewfinder (at top center of this photo) is a small monochrome screen with a magnifying eyepiece that can be rotated to different angles.*

On field cameras, in contrast, the zoom control is usually a button or a lever located on a grip handle. Focus may be automatic; if it is manual, it is set by manipulating a focusing ring on the lens.

Tally Lights

The **tally light** is a red light on top of the camera (see Figure 5.10). In studio or multicamera productions, it lights when that particular camera is on air, informing the talent and crew that the camera signal is being broadcast or

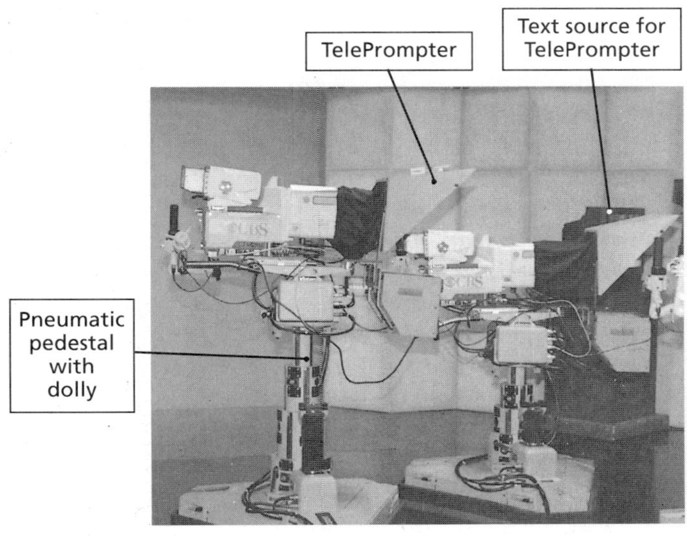

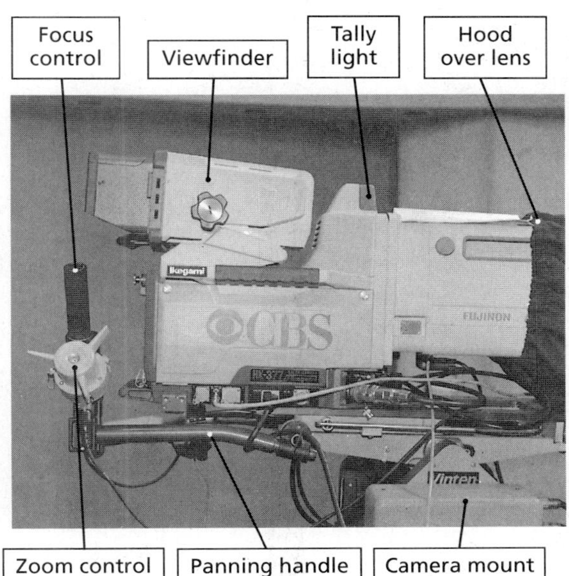

Figure 5.10 *Two views of a studio camera.*

recorded. A smaller tally light located in back of the camera lets the operator know when the camera is on air. When the tally light goes off, the operator is free to move to compose the next shot.

In field cameras, one tally light is mounted externally and another is located inside the viewfinder. They light when the camera is recording and go off when recording stops.

Intercoms and Headsets

Studio cameras are equipped with *intercom channels* or *headset connections* to carry voice communication among crew members, including production and technical staff. Headsets connected to these channels permit camera operators (and anyone else on headset) to hear the director's and other crew members' voice cues. Commercial television industry union requirements specify separate inputs for technical personnel and production personnel, even though the audio feed for these channels is the same. Some higher-end field cameras are also equipped with intercom channels.

Cue Cards

On-air talent use a variety of means to deliver scripted copy. Among the oldest is memorizing copy before show time. However, many television actors have come to rely on prompters to deliver their lines, and newscasters use such devices routinely. A *prompter* is any device used to provide script or text to on-air talent.

One of the simplest methods of prompting on-air talent is through the use of **cue cards** (Figure 5.11), large poster cards with lines of copy written on them in heavy black letters. To keep the talent oriented properly, the cue cards are usually held by a crew member right next to the camera. As lines are read, the holder moves the card to keep the appropriate text aligned with the lens.

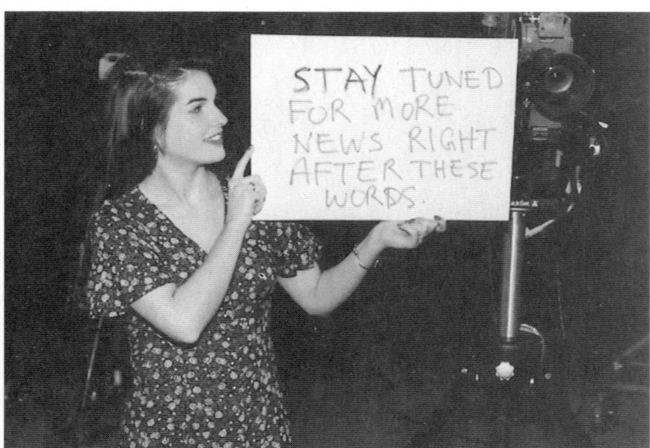

Figure 5.11 *The proper way to hold cue cards: next to the camera lens, with the line being read at the same height as the lens.*

Cards are cheap to use and immune to mechanical breakdown. However, they take time to prepare, require a floorperson to hold and change them, and are therefore useful for only relatively short amounts of copy. Further, when they are brought close to the talent (especially talent oriented directly to the camera), they may noticeably force the talent's eyes to glance off to the side of the camera lens. Conversely, when they are moved farther away from the talent to reduce this effect, they may be harder to read.

TelePrompters

The **TelePrompter** (Figure 5.12) solves many of the problems associated with cue cards. Studio TelePrompters are electronic text display devices that place copy directly over a camera lens using two mirrors. One mirror reflects text from a monitor above the camera to a second mirror placed over the camera lens, which in turn reflects the copy to the talent. By placing written copy directly over the lens, the TelePrompter eliminates the need for talent to glance off to the side of the camera to read. Instead, the talent can appear to have direct eye contact with the viewing audience.

Most studio TelePrompters use a character generator to electronically roll typed copy up or down the screen of a monitor mounted on each talent camera. Older TelePrompters use hard-copy news typewriters to provide script material that is then viewed by a television camera and forwarded to any prompter designed to carry it. For field use, available TelePrompters range from a simple paper roll attached to the camera (reminiscent of cue cards) to sophisticated, battery-powered devices similar to studio prompters.

In addition to the advantages just mentioned, TelePrompters provide identical copy to all cameras, allowing talent to look from one camera to another without losing their place. But like any other mechanical or electronic device, TelePrompters can break down. Therefore, it is wise to have a hard copy of the script available just in case. It is even better to have performers who are well prepared and familiar with their assignments.

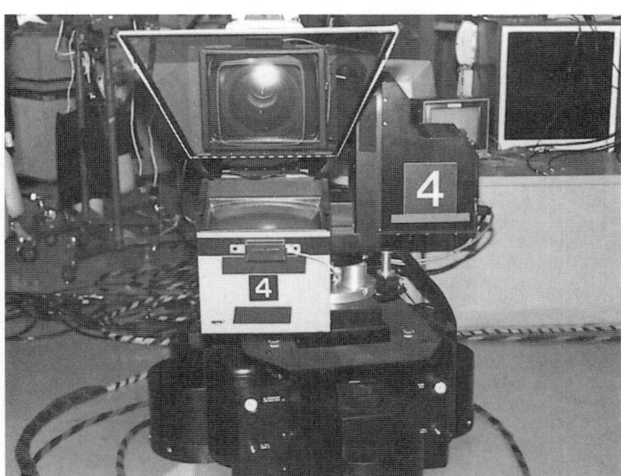

Figure 5.12 *A TelePrompter. When the device is in use, the text to be read appears right in front of the lens.*

BASIC DESIGN ELEMENTS OF PICTURE COMPOSITION

Composing quality pictures for video requires both an aesthetic and a rational sense. Picture composition should be shaped by the message you wish to communicate to and the effect you seek to create in your audience. Sometimes the goal of a program is to provide information. Sometimes it is to move the emotions. Deciding what to show the audience is essentially a rhetorical act.

In composing quality images for television, it is helpful to be familiar with formal canons of *static* picture composition, such as those used in photography, drawing, and painting. Line and shape, texture and pattern, color and contrast, depth and perspective, placement of key features—all of these design elements are important in video production.

Line and Shape

The boundaries of objects are defined by lines. Our eyes follow naturally along dominant lines to differentiate the objects that make up the television picture. By using dominant straight and curved lines, we can direct the attention of the viewer to different parts of the video image (Figure 5.13). We can also use lines dynamically to move the viewer's attention from one picture to another.

In addition to marking boundaries, lines define the shapes of objects for the viewer (see Figure 5.13). In a two-dimensional space, it is important to define shapes clearly if we want objects to be easily identified. Shape also helps to convey a convincing sense of solidity and depth, thus helping to preserve an impression of three dimensions in the video space.

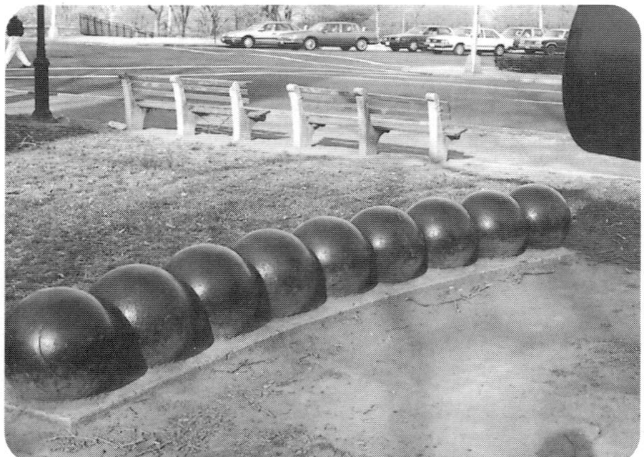

Figure 5.13   *Dominant lines define the boundaries and shapes of objects in a two-dimensional image. In video, such lines help to provide an impression of depth and solidity, and they draw the viewer's attention to important parts of the image.*

Texture

Texture refers to the surface qualities of objects: how rough or smooth they are, whether they are hard or soft to the touch, and so forth. As Figure 5.14 shows, the video image can convey a great deal of information by differentiating objects from one another on the basis of their texture. Clearly defined textures telegraph the tactile qualities of objects and provide a sense of closeness. Less defined textures make objects look farther away.

Pattern

Pattern may be defined as the repetition of a design element (Figure 5.15). Patterns tend to create rhythmic qualities and a sense of movement. Depending on how they are used, patterns can instill feelings of boredom or excitement in the viewer.

Color and Contrast

Color can differentiate objects from one another in the video image. Conversely, if objects of the same color are close together, they may tend to merge, confusing the viewer. Color arrangement on the screen is therefore an important part of program clarity.

In addition, sharper, more brightly colored objects generally appear to be closer, and less vividly colored objects appear to be more distant. (See Figure C.4 in the color insert.) Colors may also symbolize different psychological states: for instance, white is often used to symbolize purity, whereas red may indicate passion.

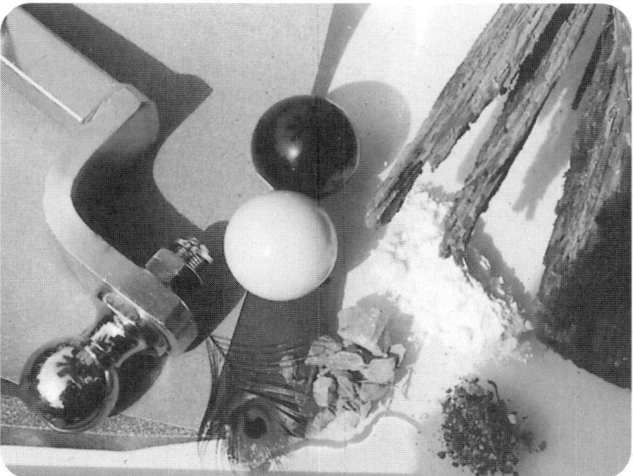

Figure 5.14 *Contrasting textures help to differentiate objects on screen. Note how much the texture helps you identify these items: a trailer hitch resting on four different grades of sandpaper, billiard balls, a peacock feather, stones, pebbles, a mound of flour, old firewood, and (can you find it?) a pool of motor oil.*

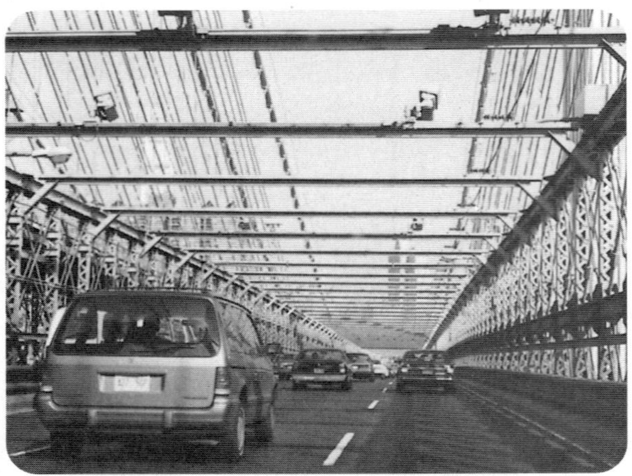

Figure 5.15   *Strong visual patterns add rhythm and a sense of movement to an image.*

Contrast—that is, differences in brightness levels—also differentiates objects from one another. Closer objects tend to exhibit greater relative contrast than those farther away.

Depth and Perspective

As mentioned in Chapter 3, *perspective* helps create the illusion of depth through the use of converging lines. Depth is also indicated by the elements of focus and image size. Images that are small and hazy appear to be far away; those that are larger, sharper, and more detailed appear to be closer.

Placement of Key Elements

As Figure 5.16 shows, the key elements of a picture should be prominently placed on the screen. Secondary picture elements should support the main idea but not dominate it. Irrelevant objects should be removed from the shot.

Balance

Balance refers to the psychological sense of stability, steadiness, and restfulness in the video image. For example, if picture elements are distributed across the screen evenly, we tend to view the picture as more balanced than if all items are on one half of the screen and the other half is empty. Unbalanced shots tend to convey a sense of tension and discomfort. Sometimes, of course, temporary discomfort is what you want to achieve.

A simple way to create imbalance is to cant the camera so that the subject looks tilted in the screen (Figure 5.17). Such canted angles are also called *Dutch angles*. This effect is often used to convey a sense of psychological stress

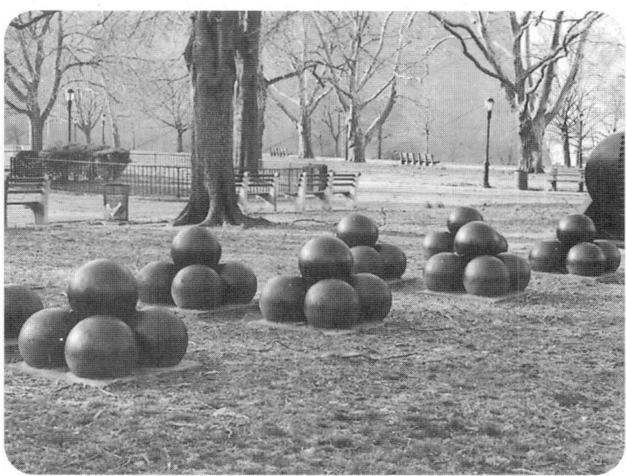

Figure 5.16 *One key to picture composition is to place the crucial elements in prominent positions on the screen. Here the cannonballs dominate the foreground and middle ground, with their importance reinforced by the strong diagonal line they form. The lesser elements—trees, benches, railings—occupy much less prominent positions.*

or loss of control in a character. On the other hand, Dutch angles can also convey a sense of power.

Achieving a sense of balance does not mean the elements must be evenly distributed. Visual interest may be enhanced through some degree of asymmetry without disturbing the overall sense of balance.

Unity and Variety

From a purely functionalist perspective, a program has *unity* when it embodies a central theme in all of its aspects and each of its elements seems to be connected in a whole. Unity also means the content contains nothing superfluous. Put another way, a program is unified if a gaping hole results when part of it is removed.

However, incorporating unity into a program does not mean it lacks *variety*. Unity does not mean harping on the same idea the same way over and over again. Such repetition usually results in boredom.

Rather, *variety in unity* is the goal. In the first movement of Beethoven's *Fifth Symphony*, the famous four-note theme is repeated several hundred times, but by different instruments, at different volumes and intensities, at different speeds, and in different registers. In the same way, video presentations should express a unified theme in a variety of ways to maximize interest and effectiveness. For example, NBC's "The Tonight Show," which has aired roughly five nights a week for more than three decades, features a monologue performed every night by the host, but the jokes are never the same.

Of course, a functionalist perspective is not the only one that may serve you. More decorative approaches may also be appropriate at times. For exam-

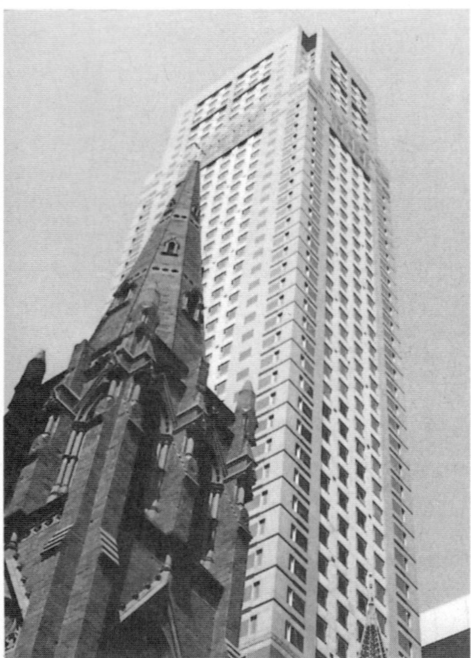

Figure 5.17   *Dutch (canted) angles, with the objects tilted on the screen, can create a sense of power, tension, or other psychological effects.*

ple, music videos lean heavily on decorative impulses for their content, as do circus spectacles. In these cases an action, shot, or sequence would not be considered superfluous simply because it was not functionally imperative. Therefore, you need to decide which aesthetic principles you are using to judge how to shape the video content.

PICTURE COMPOSITION IN VIDEO

Now that some general principles of picture composition have been introduced, we can turn to more specific applications of these concepts in video. This section presents the specific language associated with video picture composition and aesthetics, along with examples from typical program situations.

Framing

Framing defines the area of the picture. In literal terms, the frame is the perimeter around the edge of the picture that limits your field of view. The process of framing includes deciding what will be included in each shot and what will be excluded. Framing should be influenced by the principles of static picture composition that we just covered and by the dynamic needs of the program from one moment to the next.

Let's go back to the two-person, stationary-talent interview situation that we introduced in Chapter 4. Assume we start with an image of host and guest sitting together in the frame. Now, when the host asks a question, both static and dynamic considerations suggest that we frame the next shot without the host, zeroing in on the guest for a response (Figure 5.18). A close-up of just the guest is desirable at this point, for a number of reasons. First, from a static perspective, since television is a low-resolution medium, close-ups are important for presenting information clearly. Second, the close-up on the guest forces the audience's attention to the guest's answer, synchronizing the video and audio and thus increasing intensity. Third, from a dynamic standpoint, since we have just established the relationship of the host and guest in the opening shot, we may now proceed to a close-up of the guest without losing context or continuity for the audience. However, after awhile, another shot framing the host and guest together may again be needed to reestablish their relationship. Hence, framing decisions are influenced by time considerations (also called *pace*) and by shot sequences.

In addition, framing is greatly influenced by program content. For example, imagine you are shooting a cooking show in which the opening shot shows the widest view of the set, including all the ingredients and utensils laid out on a table, plus the talent and an attractive backdrop. When you frame that shot, you notice on the line monitor that the front vertical face of the table takes up a quarter of the picture. You decide that part of the table is irrelevant to the program content and ought to be excluded, or at least minimized, so you tilt the camera up. Now, however, you see that the frame includes a great deal of *dead space* above the talent's head, which also is not appropriate. As a result, you correctly decide to zoom in on the table to exclude the superfluous parts of the picture. But if this close-up also excludes some of the important items on the table, you may need to make further adjustments in the shot—for example, by moving some of the items closer together or by panning across the table. In sum, good framing generally means leaving out dead space, the superfluous, and the irrelevant and including only what is motivated and essential.

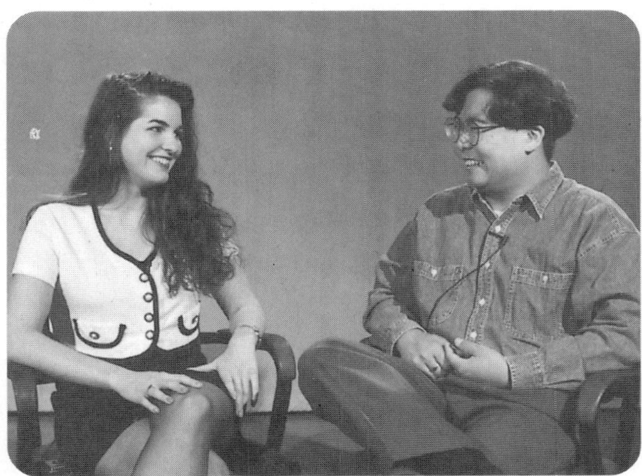 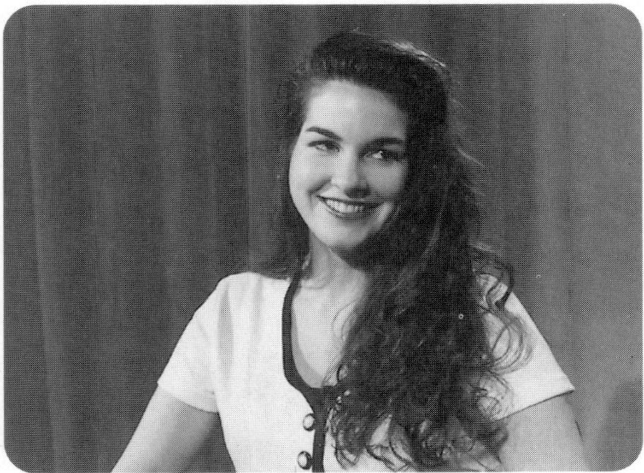

Figure 5.18   *Typical framing sequence for a two-person, stationary-talent interview program (left). The sequence begins with the host and guest together in the frame. When the host asks a question, the next shot (right) closes in on the guest alone.*

Using the z-Axis

We designate the horizontal dimension of the picture as the *x-axis*, that is, the dimension across the screen from left to right. The vertical dimension is called the *y-axis*. Also crucial is the **z-axis,** the imaginary dimension extending from foreground to background. By placing objects along the z-axis, we can create a three-dimensional sense of depth on the two-dimensional television screen.

As noted in Chapter 3, the overlapping of objects in the picture can help establish an illusion of perspective. For example, you might place a foreground object in the lower left third of the shot, overlapping a middle-ground object in the center (in a different location on the x- and y-axes) and a background object in the top right third (yet another location on the x- and y-axes). This will establish a strong sense of depth. The danger is that objects toward the foreground can blot out the others. To avoid this, check the way the objects appear when framed in the video image, not just the way they look to you in actual three-dimensional space.

Focus

As you know from Chapter 3, *focus* refers to the sharpness of the objects featured in a shot, and *depth of field* is the area in which the images are acceptably sharp. Changing focus and depth of field can shift the viewer's attention from less important parts to key aspects of the video image.

When focus is changed to shift attention from foreground to background elements of the picture, or vice versa, the change is called *racking focus* (Figure 5.19). Imagine a character in clear focus facing the camera in the left foreground of a shot. The character hears a gun being cocked. As she turns and looks over her left shoulder in the direction of the sound, her face defocuses and a pistol in the background suddenly comes into focus. In this way, racking focus from foreground to background shifts the viewer's attention from the character to the gun.

Selective focus may also be used to stress important picture elements while deemphasizing irrelevant, confusing, or distracting parts. Say you are con-

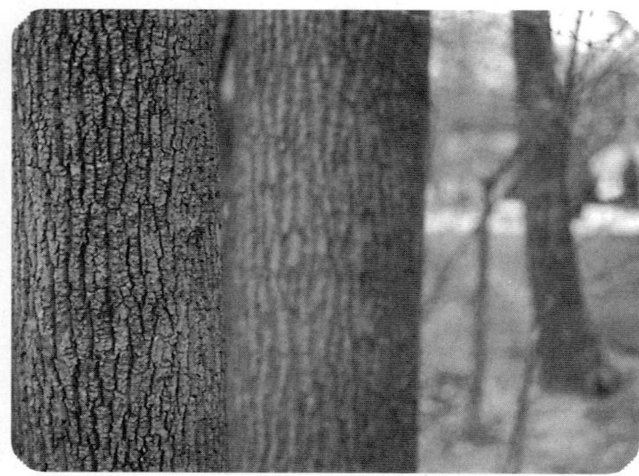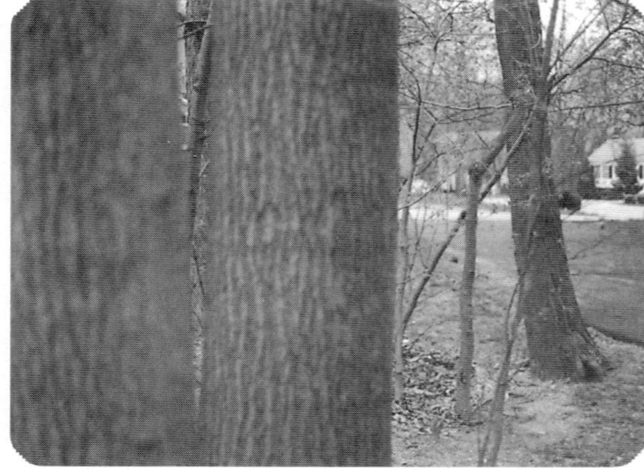

Figure 5.19   *Racking focus. In these photos, the focus shifts from the foreground to the background.*

ducting a serious interview with a nature expert about endangered species, and you have the expert on camera with specimens of elephant tusks behind him. In the viewfinder you notice that one tusk appears to be growing out of his head, making him look comical. To deal with the problem, you bring the subject closer to the camera, thus reducing depth of field. Now when you set focus on the speaker once again, the tusk is pushed out of focus, making the picture more acceptable.

Camera Angle

Camera angle can greatly influence the amount of information conveyed in a particular shot, how the subjects look, and the meaning of a shot or scene. Therefore, it is essential to consider whether the camera should be lower than, higher than, or on the same level with the subject being pictured.

When the camera lens is at the same height as the subject, the picture is said to be shot at *eye level*, or on a normal camera angle. When the camera is lower than the subject, the picture is said to be shot from a *low angle*. When shot from above, the picture is said to be shot from a *high angle*.

In our cooking demonstration, for example, it may help to ped up the camera and tilt it down onto the work surface to show the cooking process more clearly (Figure 5.20). More dramatically, if we are showing a knife-wielding villain approaching the camera, we may want to use a low angle, which will grant more strength, power, and psychological impact to the subject. Conversely, high angles generally make subjects look less threatening.

You also need to consider the impact of camera angle on shot backgrounds. Low angles usually mean you will be including more vertical height in the background portion of your shots. In studios with low ceiling height, low-angle shots risk including the lights or other studio items that are not meant to be part of the set. Therefore, for tall talent, it may be necessary to ped up rather than shoot from below eye level. Remember also that camera angle can be changed during a production to accommodate different needs. Some pedestals even permit on-air adjustment, which can create dramatic effects.

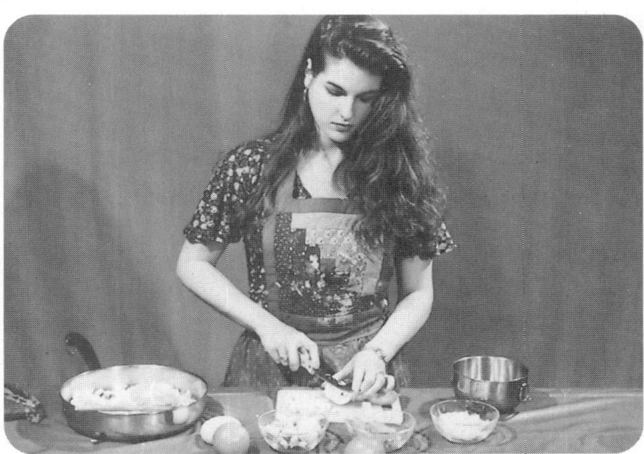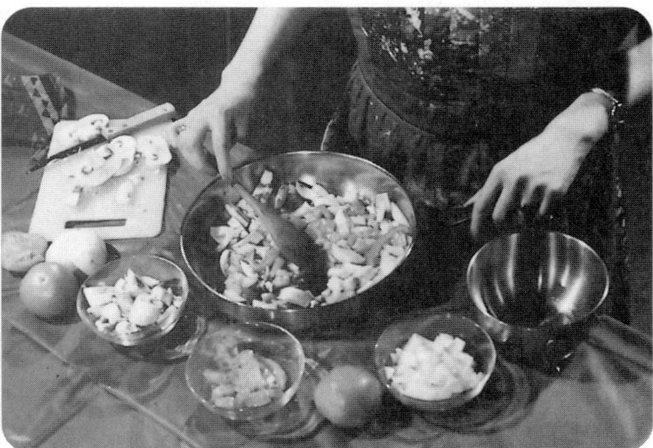

Figure 5.20 *Use of appropriate camera angles to reinforce the program's subject. After showing (left) a chef at eye level, the camera peds up and tilts down for (right) a high-angle shot of the food being prepared.*

Types of Camera Shots

Camera shots can be classified in many ways: by their function, by what they include, by the amount of screen space they allow the subject, and so on. The following sections introduce the common terminology and explain some of the reasons for choosing one type of shot rather than another.

Establishing and Master Shots   An **establishing shot** is the shot that establishes, for the audience, the relationship of talent to one another and to their setting. In our two-person interview, the opening shot that showed the guest and host seated next to each other was the establishing shot. But the establishing shot need not be the first one in a program or a sequence. Consider as an opening shot a close-up of the main character's hand about to lift a shot glass of liquor to his mouth (Figure 5.21). As he drinks, the camera zooms out to reveal that he is at a bar. The camera continues to zoom out, showing that he is alone. We see chairs upside down on tables throughout the bar. When the zoom reaches its widest angle, we notice a clock on the wall reading two o'clock near a window with a neon sign flashing to indicate that it is the middle of the night. This final shot is the establishing shot for the sequence, because it establishes the setting for the character.

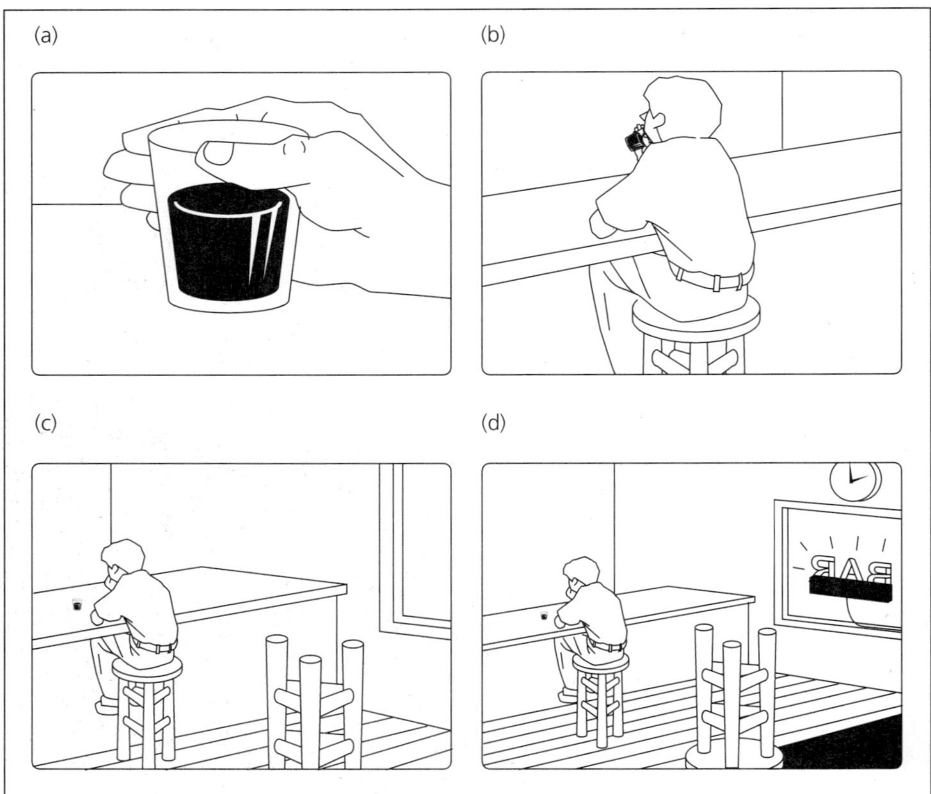

Figure 5.21   *A sequence in which the camera gradually zooms out to reveal more and more of the scene. In this case the last shot, which conveys the full setting, is the establishing shot.*

A **master shot** or **cover shot,** on the other hand, includes the widest shot of the set for a scene or program. It is important to know the extent of the master shot so that off-camera production elements (lights, microphones, booms, cue cards, crew members) can be placed into the set without inadvertently including them in the shot. In some cases a master shot can be used as an establishing shot, but often the two are different.

One-Shots and Two-Shots As noted in Chapter 4, a **one-shot** is a frame containing one person. Similarly, a **two-shot** contains two people, a **three-shot** contains three people, and so forth. Of course, the arrangement of the person or people in the frame is highly variable. The subjects can be in tight close-up, can be far away, or can be arranged in a combination of screen distances (foreground, middle, and background).

Long Shots, Medium Shots, and Close-ups Another way to classify shots, as Figure 5.22 shows, is by the proportion of total screen space a subject occupies in relation to the overall field of view. In a **long shot (LS),** the entire subject (head to toe) generally occupies significantly less than half the total screen space. If the subject takes up far less than half the screen space, the shot is called an *extreme long shot (ELS)*. When the subject occupies closer to half the total screen space, we describe the shot as a **medium shot (MS)**. Similarly, **close-ups (CUs)** and *extreme close-ups (ECUs)* describe frames in which the subject dominates.

In medium shots and close-ups, the subject's body is often truncated. For example, in a scene at a dentist's office, a medium shot might give most of the screen space to an image of the dental hygienist from the waist up. An extreme close-up could use most of the screen space to picture just the patient's mouth.

In one classical approach, directors begin with long shots to establish setting and relationships among characters and then move to progressively tighter shots to emphasize key program elements as the program unfolds. In many cases, though, other techniques may be advisable. In a murder mystery that opens with the murder scene, the director might want to hide the identity of the perpetrator. The program might start with a close-up of the murder weapon in use without letting the audience see the killer. In this example, wide shots would follow rather than precede the close-ups.

Head, Bust, and Waist Shots For shots of people, a more exact terminology may be used to indicate the portion of the body to be included. A *head shot*, obviously, is a shot of the person's head. A *bust shot* generally includes the head, shoulders, and upper part of the chest. A *waist shot* includes everything from the waist up.

In framing such shots, be aware of exactly where the body is cut off. Shots that crop a person at a joint line—for instance, exactly at the knees, waist, or chin—tend to appear awkward (Figure 5.23). It is usually better to let the body's natural cutoff lines fall either clearly within the screen or beyond its bottom edge.

Headroom, Noseroom, and Leadroom Another framing consideration involves giving human subjects a comfortable amount of room on the screen, as illustrated in Figure 5.24. For instance, you normally want to provide enough **headroom**—space between the talent's head and the top of the frame—so that the head does not appear to be squeezed or cramped. On the other hand, too much headroom can make the subject seem overwhelmed or lost in the picture.

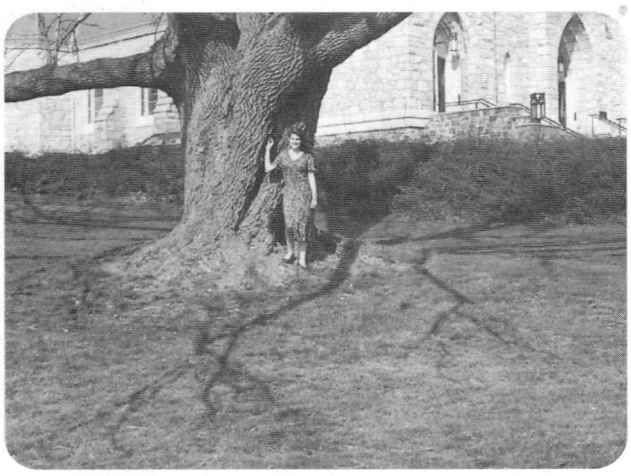

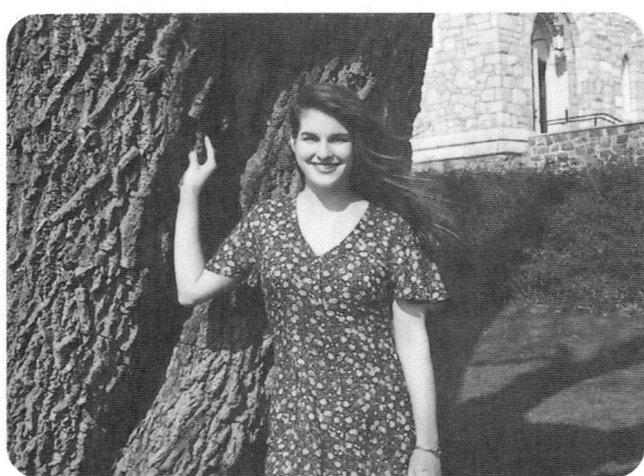

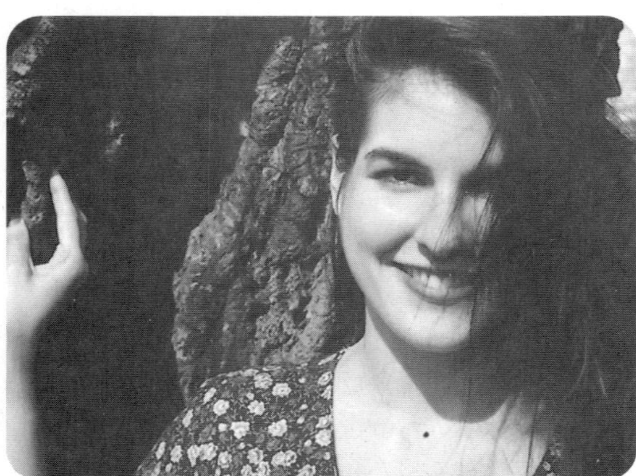

Figure 5.22   *Shots classified by the proportion of screen space the subject occupies. (top) Long shot. (center) Medium shot. (bottom) Close-up.*

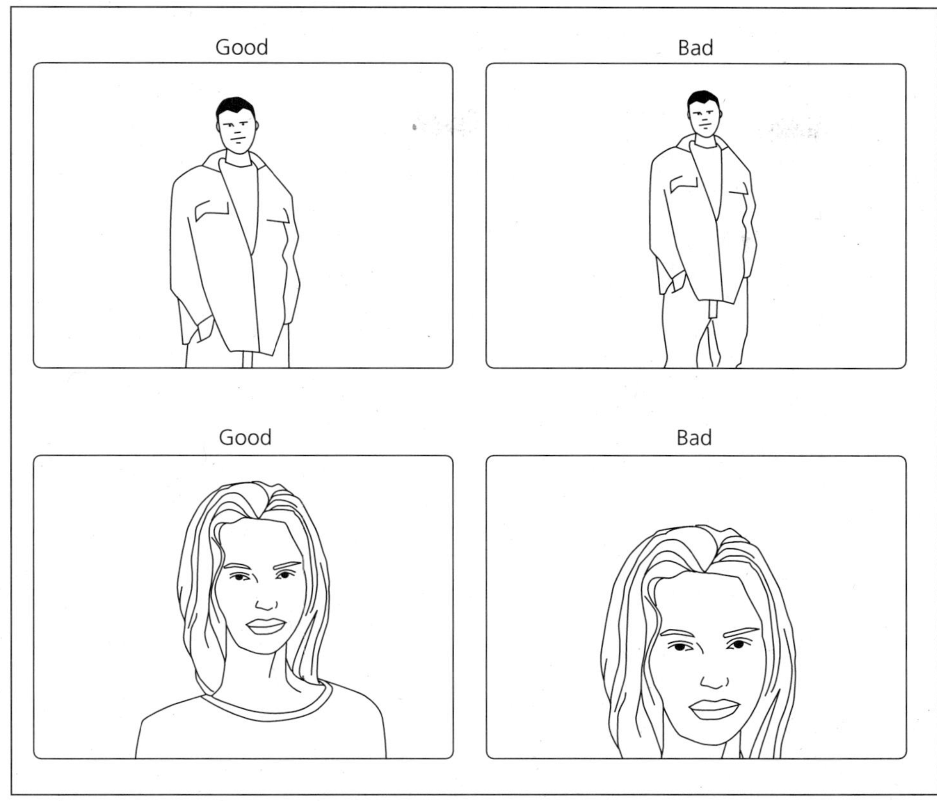

Figure 5.23   *Preferred ways to frame a person on screen. Note that cropping a person exactly at a joint tends to look awkward or uncomfortable, as though the rest of the body has literally been chopped off.*

In head shots taken from the side, **noseroom** refers to the screen space in front of the talent's face. In conventional composition, noseroom is generally greater than the screen space in back of the head. Restricting noseroom tends to create tension and a sense of entrapment of the subject. **Leadroom** is a similar concept applied to moving talent. If a horse and rider are moving from left to right on the screen, leadroom is the space to the right, in the direction the talent is moving.

You should determine how much headroom, noseroom, and leadroom to use depending on the effect you wish to create. Also, keep in mind that you may need to move the camera to maintain appropriate space around the subject as the picture changes.

Over-the-Shoulder Shots   When two subjects are facing each other, a common technique is to shoot one talent over the other's shoulder (Figure 5.25). For example, it is common for the camera to frame a *speaker* over a *listener's* shoulder. This **over-the-shoulder (OS) shot,** as it is called, establishes the relationship between the talent, presents an intimate image of the speaker's face, and removes questions the audience might have about whom the speaker is talking to.

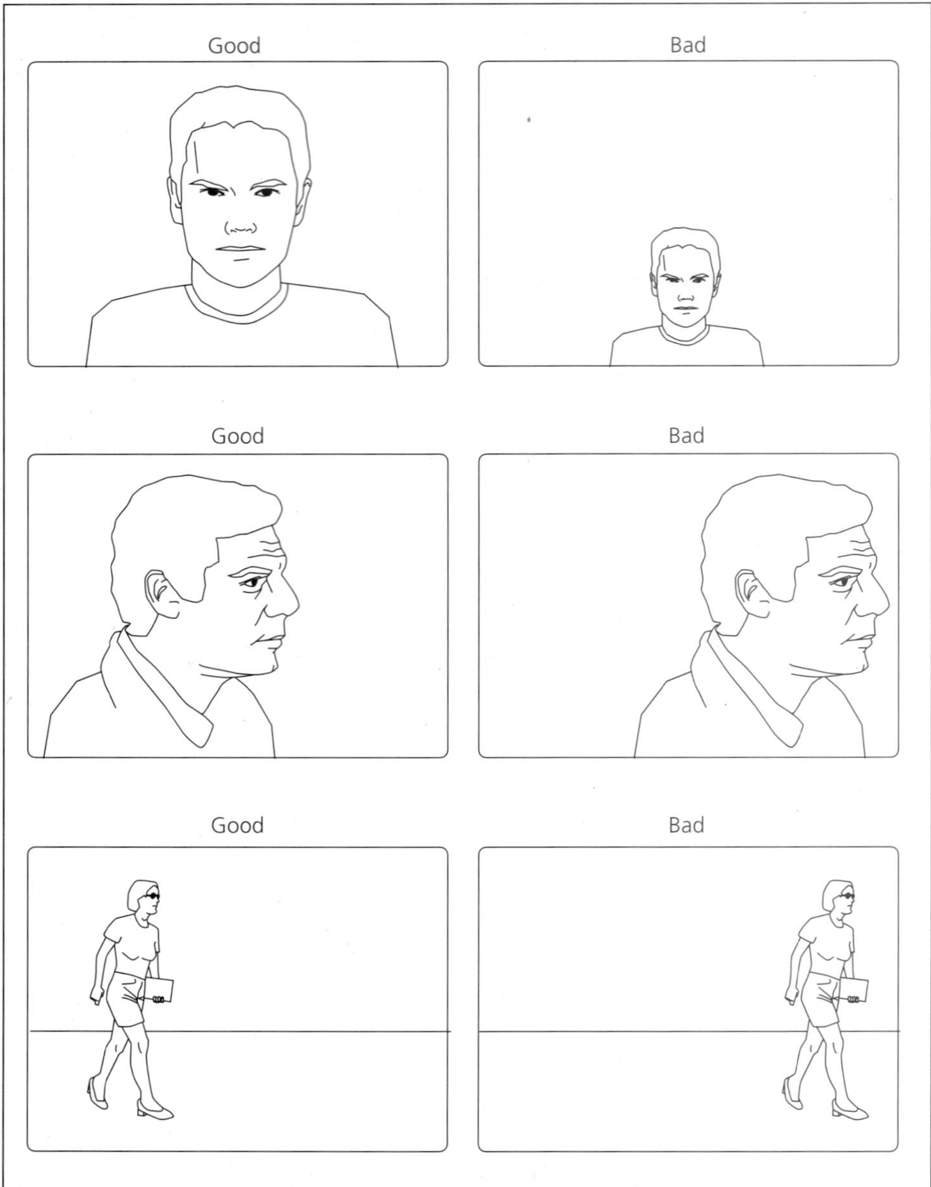

Figure 5.24   *Subjects should be framed so that they have an appropriate amount of room on screen. These drawings illustrate correct and incorrect (a) headroom, (b) noseroom, and (c) leadroom.*

In addition, the OS can easily be changed to an even more intimate close-up of the speaker by zooming in past the listener's shoulder (see Figure 5.25). After a time, the relationship can be reestablished by zooming out (on or off air) to recapture the back of the listener. This type of camera movement provides shot variety and visual interest, emphasizes relationships among talent, and preserves context and unity in the video space. Of course, a different camera can be positioned behind the speaker to carry an OS of the listener's reactions, offering additional variety and visual interest.

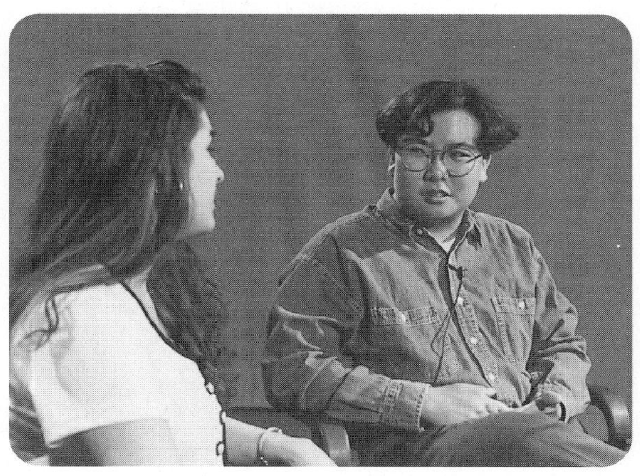
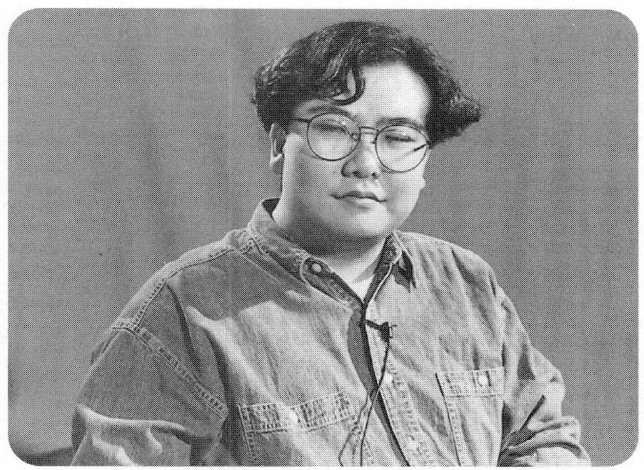

Figure 5.25   *Over-the-shoulder shots. (a) A standard OS establishing the relationship between the talent. (b) A zoom-in to a close-up.*

Reveals and Trims   To communicate important information to your audience, camera movement can reveal or isolate significant picture elements, sometimes with dramatic effects. Imagine beginning with a high-angle one-shot of your central character's face in extreme close-up (Figure 5.26a). As he delivers an especially deferential apology to someone off camera, a slow

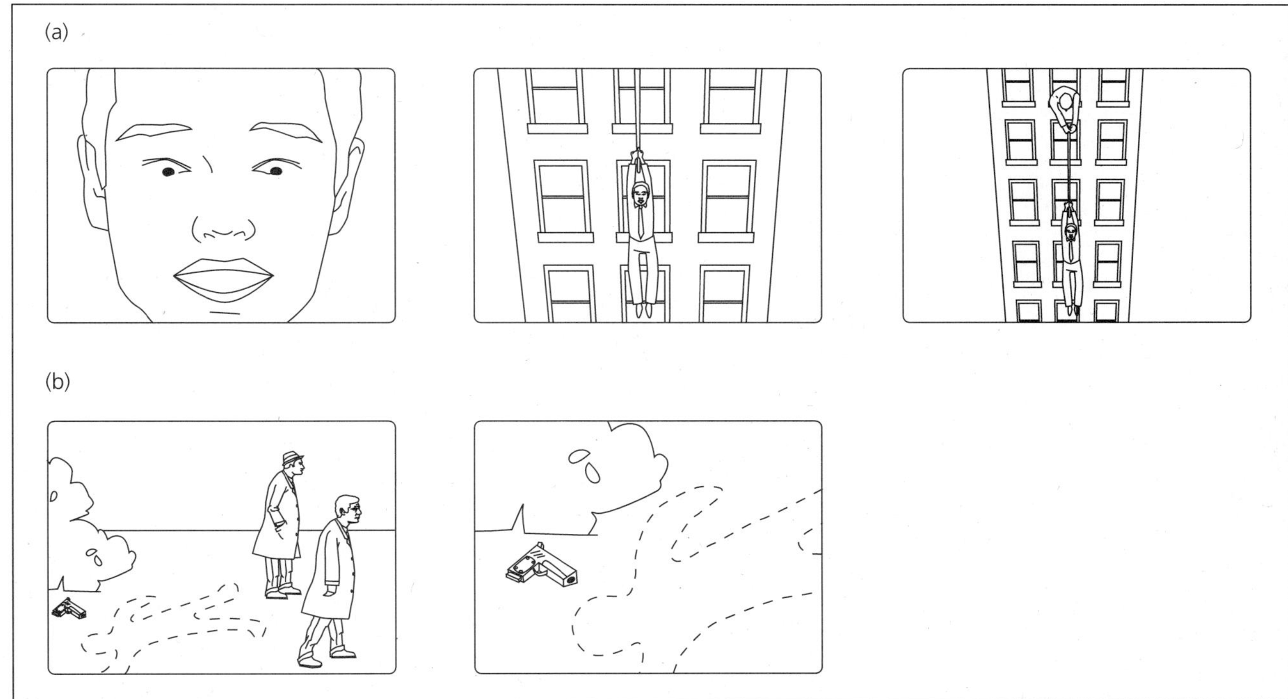

Figure 5.26   *Camera movement used to disclose or highlight important information. (a) A reveal, as the camera zooms out, gradually conveys the subject's peril. (b) A trim, as the camera zooms in, isolates a weapon the detectives have failed to notice.*

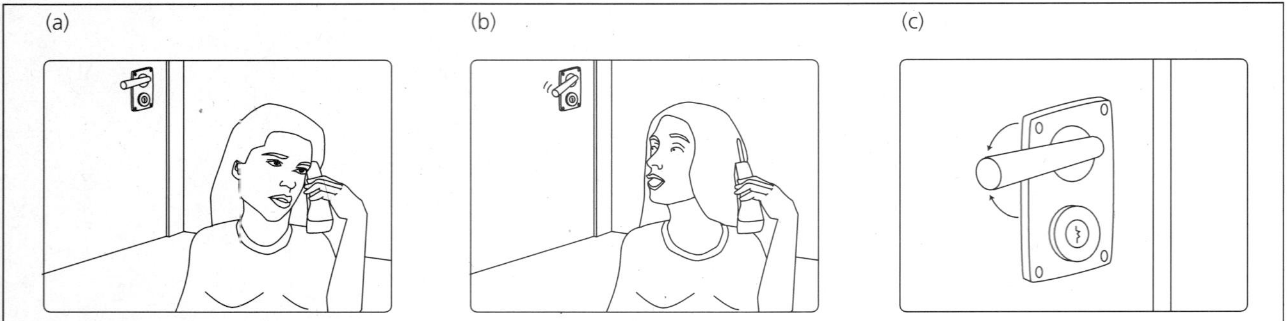

Figure 5.27   *Sequence illustrating a point-of-view (POV) shot. (a) A character gets a phone call about an escaped murderer. (b) She hears her door handle rattle and spins around to look. (c) The camera cuts to a close-up of the ominously jiggling handle from her POV.*

zoom-out reveals that he is hanging by a rope from a Manhattan high-rise, obviously in mortal danger. This **reveal** further discloses that the person to whom he is apologizing is holding the rope. Obviously, reveals can evoke horror, humor, and other reactions, depending on the subject matter and sequencing.

In similar fashion, a **trim** from a wide shot to a tighter frame can be useful when you want to emphasize a key element of the picture. In a mystery drama, you might start with a cover shot of a crime scene as detectives file out after their preliminary investigation (see Figure 5.26b). As they leave, the camera zooms in to a close-up of a key piece of evidence (say, the murder weapon) that the police have missed. By trimming the shot to isolate the key item, the audience is given clues that can explain later action. As another example, in a cooking show, a shot of the chef placing the finished dish on the table could be trimmed to a close-up of the dish itself. Trim shots can stress the important elements, eliminate irrelevant information, and use screen space optimally.

Point-of-View Shots   Shots that appear to be from the point of view of a particular actor are called **point-of-view (POV) shots**. For example, imagine a drama in which our main character is talking on the phone with the local police. They warn her that a convicted murderer has just escaped from a nearby prison. She is framed in a waist shot as she speaks on the phone (Figure 5.27). Suddenly she is startled by the sound of the front door handle being jiggled from outside and whirls around to take a look. We cut to a close-up of the door from her POV, zooming in on the door and then trimming to an extreme close-up of the handle as it moves ominously back and forth. Rather than simply recording the action from outside, this POV shot gains dramatic impact by taking the subjective stance of the main character, thus increasing the audience's identification with her.

A FINAL NOTE ABOUT PICTURE COMPOSITION

Throughout our discussion of the camera's mechanical and electrical technologies and the marvelous things people can do with them, we have come back repeatedly to the notion of program content and the fact that the program communicates with an audience. These points deserve additional emphasis before we turn to the audio component of video production in the next chapter.

Remember: when you are directing, you call the shots. You are the one arranging the camera work in a way that will make people either want to watch or tune out. Although you must know the capabilities and limitations of sophisticated camera equipment, you also need to develop your understanding of what will enlighten and move the audience. Successful programs have a purpose. They present content clearly, with an intent to entertain, inform, and/or persuade an audience. A camera that can automatically achieve that aspect of video production has yet to be invented.

PROFESSIONAL POINTERS

Working with Cameras

- Keep the floor clean. Because video makes extensive use of close-ups, small jolts to the camera, such as when it rolls over a piece of gum, may cause severe rocking of the image.
- Check your viewfinder registration. Faulty alignment between the viewfinder and the line monitor can spoil your picture composition.
- If you are using cue cards, be sure to number them. Numbers help you keep the cards in correct order. If they get out of order, they may confuse the on-air talent.
- With both cue cards and TelePrompters, find the correct distance from the talent. The prompts should be close enough to be readable but not so close that viewers notice the talent's eye movements.
- Adjust the camera angle for the needs of the program, not for the height or comfort of the operator. Check the line monitor to be sure you have framed the shot correctly.
- When framing a person, don't truncate the image at the edge of a body part. "Chopping" the person in an obvious way tends to make the audience uneasy.

KEY TERMS

tripod *(80)*
studio pedestal *(81)*
mounting head *(82)*
crane *(83)*
jib arm *(84)*
body mounts *(84)*
viewfinder *(88)*
tally light *(89)*
cue cards *(90)*
TelePrompTer *(91)*
framing *(96)*
z-axis *(98)*
establishing shot *(100)*

master (cover) shot *(101)*
one-shot, two-shot, three-shot *(101)*
long shot (LS) *(101)*
medium shot (MS) *(101)*
close-up (CU) *(101)*
headroom *(101)*
noseroom *(103)*
leadroom *(103)*
over-the-shoulder (OS) shot *(103)*
reveal *(106)*
trim *(106)*
point-of-view (POV) shot *(107)*

QUESTIONS FOR REVIEW

1. How do different camera mounts affect camera movement in both studio and field?

2. What terms do directors use to request different camera moves from the camera operator?

3. What electronic camera components help the camera operator compose video images?

4. What are the advantages and shortcomings of the various prompting devices available in video production? How should cue cards be set up for talent oriented directly to the camera?

5. What static design elements should be considered when composing pictures for the camera?

6. What dynamic considerations affect picture composition and sequence?

7. In addition to terms such as *close-up, medium shot,* and *long shot,* what other terminology might you use to specify precisely how you want shots of talent to look?

6 Understanding Sound and Microphones

Sound has always been part of the video medium. Since the 1980s, however, audio in television has improved significantly. This is because television receivers are finally being made with high-quality audio technology. Many newer sets feature multiple speakers with full-range audio, stereo sound capability, and even surround sound options. This is a great advance over the tiny single speaker that was once the norm.

But the importance of sound in television programs does not depend only on the nature of the equipment. It is also due to differences in the way we perceive sound and pictures. Think about what attracts your attention to a program when your mind is elsewhere. If you are turned away at the start of a program, it is the audio that draws your attention because your ears (unlike your eyes) can pick up signals from all directions. That is why, from a production standpoint, it is usually wise to bring the audio in either simultaneously with the video or before the video. Moreover, the ear can focus on multiple sounds simultaneously, providing a powerful means of setting a scene quickly.

Another perceptual difference between audio and video is that our eyes are much more forgiving than our ears when noise enters the signal. (Remember from Chapter 1 that noise is any unwanted material that interferes with the desired content.) Human vision is much coarser and more tolerant of defects in detail than the sense of hearing. For example, most people can detect differences in quality between analog cassette and a CD but not between 8 mm video and Beta-tape. In short, noise is more obvious to the ear than to the eye, and sound quality is therefore extremely important.

This chapter describes the physical nature of sound and our perception of it. It explains how microphones work and how they are used to capture sound for video production. The topics include:

THE NATURE OF SOUND

the sound wave • the sound envelope

MICROPHONES: BASIC TYPES AND CHARACTERISTICS

generating elements • performance characteristics • pickup patterns

MICROPHONE SELECTION

audio design principles • visible mics • off-camera mics • wireless mics • other technical considerations

STEREO AND SURROUND SOUND

problems in stereo miking • surround sound

THE NATURE OF SOUND

The Sound Wave

Sound is the vibration of molecules caused by the motion of a physical object. Most often, the vibration consists of the transfer of energy from one air molecule to another. However, sound may result from the motion of a physical object in the presence of any molecular medium that has some degree of elasticity (including gases, liquids, and solids). *Elasticity* refers to the tendency of a molecule to bounce back to its original position after bumping into one nearby. Hence, in the presence of air, a hammer hitting a concrete block, a guitar string being plucked, a violin being bowed, or vocal cords being moved by a column of air in your windpipe can all cause nearby air molecules to bump into one another, sending an air pressure wave of sound out from a source.

When sound travels through air, molecules that are disturbed do not travel the entire distance from the source to the farthest point the sound reaches; rather, they transfer energy from one molecule to the next. In this way, each molecule moves only a small distance. This process is similar to the action you observe when a cue ball strikes a row of billiard balls, leaving the cue ball near the first ball it strikes but transferring the energy from the cue ball to each ball in line, causing the last in line to move without ever coming in contact with the cue ball. It is the energy from the cue ball that moves through the row, not the cue ball itself.

As another example, when a guitar string is plucked, air molecules closest to the string are disturbed by its rapid back-and-forth motion. This movement is called *vibration* or *oscillation*. As the string vibrates, a sound wave is propagated as air molecules transfer energy to adjacent molecules. If you are in range, air molecules vibrating sympathetically with the guitar string strike your eardrums, causing you to hear the sound. The motion from one molecule to the next imitates the original vibrations of the sound source, matching the rate of motion of the string.

As the string moves, it compresses nearby air molecules in the direction the string is moving. This *compression phase* consists of a bunching together of air molecules, increasing the density of air in the direction the string is moving (Figure 6.1). As the string moves back, it sends a compression reaction in the opposite direction, causing molecules to pull farther apart in the initial direction. This thinning of air molecules in the initial direction is called a *rarefaction phase*. Alternating compressions and rarefactions can be pictured as the same kind of sine wave we described for radio energy in Chapter 2, and the wave features the same components of frequency, amplitude, wavelength, and phase. Actually, this explanation does not take into account the added vibration of the body of the guitar, which vibrates sympathetically with the guitar string. The guitar body acts as an amplifier as it vibrates a greater volume of air than the string can do alone.

Frequency and Pitch   The *frequency* of a sound wave refers to the number of times per second the sound wave completes a cycle from compression to rarefaction. Using the same notation used to describe radio waves, audio frequency is measured by the number of cycles per second (cps) or hertz (Hz). For example, if the rate of vibration of a guitar string is 50 cycles per second,

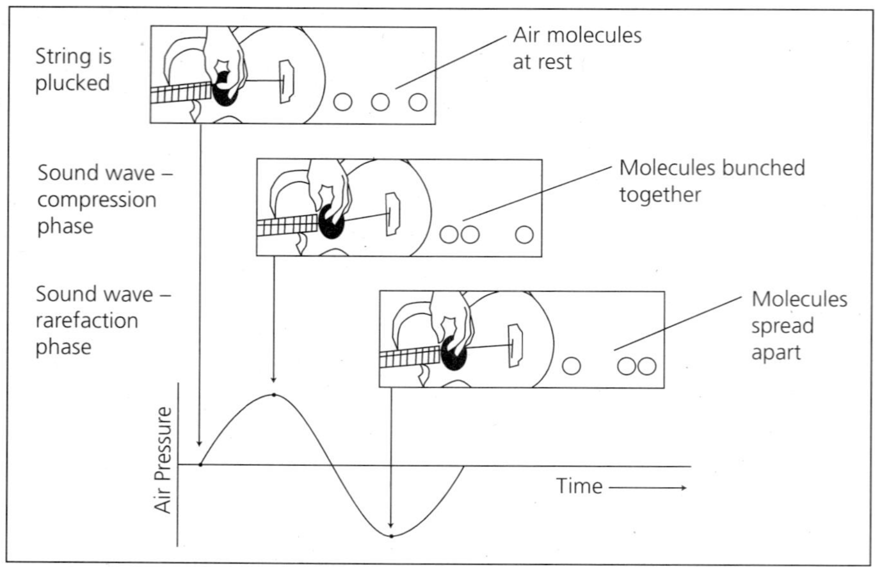

Figure 6.1   *The compression and rarefaction phases of a sound wave.*

the notation used to designate this frequency is 50 cps or 50 Hz. Similarly, we designate a sound wave of 10,000 cycles per second as 10,000 Hz, or 10 kilohertz (10 kHz).

The normal range of human hearing is roughly from 20 Hz to 16 kHz. However, our perception of sound changes so that as frequency increases, pitch increases. **Pitch** refers to the perceived highness or lowness of the tones of particular sounds. At higher frequencies, sounds are thinner and shriller; at lower frequencies, sounds are deeper. The low notes of a bass fiddle are in the range of frequencies between 20 and 100 Hz, whereas the high notes of a piccolo can go above 3,500 Hz. The faster an object vibrates, the higher the pitch it produces.

Amplitude and Loudness   When an object vibrates in any elastic medium, molecules surrounding the object also vibrate if the vibration is intense enough. The more intense the vibration, the greater the pressure exerted on the surrounding medium and the greater the number of molecules set in motion. The magnitude of intensity is known as the *amplitude*.

Human perception is such that as amplitude rises, loudness increases. For example, an opera singer singing a note of constant pitch (say, 262 Hz, known as *middle C*) can vary the intensity from loud to soft by regulating the volume of air passing across the vocal cords. By keeping the amount of air relatively small, a softer note is produced (low volume); conversely, increasing the amount of air flow increases volume, making the note louder.

We can discern a broad range of variations in loudness. For example, we can hear both a whisper and a cannon shot from a distance of ten meters, even though the cannon shot may be 10 million times louder than the whisper. We call the extent of variation in loudness that we can hear *dynamic range*. Because dynamic range is so broad, we use a logarithmic scale of values known as the *decibel scale* to measure and express differences in loudness efficiently.

Basically, the **decibel (dB)** is a unit of measurement that expresses logarithmically a ratio between two sound levels or two electrical power levels generated by different sounds. People tend to perceive that a sound has "doubled" in volume when it has increased about 3 dBs, whereas a 1 dB change is barely distinguishable. To compare volume levels of different signals, broadcasting stations use as their standard a decibel value known as the **volume unit (VU),** measured by a *VU meter*. (See Chapter 7 for a discussion of how the VU meter is used.)

Frequency and Loudness   The relationship between amplitude and our perception of loudness is complicated by another factor: our ability to hear sound across the entire frequency spectrum is not uniform. Of course, hearing may be affected by such factors as age, health, and exposure levels to sound over time. But in general we are more sensitive to sound in the middle range of audible frequencies, from about 300 Hz to 3,500 Hz, than we are to either the low end (*bass*) or high end (*treble*). Therefore, even if your audio equipment presents the full range of audio frequencies at equal volume, our subjective perception of sound as presented this way tends to favor the midrange frequencies. This means midrange sound *masks* other portions of the spectrum presented at equal volume. Because of masking, it is sometimes necessary to vary volume levels according to our sensitivity to different frequencies. When we electronically alter the volumes of certain frequencies to increase or eliminate masking, we call that process *equalization*.

Wavelength and Phase   As with light and radio waves, the wavelength of a sound is inversely related to its frequency. *Wavelength* is the distance traveled by one complete cycle of compression and rarefaction. Lower frequencies have longer wavelengths, and higher frequencies have shorter wavelengths.

Sound waves also have a *phase* component, which refers to the time relationship between two or more sound waves at a given moment in their cycles. For example, if two sound waves of identical frequency are propagated at the same instant, their compression and rarefaction phases will coincide. The two waves will be received perfectly in phase with each other at a point equidistant from their sources, assuming they travel through media of equal density. The important point is that *received waves that are in phase reinforce one another; the sound you hear gets louder*. They yield a net gain in amplitude (called *constructive interference*) equal to the sum of their individual amplitudes at a given instant. Conversely, waves of the same frequency received perfectly out of phase with one another exhibit an opposite effect called *destructive interference*, the net result of which is a canceling of their amplitudes. *The sound you hear gets lower, possibly even resulting in silence* (Figure 6.2).

Of course, neither situation occurs very often. Sound waves usually propagate from different sources at different times and at different wavelengths, so they are rarely perfectly in or out of phase with one another. Nevertheless, interference factors are critical to sound design, especially with regard to microphone placement and talent proximity and when recording in stereo.

Waveforms and Timbre   *Timbre* refers to the unique sound quality that comes from variations in frequencies that accompany most tones we hear. Timbre is that quality of auditory sensation that enables a listener to discern that two sounds of the same loudness and pitch are dissimilar. For example, although a guitar string vibrating at about 262 Hz matches the frequency of middle C on a properly tuned piano, a blindfolded listener would probably

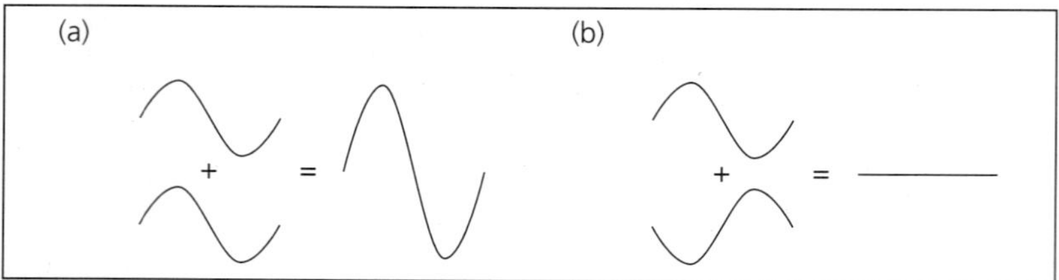

Figure 6.2   *Interference of two sound waves. (a) Constructive interference when the waves are in phase. The two waves reinforce each other, creating a louder sound. (b) Destructive interference when the waves are out of phase. The two waves cancel each other out.*

have no trouble distinguishing the guitar's C from the piano's. Why is this so? It is because most sounds are not just pure tones; rather, they contain additional frequencies generated uniquely by the source producing the original tone and, perhaps, by surrounding media with which the original tone interacts. In musical terms, these additional frequencies are known as *harmonics* (tones that are exact multiples of a fundamental tone) and *overtones* (tones that are not multiples of a fundamental tone). In fact, the richness of our auditory experience is due largely to overtones and harmonics. Recall that human hearing extends to 16 kHz. Yet the frequency of the highest piano note is only 4,186 Hz, leaving about 12 kHz just for harmonics and overtones. Obviously, harmonics and overtones account for a major segment of the audio spectrum, at least as far as music is concerned.

When a single piano note is struck, instead of generating a simple sine wave at a single frequency, the sum of harmonics and overtones generated by the fundamental tone yields a complex *waveform* (Figure 6.3) that uniquely identifies it as having come from a piano. A guitar has its own special harmonic structure that yields its own unique waveform. A clarinet has another. In short, complex harmonic structure makes notes of the same pitch have different timbres. It is timbre that makes it difficult to confuse the sound that comes from Eddie Vedder's throat with that of Luciano Pavarotti, even when they sing the same note.

The Sound Envelope

Our perception of a particular sound is also affected by its **sound envelope,** or the characteristic way it begins, sustains, and ends (Figure 6.4). The start of a sound is known as the *attack* phase. The attack phase for the sound of a violin

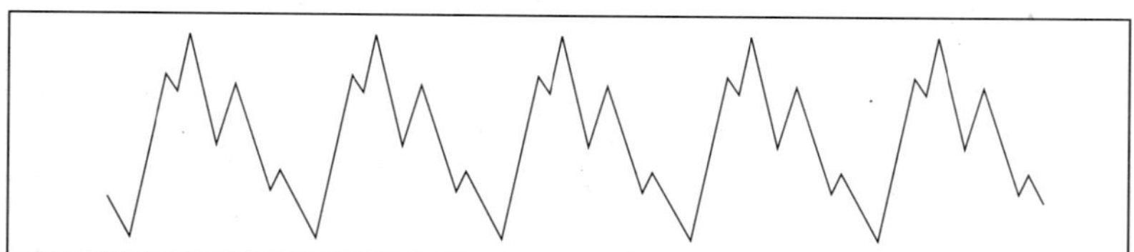

Figure 6.3   *Sample of a complex waveform as it appears on an oscilloscope. This particular form is produced by sounding the first A above middle C (frequency 440 Hz) on a standard piano keyboard.*

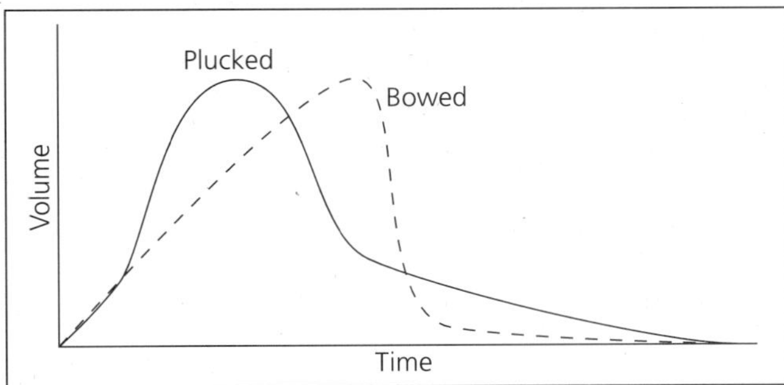

Figure 6.4 *Two sound waves of the same frequency that have very different sound envelopes. Compared to the wave shown by the solid line, the wave represented by the dashed line has a slower attack phase, a shorter sustain phase, and a sharper decay phase.*

string that is bowed is different from one that is plucked. The difference is in duration, amplitude, and timbre. The same note created by striking a bell exhibits yet a different attack than either of the methods used to produce the note on a violin.

The *sustain* phase of a sound refers to the time the sound lasts after the attack. Amplitude and duration of the sustain phase vary depending on how a sound has been generated. Some sounds take longer than others to reach a stable level after the attack phase, and some take longer to begin dying out.

The final stage is the *decay* phase. This refers to the character of a sound as it begins to become inaudible.

Sound envelopes can convey important information about the nature and origin of different sounds. Knowing about them helps you to shape sound in the editing process. For example, intentionally chopping off or reversing the attack and decay stages of a person's speech can radically alter the sounds you hear. To hear the effect of altering the sound envelope, record a person reading words backwards onto audiotape. Then play the tape in reverse. One reason the words still do not sound normal is that the attack and decay phases have been reversed.

MICROPHONES: BASIC TYPES AND CHARACTERISTICS

When sound waves are directed at a pressure-sensitive element, causing it to vibrate in the presence of either a magnetic or an electric field, a pattern of electricity matching the waveform of the original sound is generated. This is how a microphone (or *mic* for short) converts sound energy into electrical energy. Some microphone elements are shaped like a thin disk or diaphragm made of Mylar or aluminum. Others are shaped like a piece of ribbon. In all cases, sound vibrations directed at the element are changed into an analogous pattern of electrical energy.

This process is exactly the same as that described for the telephone microphone in Chapter 2, except that the telephone uses a cruder element of carbon particles to convert sound waves to a pattern of electricity. The telephone mic is considered relatively crude because it is designed only for discrimination of speech. Its pickup element responds only to audio frequencies from 300 to 3,000 Hz. In contrast, professional, broadcast-quality mics are sensitive to audio frequencies over the full range of human hearing (20 Hz to 16 kHz) and beyond.

Microphones are classified according to the types of generating elements they use, their performance characteristics, and their pickup patterns (directional characteristics). We will discuss each of these aspects of microphones in turn.

Generating Elements

In classifying microphones according to the type of generating element they use, there are three basic categories: capacitor microphones, dynamic microphones, and ribbon microphones.

Capacitor Microphones   *Capacitor microphones* (also called *condenser microphones*) use a capacitor as a generating element. A capacitor is any pair of electrical conductors (usually thin metal plates that can store an electrical charge) separated by a thin layer of air or insulating material. When one of the plates moves, a corresponding change in voltage results. In a capacitor microphone, a sound-sensitive diaphragm is used as one of the two plates (Figure 6.5a). When exposed to sound, the narrow air space between the two plates changes in direct relationship to the original sound waves, altering the amount of electricity in its circuit accordingly. The pattern of voltage variations matches that of the original sound waves.

Because capacitor mics use a capacitor instead of a magnet to induce an electric current, they must have a power supply to charge the capacitor. There is, however, a special type of capacitor mic called an *electret* microphone. These mics have their capacitor elements permanently charged during the manufacturing process and therefore do not need power to maintain a charge. Other capacitor microphones receive power either from a supply built into the body of the microphone housing or, more commonly, from a supply external to the mic. In some cases, power can be delivered to the mic over the same cable that carries the audio signal to the recording console or mixer. This is referred to as a *phantom power supply*. The required voltage is available to a capacitor mic plugged into the input but not to other types of mics that do not require power.

Since capacitor microphones (electret or otherwise) produce such a small voltage output, they must boost their electrical signals to usable levels. To do this, a *powered pre-amplifier* (or *pre-amp*) is used. The pre-amplifier is located inside the microphone housing and receives its power from either an internal battery (as is the case with most electret mics) or a remotely located power supply.

Dynamic Microphones   The *dynamic microphone,* also called a *moving coil microphone,* uses an extremely thin Mylar or aluminum diaphragm attached to a movable coil of wire suspended in the magnetic field of a permanent magnet

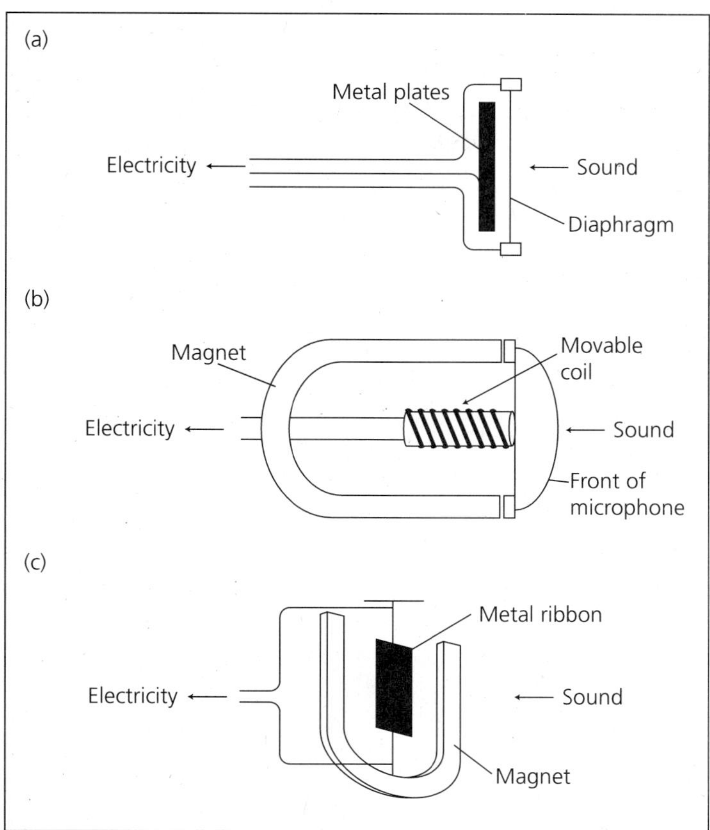

Figure 6.5 *Three different types of microphone generating element. (a) A capacitor mic. (b) A dynamic mic. (c) A ribbon mic.*

(Figure 6.5b). The diaphragm vibrates when subjected to air pressure waves supplied by a sound source. As the vibrating diaphragm moves the coil, an electrical voltage that matches the original sound pattern is induced.

Ribbon Microphones In place of a diaphragm or wire coil, *ribbon microphones* use an extremely thin, lightweight, corrugated metal strip suspended in a powerful magnetic field (Figure 6.5c). Air pressure waves vibrate the metal ribbon, inducing a voltage pattern that matches the original sound wave. Ribbon microphones are also called *velocity microphones* because the voltage generated as the ribbon cuts through the magnetic field is the electrical equivalent of the velocity of the air molecules themselves.

Performance Characteristics

The range of frequencies to which a microphone is sensitive is its *frequency response*. Manufacturers supply a specification (spec) sheet for every microphone containing a graph of the mic's response curve. The response curve plots dB gain on the vertical axis against frequency range on the horizontal axis (Figure 6.6). The flat part of the curve indicates the frequency range over which audio response for that microphone is constant. The ends of the curve indicate the limits of the mic's sensitivity. The spec sheet also shows where the mic has "peaks" or "roll-offs." Peaks are portions of the frequency spectrum over which audio sensitivities increase; roll-offs are portions over which sensitivity drops out.

Mics differ considerably in their frequency response and also in their general ruggedness and durability. Here is a brief summary of the usual performance characteristics of the three different types of mics:

- *Capacitor mics.* Because of their electrostatic design, capacitor mics generally deliver the highest-quality sound across the entire frequency spectrum. They reproduce more detailed sound than other mics and respond more quickly to sudden changes in sound (such as quick attacks and decays). However, because their generating element is made of an extremely thin sheet of metal or metal-coated plastic, they are less rugged than dynamic microphones. In addition, they are less impervious to extreme weather conditions. Nevertheless, they do have applications in field settings.
- *Dynamic mics.* Dynamic mics, by virtue of their construction, are among the most rugged and durable of all professional microphones. Their ability to withstand physical shocks and severe weather (both temperature and humidity variations) makes them a fine choice for field productions. In terms of frequency response, dynamic mics discriminate high-frequency sound more clearly than other types of mics.
- *Ribbon mics.* Ribbon mics are the most fragile of all professional microphones, and for this reason are not recommended for field use. In terms of frequency response, ribbon mics produce a warm, rich, mellow sound, which is excellent for studio announcing and musical presentations. However, because ribbon mics are activated by sound velocity, they may be especially affected by sudden sharp sounds. In fact, loud sneezes or coughs directed into a ribbon mic at close range can permanently damage it.

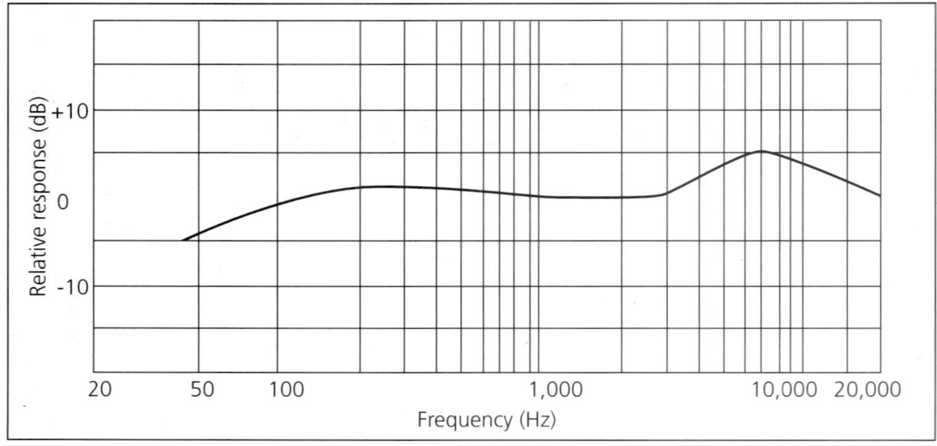

Figure 6.6   *Response curve for a typical wireless lavaliere mic.*

All mics, regardless of type, can encounter noise problems from outdoor wind. Moreover, sibilant vocal sounds, such as the letter *s,* can create a wind effect in any setting, and the consonants *b, p,* and *t* can cause "popping" noises. Some speakers are harder to mic than others because of the way they produce these letters. Proper mic technique, however, can help eliminate such problems. Popping sounds can be almost completely eliminated by keeping the talent at a proper distance of about 12 inches from the mic. Another technique involves placing the mic somewhat to the side of the speaker or slightly off-axis to the direction of speech. For close miking in studio situations, *pop filters* made of mesh are used to cover the mic head. More extreme or windy outdoor situations call for foam windscreens or other devices.

To offset sound problems caused by rough handling, some mics are designed with their diaphragms doubly insulated from their housing. In addition, a mic may be placed in a *shock mount,* which suspends the microphone in a protective housing to isolate it from mechanical vibration.

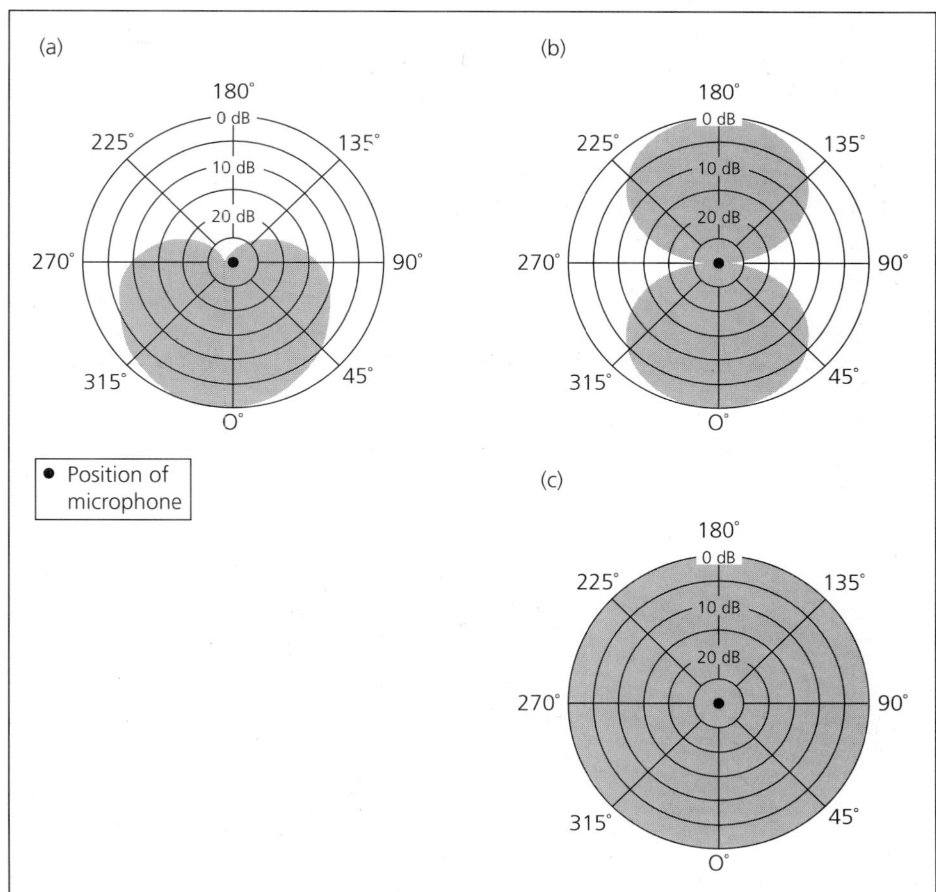

Figure 6.7 *Polar diagrams showing three basic microphone pickup patterns. (a) Unidirectional or cardioid pattern. (b) Bidirectional pattern. (c) Omnidirectional pattern.*

Pickup Patterns

The *pickup pattern* of a microphone refers to the direction(s) from which the mic is sensitive to sound. There are three basic patterns:

- If a mic is sensitive to sound from mainly one direction, its pickup pattern is *unidirectional*. Often a unidirectional mic is referred to as a **cardioid mic** because the typical pickup pattern is somewhat heart-shaped. Depending on how concentrated the pattern is, cardioid mics may be further classified as *supercardioid, hypercardioid,* or *ultracardioid,* each of which describes progressively more directional sensitivity.
- Mics with pickup patterns in two opposite directions are called *bidirectional.* Their pickup pattern resembles a figure eight.
- Finally, if a mic is equally sensitive to sound from all directions, its pickup pattern is said to be *omnidirectional* or *nondirectional.*

These types of pickup pattern are illustrated in Figure 6.7, which shows each pattern on a *polar diagram,* a concentric circle graph marked off with pie-shaped segments. The microphone is at the center of the graph, and each circle farther from the center represents a drop in microphone sensitivity. Polar diagrams can also indicate the effective distance of a microphone. Such information can help you determine how far talent can be placed from the mic and still be "on mic" or maintain "presence" (or sense of closeness).

MICROPHONE SELECTION

Selecting a microphone and positioning it for use should be based on the job you need it to do, as well as on equipment availability. For example, for a round-table discussion with several participants, simplicity might dictate a single, stationary, omnidirectional mic placed in the center of the table (Figure 6.8). If, however, you are covering a sports event with an announcer, a color commentator, and much unwanted crowd noise, you might consider using separate unidirectional mics for each speaker. These will help limit sound to

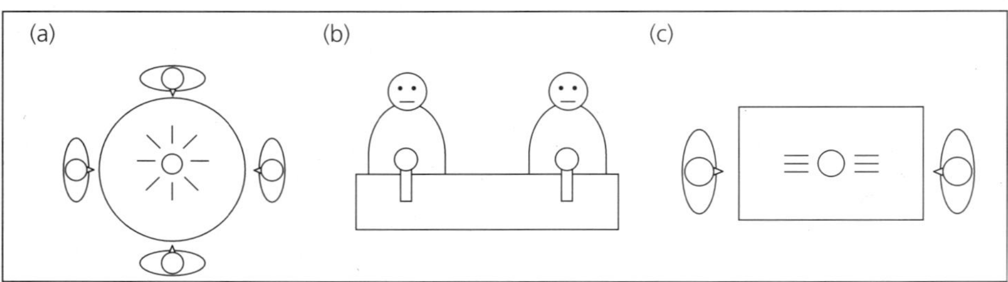

Figure 6.8   *Typical microphone selections for different situations. (a) For a round-table discussion, a single omnidirectional mic in the center of the table. (b) For two sports announcers calling a game from the booth, separate unidirectional mics. (c) For a host-guest interview with the two talents seated at opposite ends of a table, a single bidirectional mic.*

just their two voices. But if you decide to include crowd noise, you can hang some additional omnidirectional mics near the crowd. For a host-guest interview, a single bidirectional mic may be the best way to go. However, if the host's voice is significantly louder than the guest's (or vice versa), you might wish to use two cardioid mics with a great deal of separation, allowing you to set different audio levels.

From these simple examples, you can see there is much to consider in deciding what microphones to use and how to deploy them. Audio design means more than just placing mics in the vicinity of your talent and going on the air. Add the questions of whether sound will be monophonic or stereo, whether the mics should be visible or out of sight, and whether the talent is stationary or moving, and you have many issues to consider. The following sections describe strategies for selecting microphones to meet specific program objectives. Keep in mind that a particular mic may have various applications, but it is likely to be better for some jobs than for others.

Audio Design Principles

On-camera Versus Off-camera Mics Generally, when talent is oriented directly to the camera, as in news, interview, talk, game, quiz, variety, and audience participation shows, it is acceptable (though not required) for microphones to be visible on camera. It may be desirable to display a host's mic to impart authority, as is often the case with such performers as Oprah, Geraldo, Sally Jesse Raphael, or Montel.

In contrast, in programs featuring talent oriented *indirectly* to the camera, as in soap operas, dramas, and situation comedies, visible mics are often a production error, since they break the show's illusion of reality. Of course, not all on-air personnel in news look at the camera, and not all talent in sketch comedy avoid it, but most of the time talent's orientation to the camera is a strong indicator of whether it is proper for mics to be visible. From this consideration comes the first principle of audio design: decide whether or not mics should be seen on camera.

Stationary Versus Mobile Mics After deciding whether mics should be visible, determine whether they should be stationary. One factor that often influences this decision is talent movement. However, if the talent moves, that does not necessarily mean his or her mic must go along for the ride. The critical factor is: *put mics where the sounds that must be picked up are located.*

For example, if you are miking a drama and your talent moves across the entire studio floor but speaks in only one location, you may mic just that one location, perhaps with a hanging mic over the speaking area or concealed in the area where the speech takes place. On the other hand, if your talent has a frenzied scene, moving around a large set while delivering a key speech, you may need a mobile mic, perhaps a mic on a boom or a wireless mic concealed on the talent's body.

Whatever you do, base your strategy on both production values and equipment and personnel availability. Remember that in most cases, simplicity is a virtue. It is not necessary to mic every talent location, only those where sound must be picked up. Only by careful study of the script materials, floor plan, facilities, and lists of crew and talent can you determine how to proceed.

Visible Mics

If it is acceptable, desirable, or plausible for mics to be seen, you have a number of alternatives. These include personal microphones worn on the clothing (most often in the upper chest area) or held in the hand. Other visible mics include those on desks, stands, and tables.

Lavaliere (Lapel) Mics   The **lavaliere** or **lapel mic** (Figure 6.9), as the name implies, is clipped to the front of the talent's clothing, in the chest area, close to but below the talent's mouth. It may be fastened to a tie, jacket lapel, dress, or blouse. Because it is positioned well below the chin, the speaker does not talk directly into it. Therefore the lavaliere mic (or *lav* for short) is usually omnidirectional. In addition, many lavs have a built-in high-frequency boost to compensate for the loss of high frequencies, which tend to travel more in straight lines than other frequencies do.

Virtually all lavs are of the electret capacitor type. This enables them to be made quite small and unobtrusive. The battery pack and cable connector can be placed behind or to the side of the talent, out of camera range.

Lavs offer many advantages. Since they are small, they can be easily placed on the talent without getting in the way. Their small size also makes them easy to conceal under talent's clothing, so they can be used for either on-camera or off-camera applications. With the mic clipped to clothing, the talent's hands are free, and the talent can move without worrying about getting out of microphone range. Of course, lavs are great for stationary interviews as well.

However, lavs have disadvantages. Since they are designed for close, single-voice work, their frequency response is not ideal for musical performance. Moreover, their small size allows them to be easily damaged (especially their fragile mic cable), misplaced, or lost. Further, although their small size makes them easy to conceal under clothing, poor placement can result in muffled sound. Even quiet activities such as holding a book close to the chest can muffle speech or cause unacceptable noise. In addition, vigorous movement can create distracting noise because the mic can jostle and rub against clothing or bang against jewelry. Vigorous motion can even cause lavs wired to the body to pull loose.

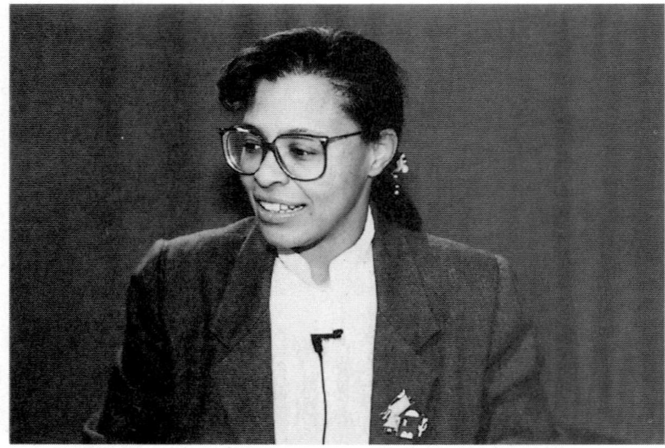

Figure 6.9   *A lavaliere mic clipped to talent's clothing.*

For these reasons, it is best to have the audio operator secure the lavs on the talent and then caution the talent about possible problems. For talent engaged in vigorous movement, it is better to avoid using wire lavs altogether. Wireless lavs are available for this type of movement, though they remain subject to many of the other problems typical of lavs. Aerobic exercise programs use wireless lavs with great success because the producers give a good deal of thought to securing them to the body properly and keeping noise problems to a minimum.

Headset Mics   Headset mics (Figure 6.10), often used by sports broadcasters and pop singers, allow the user to position the mic extremely close to the mouth while maintaining constant distance even when the head is turning. They also free up the user's hands as well as the work space in front. Headset mics may be unidirectional to limit background sound or omnidirectional to include it. Because the mouth is so close to the pickup element, the headset mic often has a built-in pop filter to reduce unwanted blowing sounds caused by breathing as well as popping sounds from consonants such as *b, p,* and *t.*

Headset mics equipped with headphones can be fed two separate audio signals, such as program audio through one ear and director's cues through the other. Being thickly padded, the headphones enable the wearer to hear production cues in the presence of crowd noise. Some models feature a "cough switch" to enable the user to turn off the mic when desired.

Hand Mics   Hand mics (Figure 6.11) afford mobility and control to program hosts who use them on audience participation programs and to news reporters in the field. Singers use hand mics to enhance their performance by bringing the mic closer to the mouth for quiet passages and moving it away during climactic parts. Some use the mic as a prop, tossing it from hand to hand, even swinging it by the cord to excite the audience, but these kinds of antics can damage the equipment.

If you use a hand mic, you carry much of the responsibility for sound quality. For example, you must be aware of the mic's ability to pick up sound sources at different distances. As distance increases, mic sensitivity decreases. Further, if the mic is the conventional wire type, you need to know the extent of the area talent can move about in without yanking and damaging the cable.

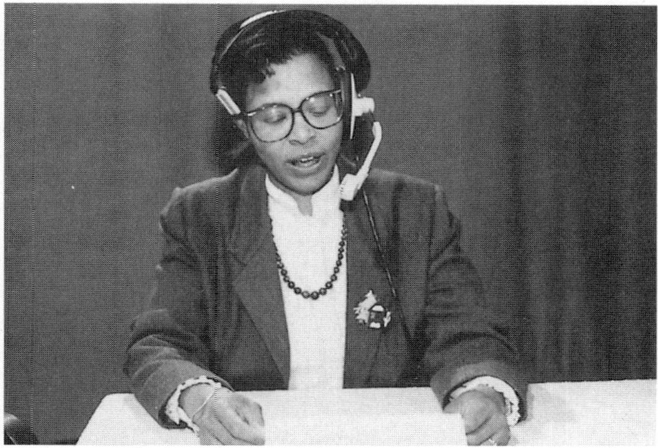

Figure 6.10   *Headset mic worn by talent.*

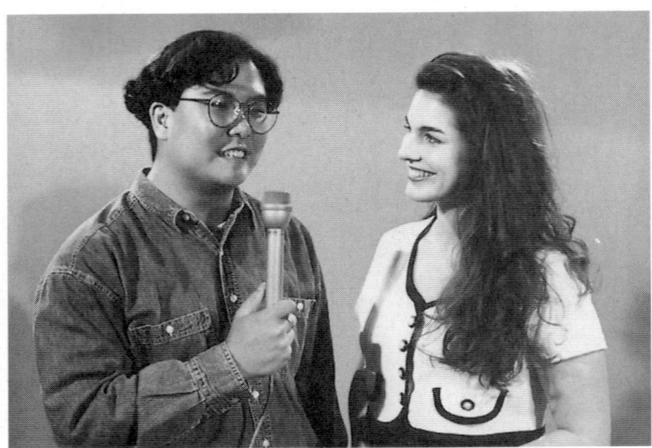

Figure 6.11   *Hand mic held by talent so that it can be easily tilted toward the other person when it is her turn to speak.*

In an interview, be aware of the volume of the interviewer's voice relative to the subject's; favor the softer voice with closer placement. Also, when alternating between speakers, try to move the mic between them at the proper time to avoid clipping opening or closing remarks. Finally, if the interviewee is easily intimidated by the microphone, you may need to comfort him or her beforehand to reduce anxiety.

A wide variety of hand mics with different performance characteristics are available. Some have omnidirectional pickup patterns; others are unidirectional. The more rugged ones for field production use dynamic pickup elements; however, electrets and ribbon hand mics are also available.

Desk Mics   Desk mics (also known as *table mics*) are often simply hand mics clipped to desk stands (Figure 6.12). They are generally visible, stationary mics suitable for talent seated at a desk or table or standing at a podium. Desk mics are frequently used in news and panel shows. Desk stands vary from

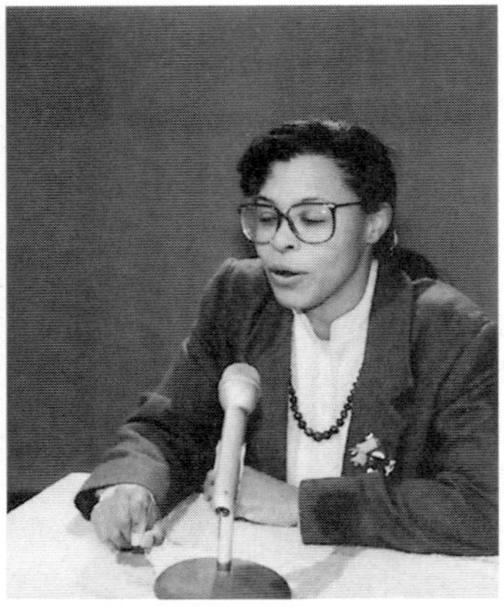

Figure 6.12   *Talent using a desk mic.*

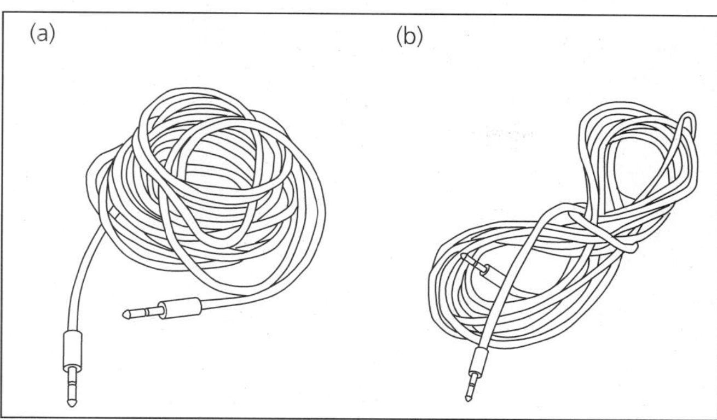

Figure 6.13   *Two ways to store cables. (a) Coiling, which often produces tangles. (b) The ribbon method, a better alternative, which eliminates tangling.*

fixed, static supports with a heavy base to the gooseneck variety with a clamp for quick, easy attachment to a podium or lectern at a news conference.

Since desk mics are most often used for speech, dynamic mics are a good choice for this function. Their ruggedness also makes them especially well suited to desk use, where they may be inadvertently knocked around.

Stand Mics   Desk and hand mics can be mounted on floor stands (Figure 6.14). This arrangement is good for stationary performers when it is appropriate for the mic to be visible (such as with singers, stand-up comics, and an-

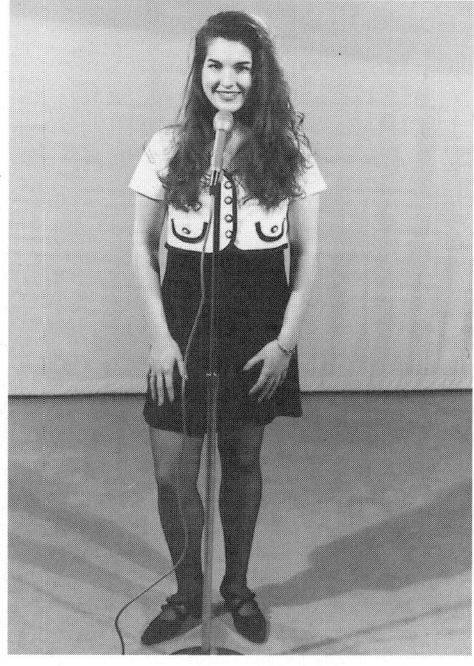

Figure 6.14   *Talent using a stand mic.*

Handling Mics

- Even ruggedly designed mics should be treated gently. Don't bang them around. Don't drop them. When you want to get one out of your hand, place it down carefully in a safe place, out of harm's way.

- To see if a mic is on, speak into it or scratch softly across the screen of the pickup element. Do not blow into it.

- Respect the integrity of the equipment, including the mic cables. If they get tangled or get caught on furniture or other obstructions, don't yank them to free them. Untangle them gently.

- When you walk with a mic in one hand, hold the cable with the other hand to avoid pulling the cable with the mic, which can damage the connector.

- After a production session, disconnect mics and store them in their boxes or pouches to protect them from loss or damage. When storing cables, use the *ribbon method* (Figure 6.13) instead of coiling the cables to minimize tangling.

- Avoid the unsightly look of a tangled mass of wires by dressing the cables. Tape the cables unobtrusively along a table edge or other convenient surface, out of camera view and out of harm's way.

- Even if a cable is concealed under a rug on the set, tape it down to minimize the chance of tripping someone.

- To keep users from disrupting desk mics, tape the mics into position if possible. Instruct the talent not to touch the mics or tap on the table or desk. If mics must be shared among speakers (not the best situation, but it does happen), shock-mount them and instruct users to pass them gently.

nouncers). The mic and mic stand can also become props for the performer, who can move them around, lean on them, tilt them, disconnect the mic from the stand, reconnect it, and so forth. The major advantage of the mic stand is that it leaves the performer's hands free while fixing the location of the mic precisely. The major disadvantage is that it can limit movement.

Mic stands come in various shapes, from straight vertical stands to those with goosenecks. There are also stands with crossbars to support additional mics. These are useful, for instance, for a single performer who plays the guitar while singing. In this case, a single stand can accommodate separate mics to pick up guitar and voice.

Off-camera Mics

Off-camera mics may be mobile or stationary, off the set or concealed on the set, hanging, attached to a boom, or held by a crew member. This section discusses the chief uses of some of the more common off-camera mics.

Hanging Mics   Any mic hung over a set, even a hand mic or a lav used in this fashion, is a **hanging mic** (Figure 6.15). The main function of such mics is to provide microphone coverage for a stationary performance area. Usually they are hung out of camera range. Examples of production situations that can be carried by hanging mics include a talent speaking while demonstrating an exercise or preparing a recipe, a writer performing a poetry reading, or two actors playing out a scene of a chance meeting at a bus stop. A major advantage of hanging mics is that they enable audio coverage of talent without mic wires. Disadvantages are that they cover sound only in a limited performance area and can cast distracting shadows on talent and set pieces.

The preferred pickup pattern for a hanging mic is cardioid rather than omnidirectional. This is because omnidirectional mics tend to pick up reflected sound from the ceiling as well as extraneous noise. To keep the mic out of camera range, hang it while watching the widest camera shot (the master shot) on the line monitor. Lower the mic into the set until you see it enter the master shot, and then pull it up high enough to remove it from the shot.

After hanging the mic, be sure to check audio levels for all personnel who will use it. Be sure the mic is sensitive enough to handle the job. For example, a tall talent close to the mic may have enough presence, but a shorter talent

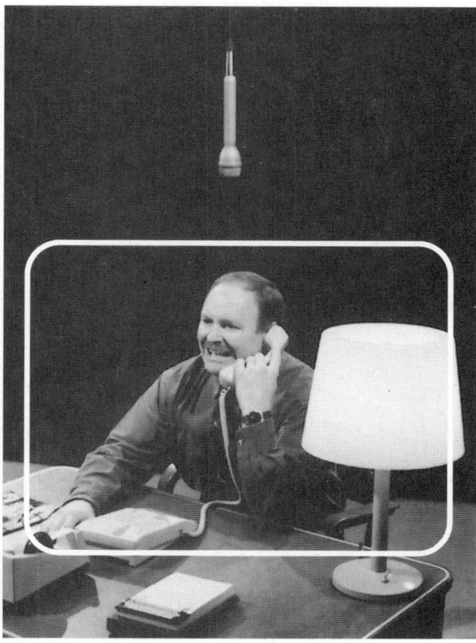

Figure 6.15   *A hanging mic. Note that the mic can be as close to the talent as necessary, as long as it is not in the shot.*

may not and may need to be miked separately. Under no circumstances should you accept an audio check from one user as proof that all is well. Check everyone's level before you approve the setup.

Concealed Mics   **Concealed mics** are hidden anywhere on the set, such as in a vase of flowers, under a sheet of newspaper, or taped to the edge of a table or to a telephone handset. Such mics are usually stationary, but a wireless lav could be transported to different locations taped to a cordless phone or worn under talent clothing. Of course, any live mic is subject to unwanted noise. If you conceal a mic under a sheet of newspaper, for instance, don't move or crumple the paper unless you want to hear a remarkably loud rustling sound!

Lavalieres make good candidates for concealed mics because they are small. If they require cables, dress them appropriately so they don't draw undue attention. Often a simple throw rug can help to hide the mic wire.

Boom Mics   Just as with lights, booms—essentially long arms—are useful for keeping the technical apparatus out of camera range. **Boom mics** can be set up in various ways, and the booms may be either mobile or stationary. In its simplest form, a boom mic may be nothing more than a stick with a mic on the end of it held by a crew member. This type, known as a *fishpole boom* (Figure 6.16), can be held above or below the talent out of camera range to augment spotty coverage of an area, and it can be moved as the sound source moves.

A *tripod boom* or *giraffe boom* (Figure 6.17) may also be used to support a mic if a crew member is not available or if it is preferable to secure the mic in a fixed position. The tripod boom is an excellent choice for covering a fixed audio source, but it cannot move along with a mobile sound source.

A more sophisticated device is the *perambulator boom* (Figure 6.18), a wheeled device that allows an operator to move a mic into and around a set from out of camera range using a set of pulleys and extenders. The mic can be rotated to cover more than one talent position, and the height of the boom arm can be adjusted. Some perambulator booms provide a platform with a seat for the boom operator. From this position, the operator can silently pan,

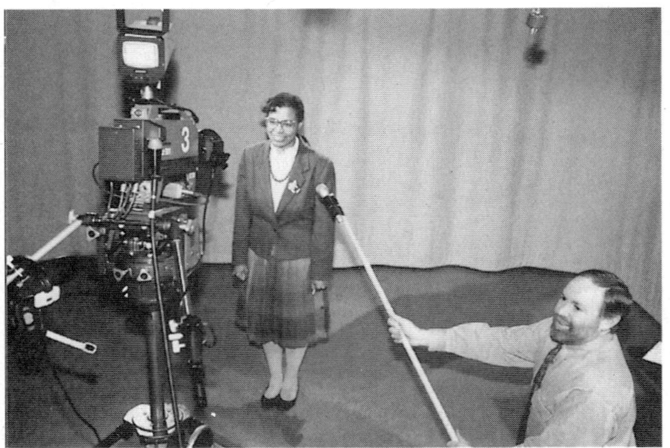

Figure 6.16   *Use of a fishpole boom. The crew member, standing well out of camera range, can use the boom to position the mic in an appropriate place near the talent.*

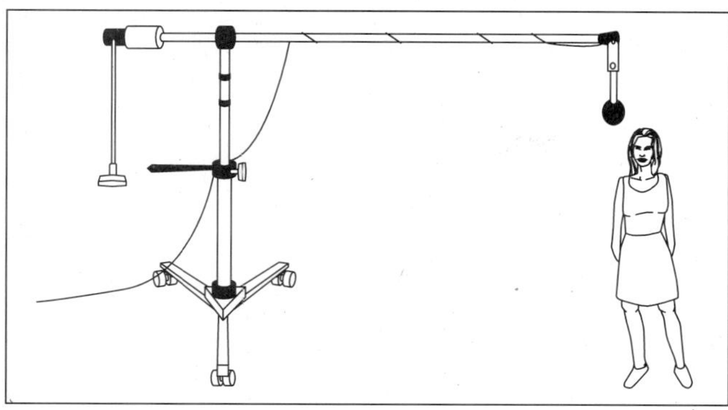

Figure 6.17 *A tripod (or giraffe) boom.*

tilt, extend, and retract the boom arm during production. Some perambulator booms provide a video monitor to help the operator judge how close the mic can get to the sound source without being seen on camera. The operator may wear a headset to hear director's cues, program audio, or even other crew members. The entire boom may be moved on its wheels by a second crew member. Perambulator booms are frequently used for drama productions such as soap operas and sitcoms.

Mobile booms provide programs with natural *sound perspective,* which refers to the agreement between video and audio distance in program content. In normal sound perspective, video close-ups are accompanied by close miking, and long shots are accompanied by more distant miking. The boom operator achieves this by bringing the mic closer to the talent when framing is tight and moving the mic away when the shot widens. Violations of sound perspective can be disconcerting to the audience.

The disadvantages of boom mics include the fact that they take up a lot of space and require up to two skilled crew members to operate them. A peren-

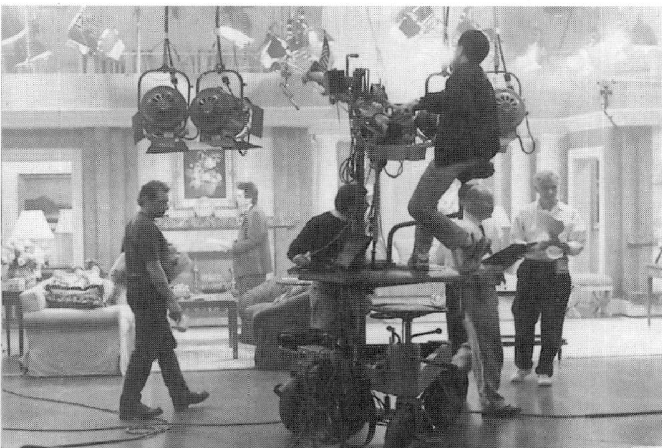

Figure 6.18 *A perambulator boom used to position a mic over the set of a soap opera. This sophisticated type of boom allows the operator to move the mic swiftly and easily during production.*

nial, more important problem with all boom mics is unwanted shadows. To avoid or minimize this problem, integrate booms from the side of the set *opposite* the key lights. This will at least avoid the harshest shadows. Above all, plan where booms will go *while* you are planning the lighting scheme. (See Table 6.1 for a brief summary of mic problems and ways to solve them.)

Long-Distance Mics   Microphones that pick up sound from a great distance offer another means of covering on-camera talent without cables. In outdoor settings, long-distance mic coverage is sometimes essential. The **shotgun mic** (Figure 6.19) has been developed to serve these needs and has applications in sports coverage and on-the-scene news coverage. The shotgun mic uses an interference tube to attenuate sound from all directions except the one in which it is aimed.

A **parabolic mic** (Figure 6.20) uses an omnidirectional or unidirectional mic at the focal point of a concave parabolic reflector, with the mic's pickup element facing the center of the reflector. The concave surface of the dish is then aimed at a sound source, and collected sound is reflected into the microphone from the direction the dish is facing. A headset can be used in conjunction with the mic to locate the optimal position.

TABLE 6.1

Some Microphone Problems and Their Common Solutions

Problem	Solution
"Popping" sounds or too much sibilance in talent's voice	Keep the talent at a distance of about 12 inches from the mic. Position the mic somewhat to the side of the speaker or at a slight angle so that the speaker talks across the mic rather than directly into it. Or, if the mic must be close to the mouth, cover the mic head with a pop filter.
Outdoor wind noise	Use a pop filter or a foam windscreen.
Audio transition difficulties when talent moves around the set	Use a boom mic, or a wireless mic attached to the talent's clothing. If you must use stationary mics, use the same model in each performance area and check that the mics provide uniform coverage.
Interference between mics in close proximity to one another	Either put the mics as close together as possible so they receive sound simultaneously from the same source (*dual redundancy setup*) or place them at least three times the distance from one another that each is from its source (the *three-to-one rule* for multiple mic setups).
Shadows from hanging mics or boom mics	Position the mics out of the direct path of lights, especially key lights with a hard directional throw. For example, put the boom mics on the opposite side of the set from the key lights. Better yet, plan ahead in preproduction so that you can integrate the mic setup with the light plot.
Hum or other noise from power lines	Hanging mics especially may pick up hum from power lines that are improperly grounded or shielded. Try running the cables away from and perpendicular to the power lines. If this problem can't be fixed, find another way to mic the area.

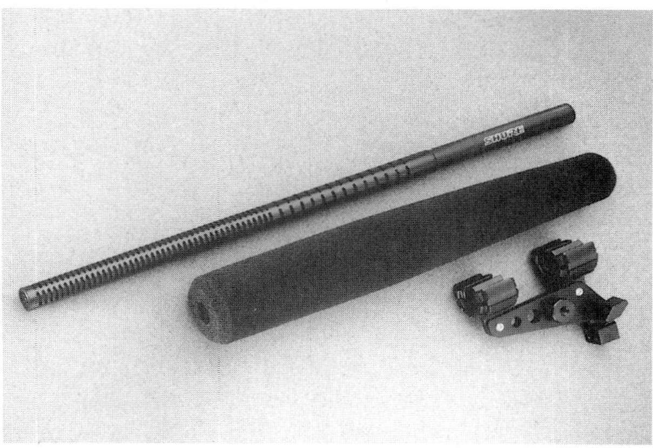

Figure 6.19 *A shotgun microphone.*

The advantage of highly directional mics is that they permit audio coverage without being visible on air and without restricting talent movement with mic cables. However, they have some drawbacks. The greater the directionality of a mic, the more compromised its frequency response tends to be. The flat part of the response curve, the part showing the frequency range over which audio response is constant, tends to be narrower than that for less directional mics.

Moreover, because of the shotgun mic's construction, it becomes less directional with lower-frequency sound, especially wavelengths that are longer

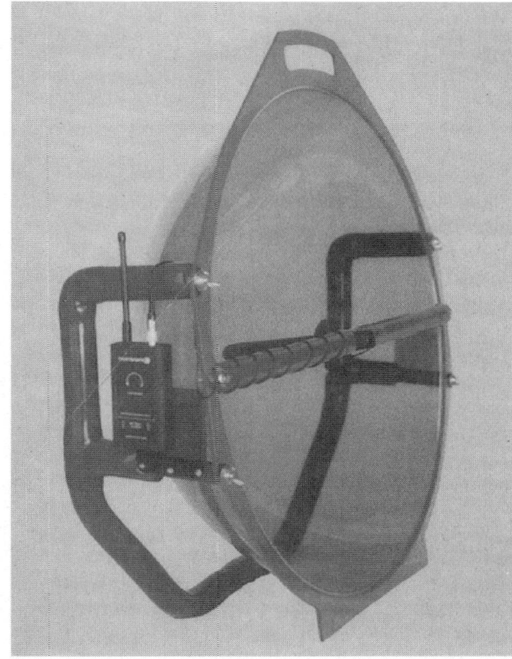

Figure 6.20 *A parabolic microphone.*

than the length of the tube. This means that at low frequencies it fails to discriminate sound from unwanted directions. The same is true for parabolic mics.

In addition, for all long-distance mics, since sound travels more slowly than light, the video of a speaker's face (especially mouth movements) can appear to be out of sync with the words spoken. For this reason, it is sometimes better to use a concealed mic near the talent.

Pressure-Zone Mics   A special type of mic designed to pick up sound from different sources at varied distances with roughly equal volume and clarity is the **pressure-zone mic (PZM),** also called a *boundary mic* (Figure 6.21). The PZM uses a pickup element (usually an electret transducer) pointed at a hard, flat, built-in plate. This makes the PZM most sensitive to sound reflected from the plate instead of from direct and indirect waves from floors, walls, and other reflective surfaces at different distances. The PZM's design reduces reverberation and muddiness caused by time differences between sounds received directly from sources and those reflected from nearby surfaces. It results in cleaner audio, since additive and subtractive phase shifts (constructive and destructive interference) are greatly reduced.

PZMs may have either omnidirectional or directional pickup patterns, with an open and spacious acoustic quality. Their pickup patterns can be altered somewhat with the use of *blocks* designed to obstruct unwanted sounds from selected areas.

The disadvantages of PZMs result from their ability to treat all received sound roughly equally. For example, if the mic is on a table, tapping with a pencil or rustling papers will be picked up just as clearly as speakers' voices.

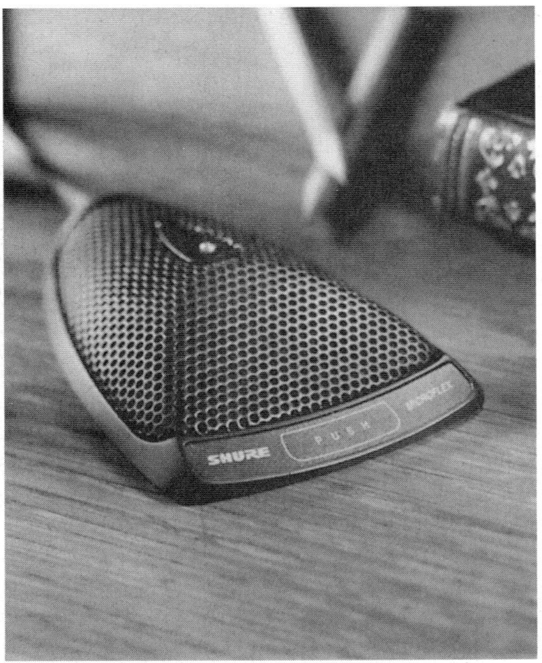

Figure 6.21   *A pressure-zone microphone (PZM).*

For these reasons, some experts restrict PZMs to specific off-camera uses—for instance, hanging them above the set merely to pick up ambiance. Others avoid PZMs entirely, finding they get better results from conventional mics in almost every situation.

Wireless Mics

Wireless mics (Figure 6.22), as we already saw, dispense with cables that can hamper a talent's movement. Performers in both indoor and outdoor settings have used wireless mics, including astronaut moon walkers, race car drivers, mountain climbers, and rock musicians. Wireless mics can be hand mics, headset mics, or any other kind, and they can be positioned either on or off camera.

The wireless mic system consists of a battery-powered FM radio transmitter (usually concealed somewhere on the talent's body) that sends a radio signal from the mic to a receiver connected to an audio console, to loudspeakers, or to any other desired output. Since wireless mics convert an audio signal into a radio signal, they are sometimes called *RF (radio frequency) mics*. When a lavaliere mic is used, the transmitter is often worn as a separate unit. In some hand mics, however, the RF transmitter may be built into the mic itself.

The receiver for the mic's RF signal can be located several hundred feet away from the transmitter. Of course, because it uses a broadcast signal, the wireless mic is subject to interference problems from surrounding conditions and from nearby transmitters. To obtain an optimal signal, it is wise to verify what radio frequencies are already in use in your area so that you can choose equipment that operates on different frequencies. Signals from wireless mics are also subject to problems from perspiration, which can shunt the signal straight to ground. To avoid this problem, keep the transmitter protected from sweat. In addition, it is helpful to keep the antenna vertical and clear of obstructions.

Tunable wireless mics allow you to tune to a frequency not already in use in the area. They are especially useful in situations where more than one wireless is being used in the same production at the same time.

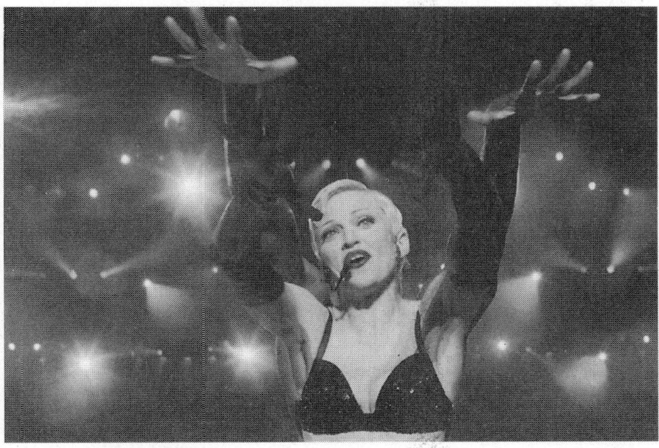

Figure 6.22 *Talent using a wireless mic.*

In recent years, wireless mics have become one of the most versatile audio instruments in the director's arsenal. Their advanced technology makes them easy to conceal. They have well-defined pickup patterns and excellent resistance to wind and clothing rustle. These advances, coupled with the fact that they can be tuned to different frequencies, make them excellent instruments in both studio and field.

Other Technical Considerations

Proximity Effect   When a sound source is placed close to a directional microphone, frequency response increases in the low portion of the audio range. This is because, compared to higher-frequency sound, low-frequency sound exhibits a greater pressure differential between sound reaching the front and sound reaching the rear of the generating element. This difference, known as a *proximity effect*, causes bass response to rise. The result can be muddiness in the bass response and masking of higher frequencies. To reduce proximity effect, many mics have a feature known as *bass roll-off*, which limits bass frequencies below a certain point.

Microphone Impedance   *Impedance* (symbolized with a Z) refers to the amount of electrical resistance in a circuit. It is measured in ohms, which are units of electrical resistance. Mics with impedance levels below 600 ohms are considered low-impedance (low-Z) mics. Those above 10,000 ohms are considered high-impedance (high-Z) mics. Low-Z microphones are less prone to noise, such as interference from fluorescent lights and static from nearby motors. Low-Z mics can also be used with longer mic cables without significant signal loss.

It is critical to match impedances between mics and the equipment to which they are connected. A mismatch may result in serious audio distortion. Mismatches can be corrected by using a matching transformer.

Balanced Versus Unbalanced Lines   Cables used to carry audio signals from microphones to their destinations (amplifiers, loudspeakers, tape recorders, consoles) may be either balanced or unbalanced. *Balanced* cable consists of two conductors and a shield that acts as a ground. *Unbalanced* cable has only one conductor and a shield that serves double duty as ground and conductor. The advantage of balanced cable is that it is less prone to electrical noise. Adapters are available to connect these two types of lines to each other if necessary.

Connectors   Microphone connectors link mics with cables or two cables to each other. Most professionals prefer a three-pin connector called an *XLR* to transmit audio signals along balanced lines. The *X* stands for the ground or shield, *L* designates the lead wire, and *R* indicates the return wire. Adapters are available to connect XLR cables to two-pin alternatives if necessary.

STEREO AND SURROUND SOUND

Our discussion so far has been limited to *monophonic* audio production. This is because in reality, all individual microphones are monophonic devices. Some "stereo mics," as they are called, contain two pickup elements oriented in dif-

<table>
<tr>
<td>

Open and Close Miking of Musical Performances

</td>
<td>

Live music requires capturing all the sounds of the ensemble. To do this, you can use one of two techniques: *open* (or *distant*) miking, in which you place mics far enough from the sound source that the same mic(s) picks up and blends most or all of the ensemble's sound, or *close* miking, in which each individual voice, instrument, or sound section is separately captured by a directional microphone placed very close to it.

The open miking approach often uses omnidirectional mics, resulting in a blended, airy sound that includes room reflections. In the close miking technique, the directional mics capture a drier, more direct sound, with little or no reflected sound and a great deal of separation among various instruments and voices. The advantage of close miking is that it gives you more control over the music because the elements are so widely separated. The disadvantage is that the lack of reflected sound reduces the sense of sonic space. Most musical recording is done with close miking techniques, but to compensate for the absence of reflected sound, reverb and pan circuits (described in Chapter 7) are often used in the postproduction mixing.

</td>
</tr>
</table>

ferent directions in the same housing. These mics provide little spatial separation and therefore offer limited stereo effect. However, more robust stereo sound can be achieved in other ways. Chapter 7 discusses multitrack stereo recordings and methods of altering sound electronically. Here we will look at ways to produce stereo effects by careful placement of regular microphones into the production environment.

Stereophonic sound exploits the fact that humans have two ears located in slightly different places, resulting in what is called *binaural hearing*. Binaural hearing enables us to experience minute differences in the intensity and arrival time of the sounds we hear. These differences add a sense of location and depth to our hearing. Stereo sound simulates binaural hearing by feeding two separate audio channels (left and right channels) with slightly different audio information to increase our sense of the *sound space* or *audiospace,* that is, the sound qualities of depth and location. Microphones may be used to create stereo effects by carefully fixing their orientation to sound sources and to one another in the audio space. The basic principle is to place mics so as to imitate binaural hearing.

One way to mic stereophonically (the *spaced technique*) is to place two omnidirectional or cardioid mics (or several if need be) perpendicular to a sound source so that sound arrives at the pickup elements at slightly different times and at slightly different volumes (Figure 6.23a). The obvious disadvantage of this method is the threat of phase problems (constructive and destructive interference) when these signals are combined in monophonic speakers in home receivers. To offset *stereo-to-mono incompatibility,* as this problem is called, the distance from mic to mic should be at least triple the distance of each audio source to its assigned mic (the *three-to-one rule*). Test equipment is available to display phase relationships so that mic placement can be adjusted to minimize interference.

To ensure stereo-to-mono compatibility, a different stereo miking technique called *X-Y miking* or *coincident miking* may be used. X-Y miking uses two

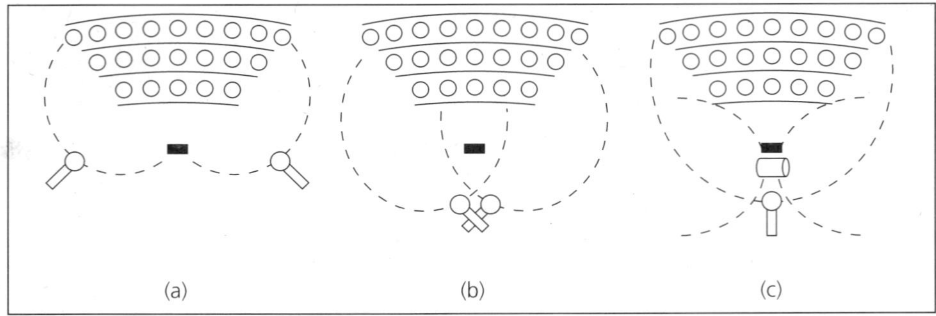

Figure 6.23   *Three techniques for stereo miking an orchestra performance. (a) The spaced technique. (b) X-Y miking. (c) Mid-side miking.*

directional mics located close to (but not touching) each other in the center of the audio space, crossed at roughly right angles, with their pickup elements on the same vertical axis (Figure 6.23b). This minimizes phase differences, but still discriminates sound intensity for left and right channels sufficiently to create stereo effects. The angle between the two mics should be dictated by the width of the sound source.

Finally, a technique called *mid-side (M-S) miking* is available for creating stereo effects (Figure 6.23c). This technique uses a bidirectional mic in close combination with a directional mic. The directional mic is aimed toward the center of the sound source. The bidirectional mic is placed parallel to the sound, covering the left and right sides of the audio space. Audio outputs from this arrangement are then fed to an audio circuit designed to use phase differences to feed left and right channels.

Problems in Stereo Miking

Recall from Chapter 4 that flat lighting was developed to minimize jarring light level differences when cutting between cameras shooting the same scene from highly discrepant positions. Similarly, it is important to design audio to minimize sonic discrepancies when cutting between cameras. This means not violating viewer expectations regarding the locations of sound sources.

Matching sounds to pictures can be particularly difficult when using stereophonic miking formats. For example, imagine an opening shot of a character appearing in the left half of the screen. Initially, it makes sense for that character's voice to come predominantly from the left channel. However, if the next shot captures the same character in the right half, does that mean the voice must now come from the right channel, or will that switch be too jarring? Do all camera angle changes require similar changes in sound perspective? Many experts would say no. The question then becomes: when should stereo sound perspective change? The answer is: when it makes sense, when it is plausible, and when it enhances the program purpose. Such change should be avoided if it will destroy continuity.

To avoid the problem altogether, it may be wise to feature voices of main characters in the center of the audio space regardless of where they appear on screen. This is, in fact, what is done with dialogue in surround sound audio.

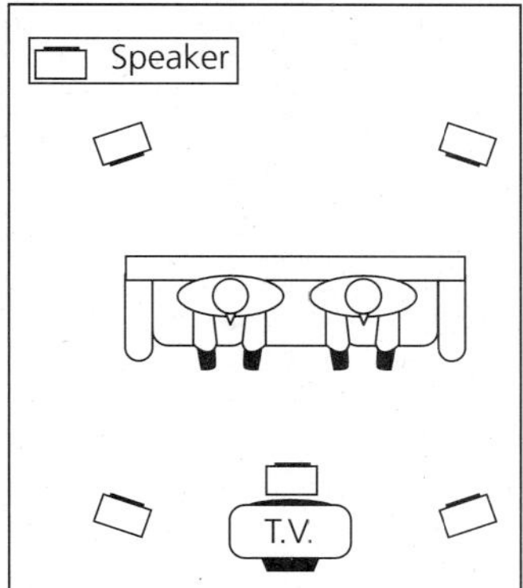

Figure 6.24 *A five-speaker surround sound layout.*

Surround Sound

A recent audio development for television home receivers is surround sound. **Surround sound** uses four or five speakers to create a 360-degree sound perspective (Figure 6.24). The five-speaker layout uses three speakers in the front, including one centered on or under the television set to carry dialogue and one on either side to provide normal left- and right-channel stereo for music, sound effects, and ambiance/environment. Two rear speakers may carry low-power, simulated, reflected sound, or they may be fully powered equivalents of the front speakers. A system called the *Dolby MP (Motion Picture) matrix process* feeds a four-channel soundtrack to the speakers to create the overall effect. Three channels feed the front (left, center, and right), and one feeds the rear.

Stereo miking techniques are used to record in surround sound. When surround sound audio is delivered to a normal two-speaker stereo system, the four channels are collapsed into two. Of course, the audio perspective problems outlined for stereo miking remain an issue for surround sound audio.

KEY TERMS

pitch *(112)*

decibel (dB) *(113)*

volume unit (VU) *(113)*

sound envelope *(114)*

cardioid mic *(120)*

lavaliere (lapel) mic *(122)*

headset mic *(123)*

hand mic *(123)*

desk mic *(124)*

hanging mic *(127)*

concealed mic *(128)*

boom mic *(128)*

shotgun mic *(130)*

parabolic mic *(130)*

pressure-zone mic (PZM) *(132)*

surround sound *(137)*

QUESTIONS FOR REVIEW

1. What are some similarities and differences among sound, light, and radio waves?

2. How are frequency and loudness related?

3. What are the effects of constructive and destructive interference, and how do they affect microphone placement in television production?

4. When mics are classified by generating element, what are the three basic types, and what are the typical performance characteristics of each type?

5. What are some of the main differences between miking a soap opera and miking a talk show?

6. What factors might you consider in miking an aerobics video on location at the beach? Assume you have a host (exercising vigorously), several other performers behind the host, music, a live audience, and natural beach sound in the background. What mics might you use, and where would you put them?

7 Audio Processing and Aesthetics

With the understanding of microphones we obtained in Chapter 6, we can move on to the storing and mixing of sound elements for video production. In addition to the sounds picked up by microphones, television often incorporates prerecorded audio material from records, magnetic tape, compact disks, digital video disks, computer disks, and even film and videotape soundtracks. Remember that audio is often the first part of a program to catch the audience's attention. Think about how often the opening audio is a piece of music, such as a theme song, and how immediately it identifies the program.

The content available from audio storage sources includes speech, music, and sound effects. A vast archive of sound, spanning many decades of audio technology, is available for use in video programs. This chapter describes the major audio storage media used in video production, briefly explains how they work, and suggests how to use them efficiently. The topics include:

USING SOUND FROM AUDIO SOURCES

phonograph records • optical disks • analog audiotape • digital audiotape • computer disk and chip technologies • videotape and film soundtracks

AUDIO CONSOLES

input levels and channels • volume and gain controls • cueing and monitoring sources • other functions of the audio console • patching

SOUND EFFECTS

APPLYING AUDIO PRINCIPLES IN ACTUAL SETTINGS

a multicamera studio setting • a single-camera field setting

USING SOUND FROM AUDIO SOURCES

Audio storage media used in television production include compact disks, audiotape (both analog and digital), computer technologies, and videotape and film soundtracks. Each is discussed in the following sections, but to illustrate some of the basic principles of audio technology we begin with an old-fashioned medium: the phonograph record.

Phonograph Records

Though phonograph records are scarcely manufactured anymore, hardly any radio or television station does not still have a pair of turntables and a record collection on the premises. And despite the conversion of many old recordings into compact disks, you may find that the audio material you need is available only on a record.

Records are made by spinning a disk with a lacquer surface in contact with a cutting stylus, which vibrates in accordance with sound waves produced by an audio source. Louder sounds result in deeper cuts into the lacquer and softer sounds in shallower cuts. An analog impression of the original sound is thus preserved by the pattern in the lacquer. When a record is played on a turntable, a needle drags lightly along the groove left by the cutting stylus, and its vibrations are converted into electrical energy and then into a replica of the original sound.

Broadcast-quality turntables enable you to cue records accurately. **Cueing** a record (or other audio source) means finding the beginning of the program material you wish to use. Further, to avoid *dead air,* or periods of silence, you need to start the program material almost as soon as the director calls for it. To do this, after you find the spot on the record where the program material begins, stop the record's spinning by placing your finger lightly on the record's edge. Turn off the turntable, leaving the stylus in place on the record. Then rotate the record backward about one-eighth of a turn. When the director cues you, engage the turntable and open the audio channel when the record reaches speed. Since the turntable takes time to reach its proper speed, you must be sure it is at speed before putting the program material on the air. Otherwise noticeable variations in pitch, known as *wowing,* will result.

Optical Disks

Optical disks include both compact disks (CDs) and digital video disks (DVDs). Unlike the phonograph record, the optical disk uses a laser beam instead of a needle to retrieve audio content. Hence there is virtually no wear when playing an optical disk, and such disks can retain their high-quality sound almost indefinitely. They also have less background noise and a better dynamic range than phonograph records do. But perhaps the most significant difference is that whereas phonograph records use an analog method of sound recording, optical disks use a digital code, meaning the original sound is converted into a discontinuous series of pits corresponding to zeros and ones based on discrete samples of the original sound. When the optical disk is played, the laser beam illuminates the pits as the disk spins, the reflected light is modified by the digital signal, and discrete values are read by a photoelectric cell. The stream of zeros and ones is then used to produce a replica of the original sound.

By assigning a discrete value to each bit of sound sampled rather than using a continuous stream of analog signals, this system of digital coding allows rapid access and retrieval of selected audio moments. It can be likened to an efficient filing system in which every sampled sound has its own unique address, making it possible to locate sounds quickly and easily.

Another major consequence of this change from analog to digital coding is that it makes the recording compatible with computer technology. What are the implications? One result is that audio information can be reproduced (*dubbed* or *duped*) ad infinitum without losing signal quality. Another is that audio data can be transmitted over telecommunications lines to multiple locations simultaneously, arriving in pristine condition. Neither of these things can be done with analog recordings.

Digital coding offers additional advantages. For example, a player can be programmed to play selections in random order. In addition, cueing selections is simple and requires little attention or skill. Many players offer a *pause* feature, which holds a selection in cue before playing it; a *review* feature, which lets you check the start and end times of a given selection; an *auto space* feature, which allows you to add seconds of silence between selections; and a *volume check* feature, which lets you locate the loudest span of a few seconds on the entire disk (useful for setting sound levels during a broadcast or dubbing operation). Further, there is no wow or flutter (high-frequency distortion caused by variations in disk speed) and virtually no start-up time before selections begin. You can achieve a noiseless transition between program choices.

Analog Audiotape

Most CDs, like phonograph records, store audio content but do not allow you to change what is on them. Though recordable CD systems are becoming more common, standard CDs are "read-only," meaning the user cannot re-record on them. In contrast, magnetic tape permits you to store *and* record audio content. Audiotape recorders have an erase head that comes in contact

> ### PROFESSIONAL POINTERS
>
> #### *Bulk-Erasing Magnetic Tape*
>
> - The bulk eraser is a large electromagnet, also known as a *degausser*. Keep tapes (and your watch) away from the bulk eraser when it is first turned on to avoid surges that are difficult to neutralize.
> - Move the tape *slowly* while it is on the eraser, at about 2 inches per second.
> - To avoid putting low-frequency thumps on the entire reel, withdraw the tape slowly while continuing to rotate it.
> - Be sure to treat both sides of the reel or cassette.
> - To avoid burning out the machine, do not leave it on too long.

with the tape right before it is exposed to a new recording signal. This neutralizes prior signals so that the tape may be used over and over again. There is also a separate piece of equipment called a *bulk eraser* that can quickly erase an entire reel, cassette, or cart of magnetic tape (audio or video). The ability to erase and/or re-record audio material on the same tape makes audiotape very useful for video production and for broadcasting in general.

Audiotape comes in both analog and digital varieties. Analog tape, which we will consider first, is a thin plastic strip coated with magnetic oxide particles that can store magnetic energy supplied by an electric current (produced, for example, by a microphone). As the tape travels past the head of a tape recorder, the tape's magnetic particles are arranged into patterns that match the strength of the magnetic field passing through the tape head. In this way, the tape preserves a magnetic analog of the pattern of electricity created by the original sound.

Many sizes and grades of analog audiotape are available. The major recording formats include open-reel tape, cartridges, and cassettes.

Open-Reel Tape   **Open-reel tape** (also called *reel-to-reel* tape) is a thin ribbon of magnetic tape wound onto a spool (Figure 7.1a). The tape can vary in length and width. For single-track monaural and two-track stereo recording, the standard tape width is usually ¼ inch. But for multitrack recording, ½-inch, 1-inch, and even 2-inch tape may be used.

Tape thickness also varies. In general, the thicker the tape, the less susceptible it is to stretching, breaking, or creasing. Thicker tape is also less subject to *print-through,* or the tendency for a signal from one layer to transfer a copy of itself to an adjacent layer on the spool. However, the thicker the tape, the shorter the length that can fit on a reel and the shorter the playing time. Tape thickness is measured in *mils,* or thousandths of an inch. So, for example, a 7-inch reel can hold 1,200 feet of ½-mil tape, providing about 30 minutes of recording time, assuming the tape travels through the tape head at 7½ inches per second. Broadcasters generally use 1½-mil tape to minimize problems from breaking and stretching.

Tape speeds also vary according to production needs. Machines can move the tape at speeds of 3¾ inches per second (ips), 7½ ips, 15 ips, or 30 ips. The

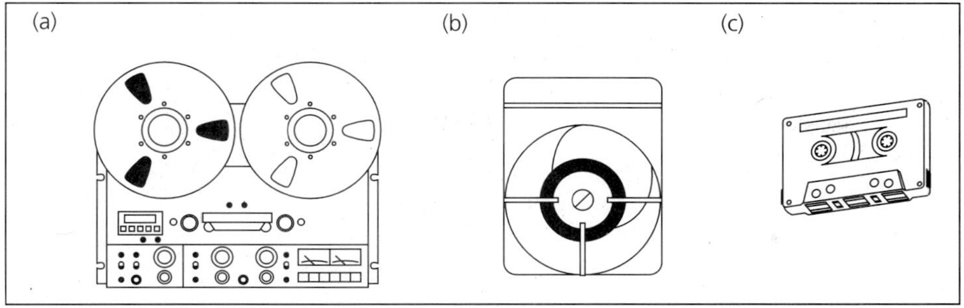

Figure 7.1   *Three formats for audiotape. (a) Open-reel audiotape machine. (b) Cartridge. (c) Cassette.*

faster the tape speed, the better the fidelity and the better the **signal-to-noise (S/N) ratio,** or the strength of the intended signal relative to the amount of interference that accompanies it. Improved S/N ratio means lower-level signals can be recorded, providing a greater dynamic range. S/N ratio improves with tape speed because the faster the tape moves, the more tape area there is to handle a given quantity of audio content. In other words, the audio information is more spread out on the tape at higher speeds, making it easier for pickup devices to read the signal. Video studios normally require tape speeds of 7½ or 15 ips.

Cartridge Tape   For shorter audio segments, cartridge tape machines are preferred over the open-reel format. **Cartridge tapes** (or *carts*) contain a continuous *loop* of ¼-inch tape inside a plastic container (Figure 7.1b). Tape lengths vary from a few seconds to about ten minutes. Carts are widely used for commercials, program themes, fanfares, and bridges (musical transitions between program segments). Short narrations are also carried on carts.

The big advantage of carts, aside from the fact that their tape is protected from dirt and rough handling by the housing, is that they can be marked with *cue tones* on a separate audio track (usually right before the program material), enabling the cart to recue automatically after each playing so that it is immediately ready for another play. Therefore, if audio content time and tape time are made equal, cueing can be kept extremely tight. In addition, carts get up to speed almost immediately, making it possible to use them with virtually no wowing.

Carts are especially useful for providing the same sound effect, such as audience applause, repeatedly. In fact, by eliminating the cue tone, you can run the cart continuously, making it possible to fade a sound in and out without having to restart the tape.

Cassettes   Tape **cassettes** (Figure 7.1c) differ from carts in several ways. First, although both have plastic containers that protect the tape from dirt and rough handling, carts contain a continuous loop on a single reel, whereas cassettes use a reel-to-reel format. Cassettes are also smaller than carts, but can hold a longer ribbon of tape. Their reel-to-reel format permits record and playback functions in two directions (unlike carts, cassettes have a flip-side). Cassettes also use narrower tape (about ⅛ inch wide) and slower speed (1⅞ ips), yielding longer playing time than carts. Cassette recorder-players work from batteries as well as normal AC wall current. All of these features make them ideal for field production.

Depending on tape length, cassette playing time commonly varies from 30 to 60 to 90 to 120 minutes. The 120-minute length is not recommended, however, because of unreliable tape speed; the thinness of this tape (¼ mil) makes it difficult for the tape to be grabbed by a dirty pinch roller. Such cassettes also exhibit sluggish fast-forward and tape rewind, and the tape's thinness makes it fragile and more subject to damage. Shorter-length tapes use ½-mil tape, which avoids many of these problems.

Small size, light weight, portability, and ease of handling make cassettes useful almost anywhere, but they do have drawbacks. Since the cassette system encases the tape inside a small reel-to-reel cartridge, it makes program material difficult to cue and edit. Moreover, the narrow gauge and slow speed of cassettes can limit their ability to produce broadcast-quality sound. Analog cassettes have poorer fidelity and S/N ratio than other tape formats. Hence, when initially introduced, they were not considered a serious choice for broadcast-quality or high-fidelity recording. However, the Dolby noise reduction system improved their performance.

Improvements in tape coatings have also made cassettes more useful in professional applications. Four grades of magnetic coating are available for

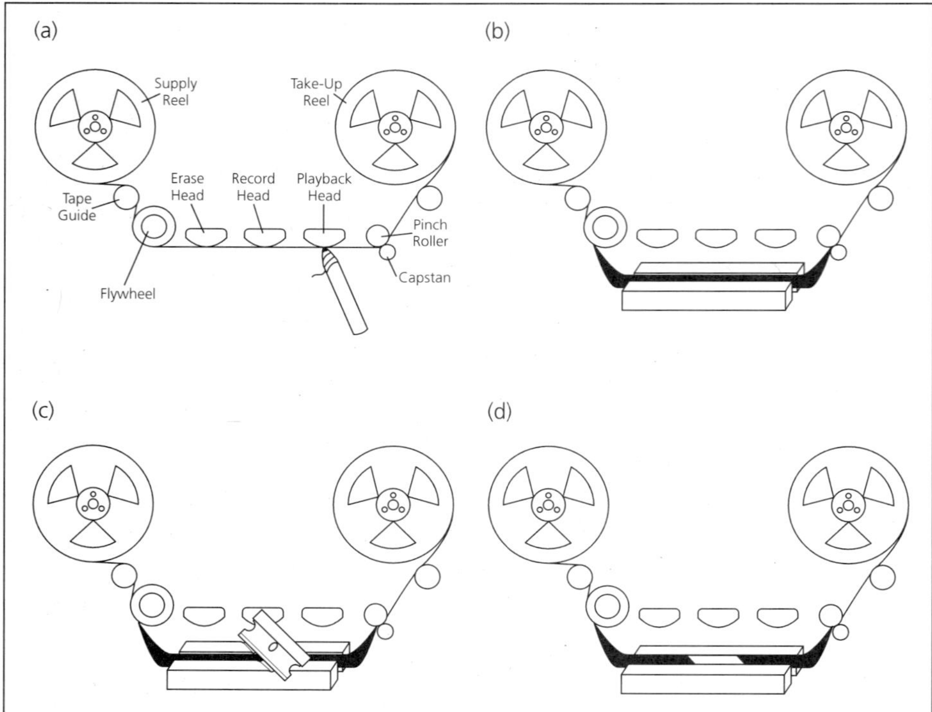

Figure 7.2   *Steps in mechanical editing of open-reel audiotape. (a) After jogging the tape back and forth to locate the precise points you want to edit, mark those points at the playback head with a grease pencil. (b) Unspool enough tape so that you can lay the portion you want to edit in the groove of an editing block. Position the tape so that one of the points you have marked is exactly at the diagonal slot in the block. (c) Draw a single-edge razor blade through the slot to cut the tape. Reposition the tape in the block for each such cut you need to make. (d) Join cut tape ends together with splicing tape.*

Procedure for Mechanical Editing of Open-Reel Audiotape

For mechanical editing, you need the following tools: a grease pencil, an editing block, a single-edge razor blade, and a roll of splicing tape.

Once you are near the content where you want to perform an edit, manually jog the reels back and forth past the *playback* head of the recorder, listening to locate the exact edit points. (Most machines arrange the erase, record, and playback heads in left-to-right order.) When you have found the exact points, mark them with the grease pencil on the back (shiny side) of the tape.

Then unspool the portion of tape to be edited, place it in the editing block, cut it on the marks you have made, and remove the unwanted segment. You may then join the tape ends together with the splicing tape or insert another segment of tape.

audiocassettes. In order of increasing quality, these are ferric oxide (type I, also called *normal* tape), chromium tape (type II, or *high bias*), ferrichrome (type III, or *intermediate bias,* a rarely used variety), and metal particle (type IV, also known as *high bias*). *For best recording quality, be sure to set the recorder's tape-select switch to match the type of tape you are using.*

Editing Audiotape Editing audiotape allows you to remove sound, add sound, or shuffle program material into any order you want. Of course, significant ethical issues are associated with this practice, especially in broadcast journalism and documentary production. Either accidentally or on purpose, editing may put words and meanings into subjects' mouths that they did not say or mean. It is the ethical obligation of the producer to edit carefully and in good faith to avoid this type of misrepresentation.

Editing can be done either mechanically (on open-reel tape) or electronically. Mechanical editing involves physically cutting the tape with a razor blade or other sharp device made especially for such work, rearranging the tape segments into the order you want, and reconnecting them with splicing tape (Figure 7.2). Electronic tape editing uses machines equipped with *total transport logic (TTL)*. These machines measure tape in terms of hours, minutes, and seconds, permitting you to mark edit points electronically. You can then edit easily without physically cutting the tape by dubbing desired segments to a blank tape on another recording machine.

In electronic editing, since no physical cuts are made in the tape, once segments are assembled (usually in order from start to finish), you cannot add or delete portions, except perhaps at the beginning or end, without either blotting out other material or leaving dead air. Therefore, it is strongly recommended that you study a log of all audio items carefully and finalize all editing decisions before editing begins. In an emergency, to save time, it is possible to redub the edited material onto yet another blank tape, inserting or deleting the crucial elements that were overlooked the first time. But this solution has a price: each time you dub a copy onto another blank tape, you lose signal quality.

The most precise electronic editing process uses a time code developed by the Society of Motion Pictures and Television Engineers (SMPTE) recorded onto a separate tape track. **SMPTE time code** editing records an eight-digit number every 1/30 of a second, with the first pair of digits marking hours, the second pair marking minutes, the third marking seconds, and the fourth marking frames. This code is read by a separate playback head on the recorder. You may recall that video frames also change every 1/30 of a second. By marking audio content in this way, SMPTE time code permits you to match sound to any video image with complete accuracy. Time code also makes it easy to edit with the use of a computer.

Digital Audiotape

Digital audiotape is a superior form of recording over analog tapes because it offers a much wider dynamic range and much lower noise level on playback. Several formats are available, each incompatible with the others. One records and plays back using a stationary tape head on a reel-to-reel machine. Another uses a cassette with a rotating head. Yet another uses videocassettes for audio-only recording. Even within these categories, incompatible formats abound. The following sections briefly describe each category's major features as well as its advantages and disadvantages.

Stationary Head Audio Recording   Within this category of digital audio recorders are two formats: *Digital Audio Stationary Head (DASH)* and *Professional Digital (PD* or *Pro Digi)*. Both recorders use ¼-inch and ½-inch open-reel tape to record digital audio and an analog reference track for editing purposes. Unlike DASH, the PD format also uses 1-inch tape for thirty-two-channel recording. Mechanical editing may be done with an analog track that accompanies the digital version of the audio content. Electronic editing with time code is also possible.

There is also a type of digital audiotape called *Digital Compact Cassette (DCC)*. This system is unique in that it encodes data by breaking the frequency spectrum into thirty-two separate bands; it also includes time code for editing purposes and index information such as song titles and artist names. One nice feature of DCC machines is that they use standard audiocassettes, and they are able to play back both analog and digital tapes (though they record only in digital format).

Rotary Head Audio Recording   Instead of a stationary head, rotary digital audio recording (R-DAT, or DAT for short) uses a recording head spinning at about 2,000 rpm to record audio content. The head's rotation increases the capacity of the tape to encode information in a limited space, making it possible to produce digital stereo at lower cost using smaller cassettes. Unfortunately, this renders the R-DAT format incompatible with all stationary formats. In addition, because of the rotating head technology, R-DAT tape at this time can only be edited electronically.

R-DAT tape is of the metal particle variety, and it is the same width and thickness as that used in standard cassettes. However, the length of the tape is generally half that contained in analog cassettes, and the cassettes themselves are smaller.

Two rotary formats called *Modular Digital Multitrack (MDM)* use videotape to store digital audio. One such system, Alesis ADAT, uses S-VHS videocassettes to record audio material; the other, Tascam DA-88, uses 8 mm videotape. The advantage of using MDM systems is that the machines are capable of recording eight tracks of audio information and can be interconnected to yield up to 128 tracks. Also, the audio can be integrated with videotapes with virtually no signal loss. Unfortunately, the various MDM systems use different videocassettes that are incompatible with one another and with other formats.

Computer Disk and Chip Technologies

One of the biggest drawbacks of tape-based technologies stems from the fact that the data are stored on a linear ribbon. Locating and accessing specific moments of program material can take a great deal of time. This problem is virtually eliminated by using computer disks with random access capability.

Computer Disk Recording   Like CD technology, computer disk recording transforms sound into computer data by assigning discrete values to samples of the sound stream. Once data are entered into the computer's hard drive or floppy disk, they can be randomly accessed, retrieved, and even altered at a *digital audio workstation (DAW)* (Figure 7.3).

The options available on the DAW allow you to change the order of sound segments as in conventional editing. Many programs also enable you to reverse sounds so that they can be played backwards. Sounds from disparate sources can be combined. Sound envelopes can be altered. Volume and frequency can be changed. Tempo can be altered without changing frequency. The possibilities are endless.

Perhaps the greatest advantage of computer-based recording is the ability to perform **nonlinear editing**. Instead of having to assemble edits in linear

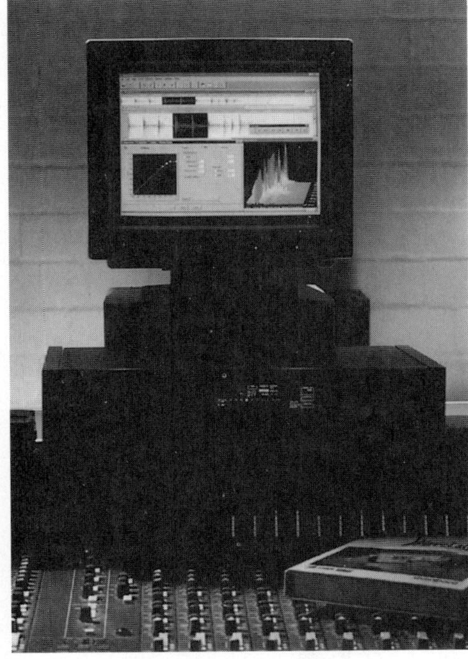

Figure 7.3   *A digital audio workstation (DAW).*

fashion as with tape-based media, the editor can manipulate segments with the same flexibility now available for word processing. In short, if you wish to insert a sound somewhere in the sound stream, the computer shifts the location of edited segments to make room for it without disturbing the overall order or obliterating any sounds already laid down, just as word processing programs do for text. In addition, when such changes are made, signal quality does not suffer, since all the audio information is digitally coded.

However, no technology is without limitations, and computer-based systems are no exception. The drawback here is the demand for large-capacity (multigigabyte) hard drives to handle the workload. For example, the amount of disk space required for twenty minutes of monaural sound (or ten minutes of stereo) is about 150 megabytes. This means several disks may be needed for longer productions.

Musical Instrument Digital Interface   With computers and electronic keyboards, synthesized sound can now be precisely fitted to video content with a digital system called a *musical instrument digital interface* or *MIDI* (Figure 7.4). Using a computer as a central control, MIDI permits you to manipulate collected samples of sounds (music, speech, or sound effects) with an array of keyboards. When processed through a MIDI system, each sound's qualities, such as frequency, volume, duration, timbre, tempo, attack, and decay, become almost infinitely manipulable. Because MIDI technology may be integrated with SMPTE time code, layers of sound can be made to fit precisely with video footage.

Even beyond the use of music, MIDI has had a significant impact on television production. For example, MIDI can assign vocal and sound effect samples to individual keys of a keyboard. By playing those keys, you can provide a customized soundtrack for any program.

MIDI's advantages include ease of editing. If you don't like a particular sound, you can change it back to its prior voicing (arrangement) with a keystroke. In addition, all of the sonic image can be produced in pristine condition repeatedly without signal loss. Further, you don't waste time rewinding tape, since all sound is stored using microprocessors. The computer memory offers instant random access.

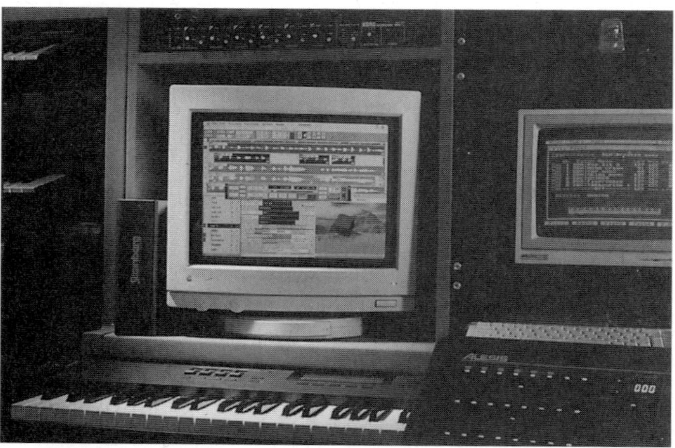

Figure 7.4   *A musical instrument digital interface (MIDI).*

MIDI's disadvantages include the need to provide separate synthesizers for each voice in some cases. Some MIDIs suffer from "lag" problems (or a loss of precise time agreement among sources) if too much data are processed at once. Further, fast playing can sometimes outrun the MIDI's ability to keep up with the sound stream. Finally, MIDIs can be expensive.

Videotape and Film Soundtracks

Traditional videotape and film contain soundtracks that can be used in television production. Analog videotape comes in several formats, including 1-inch open reel (type B and type C), ¾-inch and ½-inch cassette, and 8 mm varieties, among others.

In general, analog videotape features several tracks of information: a video track, one or more audio tracks, and a control track that acts like the sprocket holes on a reel of film to synchronize playback. In some cases, there may also be tracks for time code and video scanning information.

One-inch tape has three audio tracks. Two are used for program sound, and one may be used for time code and/or other functions, such as director's cues. The ¾-inch cassette format (also called *U-matic*) has two medium-quality audio tracks. Three ½-inch formats with four audio tracks include *Betacam SP*, *MII*, and *Super VHS*. Two of the audio tracks are used for sound, and the others may be used for control track and time code. Finally, the 8 mm format has two audio tracks and an FM-carrier audio track in the video waveform for time code.

Film has two basic sound formats: optical and magnetic soundtracks. In the *optical* format, a light shines through a track near the edge of the film, at varying intensities in accordance with an original sound stream. During playback on a projector, as the film streams by, the modulated light that gets through the film is converted into electrical energy by a photoelectric cell. Then a loudspeaker reconverts it to sound. In the *magnetic* format, on the other hand, the soundtrack is recorded on a magnetic stripe running along one edge of the film. The magnetic stripe acts essentially like audiotape. During playback, a soundhead reads the sound.

AUDIO CONSOLES

With so many audio sources available for television production, how are they made ready for use at the instant they are needed, and how are they blended seamlessly into an air-quality sound track? These are the functions of the **audio console** (Figure 7.5). Audio consoles (also called *audio mixing boards* or *audio mixers*) serve several purposes:

- Providing an appropriate interface for each audio source
- Providing volume control
- Allowing you to cue and monitor each source before it is used
- Allowing you to modify each source during production
- Blending and mixing various sources during production
- Providing an air-quality output signal

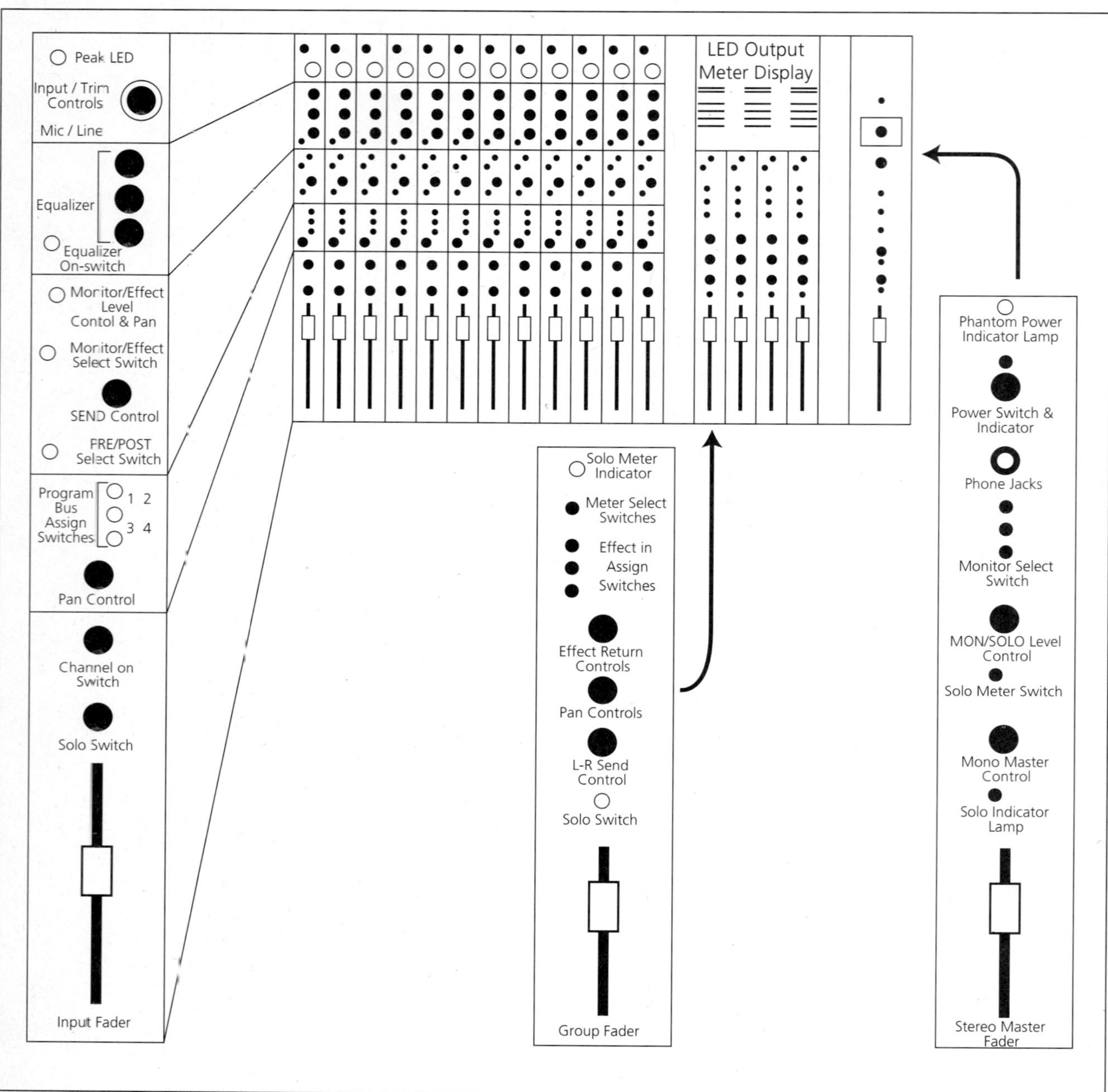

Figure 7.5   *An audio console.*

Input Levels and Channels

Audio consoles receive input signals from numerous sources. As it turns out, all sources fall into two categories of voltage level. Microphones, which deliver signals measured in millivolts (mV), or thousandths of a volt, are designated as *low-level* sources. Other audio sources, which deliver much stronger signals (in the range of one volt or more), are *high-level* sources.

This distinction is important because audio consoles accommodate the two categories differently. Microphone signals must be boosted by a pre-amplifier; other sources do not need such a boost. Therefore, you need to assign the input sources to the proper channels on the console.

Some consoles permit you to use any channel for any type of source as long as you choose the input level properly. Usually you identify a source as either high or low level by flipping a switch, clicking a dial, or pushing a button. The switch, dial, or button that does this is usually labeled *mic/line* (see Figure 7.5). By selecting the "mic" position, you tell the audio console to engage a pre-amplifier for that channel. By selecting "line," you tell the console to bypass that channel's pre-amp.

Other consoles hard-wire specific channels to accept only one type of source. (*Hard-wiring* means making a permanent connection between two terminals inside a piece of equipment.) For these consoles, a "mic" channel is always a mic channel and must not be used for anything else. Similarly, a "line" channel is solely for nonmicrophone audio sources.

If by mistake a mic is connected to a line channel, a low signal will result, and this will not become air quality no matter how high the volume is set for that mic. Conversely, if a non-mic audio source is mistakenly *patched* (temporarily connected) into a mic channel, it will be too loud, and distortion will result.

To make life easy, especially if you have a lot of sources, you can use masking tape to label each channel. Write on the tape the name of the talent (for instance, *H* for "host," *G* for "guest"). Then you can locate the correct channel for each source at a glance. To simplify further, you can arrange the assignment of channels in the same left-to-right order in which the talent appear on screen.

Volume and Gain Controls

When each audio signal is connected to the console through an individual input channel, you can control the sources with either a switch or a fader. *Switches* allow you to open or close channels *instantly*, like turning a light on or off. For instance, you can use a switch to "hit" music or "kill" mics. In contrast, *faders* allow you to gain or lose volume *gradually*, like slowly raising or dimming the house lights in a theater. Thus, faders enable you to "sneak music under" or "crossfade mics." Faders are also used to set volume levels, stabilizing the volume for each source at any level you want. During a program, the faders can also adjust levels as needed.

Faders may be adjusted to increase or decrease output volume to cancel out respective decreases or increases in source volume. This is known as *riding the gain*. Riding the gain may become necessary if a talent suddenly shouts or whispers during a speech, or when music has great dynamic range and features both soft and loud passages. Of course, there may be times when you wish to feature such dynamic changes rather than canceling them out; it all depends on the purpose of the program and the wishes of the director.

To monitor volume, consoles are equipped with a visual volume display, the **VU (volume unit) meter** (Figure 7.6). The VU meter displays volume intensity leaving the console. It is calibrated in both volume units (decibels) and (in the case of a transmitter) *percentage of modulation*. It is important to avoid overmodulation, which can cause distortion in the overall audio signal. Of course, it is also important to keep levels from getting too low. Generally, levels of 80 to 100

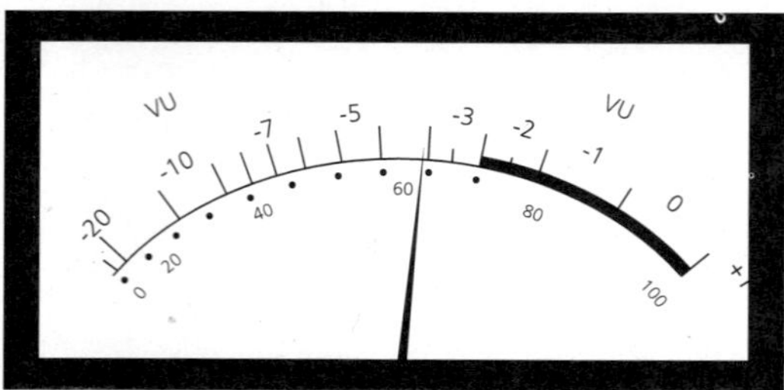

Figure 7.6 *A VU meter. Note that it shows both volume units (the outer scale) and percentage of modulation (the inner scale). Zero on the VU scale equals 100 percent modulation, which means the maximum amount the sound system can properly handle. The audio level should generally be kept between 80 and 100 percent modulation, that is, between −2 and 0 VU.*

percent modulation are considered to be *up full* and acceptable. Levels of 40 to 60 percent modulation are *under* and are still audible (good for background music). Levels below that may be *in the mud,* or too low for broadcast quality.

If the VU meter's needle occasionally goes into the red zone, above 100 percent modulation, there is no need to worry. However, if it remains there or enters the red zone the majority of the time, you may be *pinning the needle* or *peaking the volume,* either of which can cause distortion and overmodulation. To avoid these problems, you should lower the level.

But if you are feeding the console more than one source, how do you know which one is too high? The VU meter is not sensitive to *which* source may be the culprit; it only knows that the overall sound is too loud. One way to find out is to take levels for each audio source individually. To do this, turn off every source except the one for which you wish to set a level and, with the source playing, set that level. Then turn off that source and set a level for the next one. Do this for each source separately. Then, during production, ride the gain to maintain desired levels.

Though consoles provide a separate channel for every source, all the sources are eventually sent to a master gain control. The *master gain control* is simply a single fader that controls every source flowing through the board simultaneously and sets the volume level for the overall output. Some consoles also have *submaster gain controls* (or several submasters), which are used to control a subgroup of sources with a single fader. For example, if you have three mics covering the string section of an orchestra and three others on the percussion, you can feed each group of mics to a separate submaster, allowing you to balance the strings with the percussion by using just two faders.

Cueing and Monitoring Sources

Aside from monitoring sources when they are *on* air, you may also need to locate audio material, take levels, and cue sources during a show but *off* the air, so they will be ready to use when the director asks for them. To do this, you choose the cue setting for the appropriate channels. The *cue setting* shunts the channel's feed to an audition loudspeaker and away from the program audio. To *audition* an audio source means to listen to it without putting it on the air. Actually, the audition loudspeaker may not be all that loud. It may be a puny control room audio monitor or even a speaker inside your headset. But in cue mode, you can safely listen for the program material you wish to use and then set its volume level with the fader and VU meter.

Other Functions of the Audio Console

In addition to the system functions already discussed, many consoles have a *talkback* channel that enables control room personnel to communicate with the studio and other locations. Talkback is also called the *PA* or *SA,* which stand for *public address* and *studio address,* respectively.

TABLE 7.1

Some Common Features of Audio Consoles

Input System Functions

1. *Trim:* provides gain control to boost low-level sources to make them more usable; attenuates high-level sources to avoid distortion.

2. *Pad:* lowers or prevents power overload when trim is not enough. Useful especially with mics in the vicinity of very loud sources (for instance, in front of a rock group's guitar amp).

3. *Phantom power:* provides power to condenser mics, eliminating the need for a battery or external power supply.

4. *Equalizer:* alters volume level for portions of the frequency range to eliminate or enhance the psycho-acoustic quality of the sound.

5. *Filter:* a specialized equalizer used to eliminate certain portions of the frequency range.

6. *Pan:* varies relative loudness delivered to selected output channels to change the "locations" of sound sources in the audio space.

7. *Reverb:* sends a source to a signal processor, which returns it a short time later to be fed through the console again, creating an echolike effect.

8. *Pre-fader listen (PFL or SOLO):* enables you to hear one channel without having to shut off others. Using PFL turns off other channels feeding the monitor without affecting the actual output of the program.

Output System Functions

1. *Bus:* a board component that isolates or groups a signal with several others before sending them out of the board. This may be done to set a final volume level on a source before combining it with others. For instance, drums and backup vocals may be mixed together before they are mixed with lead vocal.

2. *Pan pot:* varies relative levels being fed to two output buses.

3. *Reverb:* adds an echolike effect to the output audio.

A *foldback* feature is also provided to send selected audio feeds to either small, specialized monitor speakers located very near the performer or to headphones or earpieces worn by the talent. This allows studio talent to hear important audio material, such as sound effects cues or music for lip synching. This feature is necessary because studio speakers are normally turned off whenever there are live mics in the studio. If left on, the studio speakers might create **feedback,** the loud, squealing noise that results when mics pick up their own signals from nearby speakers.

Consoles also offer a built-in *oscillator tone* that may be used to generate pure tones (sine waves) for recording test audio on videotape and for calibrating console meters. Table 7.1 lists some other features commonly found on audio consoles. Each type of console, however, has its own particular arrangements.

Patching

In the event of equipment failure or the need to use equipment not normally connected to the console (such as an additional CD player), you may use a system of *patching* to bypass, replace, or add to regular audio signal routes. A **patch panel** (also known as a *patch bay* or *patchboard*) is a board that allows you to connect any audio source with any output destination (Figure 7.7). Traditionally the connections were made with patch cords, which resemble telephone switchboard cables with identical jacks on each end. Today a computer (called a *router*) is often used instead. Either way, the concept is the same, namely, to provide a flexible means of connecting any audio source to any destination.

Three types of patch panels are used: open, normalled, and half-normalled. In an *open* system, no outputs are connected to any inputs until they are patched with a patch cord. In contrast, a *normalled* system hard-wires specific outputs to specific inputs without patch cords. In a normalled system, inserting a patch cord disconnects the normal connection. The advantage of the normalled arrangement is that it saves time in cases where the work schedule is predictable and regular but still allows flexibility in an emergency.

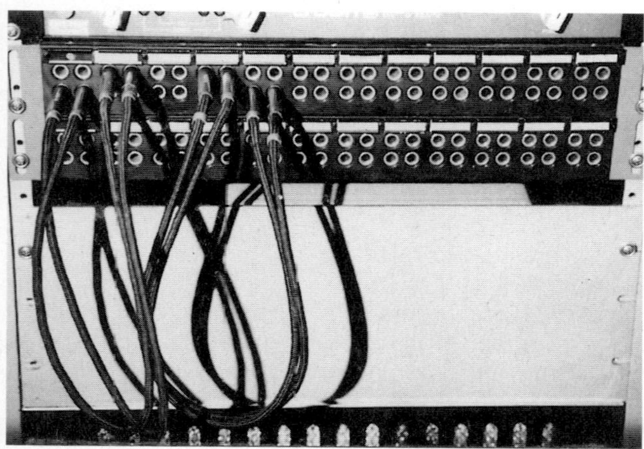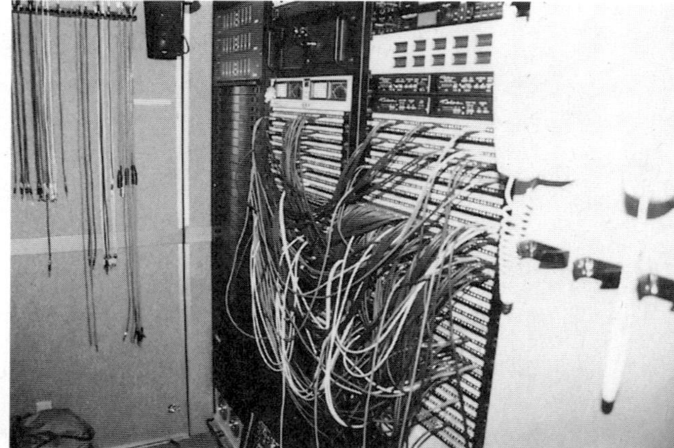

Figure 7.7   *Patch panels, small and large.*

PROFESSIONAL POINTERS

Using Patch Panels

- Set microphone faders to zero while patching to avoid surges that can damage equipment.
- Don't patch input to input or output to output.
- Avoid patching mic-level signals into line-level holes, which will result in a weak signal. Conversely, don't patch line-level signals into mic-level holes, or distortion will result.
- To avoid feedback, don't patch an output back into its own input.
- When setting up a patch panel, arrange sources on the panel by function to avoid confusion. For example, put all mics in one section of the board, all tapes in a second section, all film soundtracks in a third section, and so forth.

In a *half-normalled* system, hard-wire connections are interrupted only by inserting a jack into the *input* (bottom) hole. Inserting a jack into the output hole does not disturb the normal connection. The advantage of the half-normalled arrangement is that it allows you to feed a signal to an additional external source (such as a tape recorder) without disturbing its normal destination.

In addition to enabling you to bypass, replace, or add to regular equipment, patch panels permit you to arrange sources conveniently on the audio console. For example, patching makes it easy to match the left-to-right order of microphone faders on the audio console to the left-to-right order of talent on screen, thus making the job of monitoring their levels easier.

SOUND EFFECTS

Sounds that are not music or speech may be called **sound effects** (abbreviated SFX). Sound effects enhance video programs by providing clues about location and context. They add valuable description. They also identify, intensify, and emphasize key action, supply transitional devices, and provide symbolic meaning.

For example, to establish locale, there is hardly a better way to convey a cityscape than to inject sounds of car horns, police sirens, and the general din of traffic noise into the soundtrack of a program. Similarly, a tropical setting can be conveyed easily, cheaply, and quickly with sounds of birds and monkeys.

Sound that supports action on screen provides context. For example, when we see a window close, the sound of the window moving through the sash, followed by the contact it makes when it completes its motion, confirms the

PROFESSIONAL **POINTERS**

Creating Simple Sound Effects

- *Airplane:* For a propeller aircraft, hold cardboard in contact with the blades of an electric fan. For a jet, place a mic at the end of a vacuum cleaner tube.
- *Breaking bones:* Chew hard candy close to a mic, or break the candy with pliers.
- *Babbling brook:* Blow bubbles through a straw into a glass of water.
- *Engines and electric motors:* Run a vacuum cleaner, a blender, or a Cuisinart near the mic.
- *Fire:* Crumple cellophane. For crackling, crush heavier paper abruptly.
- *Footsteps:* Fill a trough with gravel and walk in it.
- *Ice cracking:* Crumple a Styrofoam cup.
- *Rain:* Pour sand through a tube onto cellophane.
- *Snow crunching:* Squeeze cornstarch.
- *Thunder:* Rattle sheet metal.

action. Moreover, the loudness of these noises adds meaning to the action, perhaps communicating important information about the emotional state of the character closing the window.

Sound effects libraries can be purchased on disk or tape. When such sound effects are prepared for use, they are often transferred to carts so they may be tightly cued. However, sounds may also be mechanically created in the studio or taped live on location, and these custom sound effects often allow a closer match to the action on camera. With the use of MIDIs, custom sound effects can be sampled and modified to fit specific needs. Files of effects can be stored and used over and over again during production and postproduction. Experienced television and radio sound technicians have come up with an amazing variety of ways to create sound effects with simple materials; the accompanying Professional Pointers feature lists a few of them. Even in our digital age, you'll find that these tricks will come in handy.

APPLYING AUDIO PRINCIPLES IN ACTUAL SETTINGS

Let's apply the audio principles we have learned to multicamera and single-camera productions in studio and field, respectively. The sample productions we will use illustrate some of the audio design elements considered during preproduction, production, and postproduction. The discussion is not intended to be exhaustive, since every project has its own unique objectives and challenges. But the sample productions will give you a sense of how to approach the projects you will encounter.

A Multicamera Studio Setting

Multicamera recording feeds signals from several mics and cameras to a single **videotape recorder (VTR)** (Figure 7.8). Programs broadcast live and recorded live-on-tape use this format. Such programs include sports, awards shows, public affairs broadcasts, talk and interview shows, sitcoms and soaps, and specials such as pageants and parades.

Let's consider the simple, two-person stationary talent interview we have been using as a running example. Recall that we were using two cameras for this program, which features a host and a guest. Now we'll consider a variation of the show that includes an off-camera announcer.

Preproduction   To execute the audio portion of this program, we begin by studying the script, floor plan, light plot, set, and other materials (including the facilities or "fax" lists) to determine what sounds we will need and how they will be supplied. With two talent (host and guest), it is possible to use a variety of mics to capture their words. Assume we decide to use two lavaliere mics, one clipped to the host and one to the guest. For the off-camera announcer, we decide to use a mic on a stand. This provides a means of picking up the announcer's voice, but does so without requiring the announcer to touch the mic, thus minimizing unnecessary noise that might result from handling. It also frees up the announcer's hands to hold script materials.

We also decide we want opening theme music (*intro* music) and different closing music (*outro* music). From the fax list, we see that we have at our disposal both a CD player and a cassette player. To avoid having to change CDs or tapes during the show, when we may have plenty of other matters to attend to, we decide to use the CD player for the opening and the cassette player for the close.

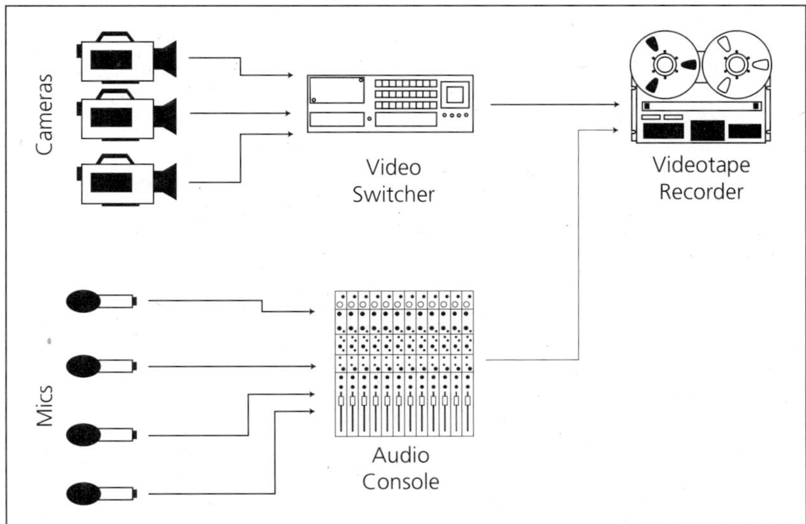

Figure 7.8   *Arrangement for a multicamera studio setting. Multiple cameras feed into a video switcher and multiple mics into an audio console. Then the audio and video signals are recorded together on the videotape recorder.*

We notice that the script features an announcer **voice-over (VO)** during the opening music, giving the guest's name and explaining the program topic. The director informs us that she does not want the music to interfere with the announcer, so we decide to bring the music up full at first and then lower it under the announcer's voice.

As soon as the announcer finishes the opening speech, we will kill his mic and then bring the music up full again, right before opening the host's mic for his opening speech. Then, just as the host begins to speak, we will slowly fade the music out. This will give shape to the audio portion of the program, avoid dead air, and telescope the sound from the host with that of the music. *Telescoping* means slightly overlapping audio sources to eliminate dead spots in the audio track, giving a seamless quality to the program audio.

As for closing music, the director asks that we make it coincide exactly with the closing credits so that as the credits begin and end, so does the music. To do this, we time the closing credits during rehearsal, determining that they last twenty seconds. We select a thirty-second piece of music. To have the music end exactly as the last credit rolls off and the screen fades to black, we backtime the music. In this case, **backtiming** means subtracting the time of an audio or video segment from the total program time so that the segment can be made to end exactly when the program ends. For our purposes, we will roll the music thirty seconds before the end of the show, but we will not fade up the channel carrying the music on the audio console until the credits begin. This strategy is known as *deadpotting*. As soon as the credits appear, we will fade up the music—gradually so that it won't sound odd if it becomes audible in the middle of a musical phrase.

Finally, we consider Murphy's Law, namely: What will we do on the day of production if the CD player suddenly breaks down? To fortify our program against the possibility of equipment failure (or unavailability), we decide to dub the CD material we have selected onto a cassette. In fact, we further decide to get both intro and outro music on CD, if possible, in case the cassette player fails. We obtain several backup mics and cables for the same reason. We hold a final preproduction meeting to make sure all personnel are familiar with their assignments.

Production   We seat the host and guest and place mics on them. Since the guest sits to the host's right and therefore looks to the left to see him, we cheat the lav slightly to the guest's left side to favor the orientation of her mouth. Conversely, since the host looks to the right, we clip his lav to his right. After the mics are in place, we have each talent speak into his or her live mic while all other mics are turned off. As each talent speaks, the audio operator takes a level. When each level is set, the audio operator places a piece of masking tape on each channel and labels it to identify the talent delegated to that source.

Each non-microphone audio source is then cued, levels are taken, and the appropriate channels on the audio board are tagged. When this process is completed for all audio sources, the audio operator fades the master gain to zero (to cut the noise level in the control room during other preparation), informs the director that audio is ready, and stands by.

During run-throughs, we check that everything is working properly. If all is well, then either the show is broadcast or it is put on tape for later use.

Postproduction   After the show is completed, it is time to strike the studio. This means all equipment must be put away. Audio cables are disconnected, tape labels are pulled up from the audio board, faders are placed at zero, audio sources are removed from their players, mics are put away, and the studio is cleared of all production materials. If the program has been taped for later use, the strike begins only after the director has made sure that a videotape has in fact been recorded and that it is of acceptable quality.

Some postproduction work may be needed to modify an audio portion of the program, a process known as **audio sweetening**. In our example, this might include adding a voice-over announcing a change in guest for the following week's program. If this type of additional material will be needed, it is often wise to prepare it during production. For example, you may want to use the same announcer to preserve a seamless quality to the voice-over.

A Single-Camera Field Setting

Elaborate field configurations feed video signals from several cameras and audio signals from several microphones to several VTRs. However, many programs shot in the field use a single-camera production style. In the single-camera style, the lone video signal, along with one or more audio signals from one or more microphones, is fed to a single VTR. When several mics are used with a single camera, an audio mixer (Figure 7.9) is needed to feed the audio signals to the VTR. In the field, the audio mixer takes the place of the studio console.

Let's consider a program that features a series of one-on-one interviews with several subjects, as well as a panel discussion with five people sitting in a semicircle around a table. We will use one camera and several microphones, and we have an audio mixer at our disposal to feed all of the mics into one VTR.

Preproduction In general, the single-camera production format differs from the multicamera approach in that various angles and perspectives at given locations are shot in separate takes. For instance, in our interview segments, video images of the host's reactions to subjects' comments will probably be shot separately to be used as "cutaways," which provide smooth transitions between segments. Assuming we will have a number of such interrupted starts and stops in the video and sound recording, consistency of sound will be especially important because it can help unify the entire program.

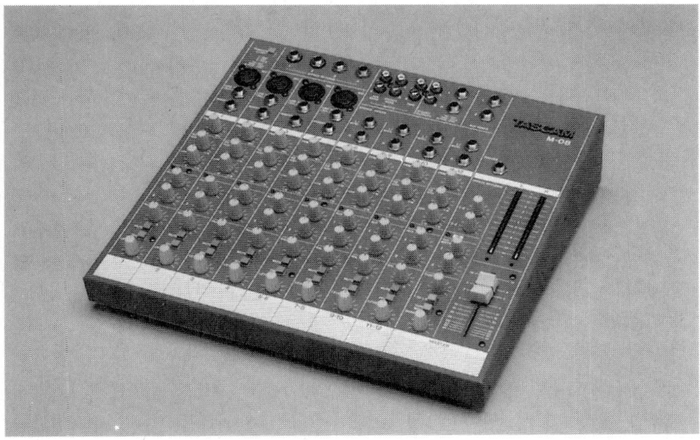

Figure 7.9 *An audio mixer small enough for field use.*

Marc Wiener

Director, Electronic News Gathering (ENG),
WCBS-TV, New York

Q: Tell the readers about your background. How did you get into this field?

A: I found my way into the field by liking music. When I went to Hofstra University, I joined up with the radio station people. They had technology, politics, music—everything I was interested in—embodied in one place, the college radio station. I became the music director and station manager.

It turned out that the Hofstra radio station, WVHC, was run very professionally, and it was a good place to prepare to go into the professional field. I went from college right into CBS-FM radio as a board engineer. I sat opposite the disk jockey, played the records, and took the phone calls. Since I could do the technical stuff, I worked my way into that also. I also produced commercials, jingles, and promotional announcements. So for thirteen years I was responsible for the sound of CBS-FM technically, for all announcements and records.

Q: So how did you wind up here in television?

A: Well, radio started getting rid of many engineers when disk jockeys began running the boards themselves. That was in 1983. Luckily, right at that time, Channel 2, where we are right now, decided they wanted to produce their own syndicated program, a feature called "Two on the Town," intended to highlight New York City. They needed an audio engineer, and I was picked to record concerts and interviews with recording artists at jazz clubs. "Two on the Town" lasted two years. It was very expensive to produce, and they did not have an aggressive marketing staff to syndicate it successfully. But rather than fire people at that point, they expanded the local newscast to an hour, and all the staff was folded into the news department. I became the audio ENG guy in local news.

Q: That's an interesting transition.

A: Well, "Two on the Town" was done like ENG. You know, it had portable lighting and wireless microphones. Eventually, I became a supervisor for the ENG crews. I picked up on the ENG equipment very quickly.

Q: What was the major difference in the transition from radio to television?

A: I like to say that television is much heavier. TV has hundreds of pounds of equipment to do the same job you could do on radio with a microphone. You also have a lot more people involved in getting something on the air. In radio, if it's a music program, you have the disk jockey and the engineer, and that's it. If it's news, you have the reporters, who can record the material themselves, bring it back to the studio, edit it on tape, and go on the air. But in television, you have the reporter in the field, the camera person, and sometimes an audio person. Then you've got writers and producers, the news director, the editor, people in the control room, and on and on.

Of course, technically things are changing. Equipment in television is becoming less bulky. Earlier you had to have specialists to make sure the equipment worked properly, but solid-state technology has made the equipment more reliable, so you don't need as many engineers anymore. Also, wireless mics have eliminated the need for a mic person. Now you just give the mic to the reporter and he or she holds it. Solid-state circuitry and AGC [automatic gain control] circuits have improved, so the mic on the head of the camera does an unbelievable job of picking up audio that you would have thought was completely unusable only a few years ago.

Q: You've started answering the question I ask all the time, which is how things have changed since you entered the business.

A: And things will change even more. We're going to networked PC computers, so the writer working on the script will be able to bring up video that has been put onto a computer hard drive. The writer links the actual piece of video to words digitally. Then the script is automatically fed to the teleprompter, and type is set along with it for text graphics. When you go to air, the person in the control room just pushes a code number for that piece and it's automatically played back from the hard drive. The script and the fonts automatically roll with the story. And just one writer-editor does the whole thing.

Q: Is the net effect of this change an increase or decrease in the need for people to work in this field? After all, we are also seeing a proliferation of channels, right?

A: The net effect is both an increase and a decrease. We need fewer individuals at each place, but we have more places to work. For people who want to get into the video field, this is a great time to get involved. The jobs are expanding tremendously because it's not only television anymore. What about video on computer games? What about industrial videos needed by corporations? What about weddings and bar mitzvahs? What about real estate sales on the Internet, where a home buyer can take a virtual tour through a house? Computer graphics people also use video. It's basically unlimited, and where it's going nobody knows.

Q: What is the deadline pressure like in this brave new world?

A: I'll give you an example of my week from hell two weeks ago: the New York Yankees. We didn't know whether they were going to win the [1996] World Series or not. However, my news director says, "The very second that the Yankees win the World Series, I want to be on the air live from the field and from sports bars around the city, and we're going to do the ticker tape parade that the city is going to throw."

Q: Did you have to use all your equipment?

A: I was maxed out! And I had to convince my people that they wanted to work both of their days off. If the Yankees lost game 6, we would have had to do this all over again on Sunday for game 7. We got a break: They won on Saturday.

Then I had to put together parade coverage from lower Broadway to City Hall, get everyone in place, and fight with the city for truck positions. This meant I had to be downtown and here at the same time to work all this out.

Q: How did you do that?

A: [Pause] I don't know. . . . A lot of phone calling. I got hold of every other station in the city to convince them that we needed to pool resources. From Monday to Tuesday, I did not go home and I did not go to sleep.

We needed to install telephone lines at City Hall so we could talk to the individual correspondents in their ears. The phone company said at 2:00 p.m. on Monday, "We don't have any more trunk lines available at City Hall. We're going to try to put another trunk line in, but there may not be any phones down there." And people always say, "Well, why don't you just use cell phones?" But when you've got 3½ million people in one cell area with so many trying to use their cell phones, cell phone technology is just out the window. You can't get a line. And in fact, it was completely unusable.

You need land-line phones. One of the secret, unseen spinal cords of television is still plain old telephone service with a hard-wire copper trunk line. If you can't talk reliably to reporters in their ears with some form of voice communication, you can't produce the show. They have no idea when they are on. They have no idea if they're talking to someone in the studio or someone at another remote location. The reporter must be able to communicate with the control room.

So the World Series was a high-pressure moment when we had no time to plan, and we still had to go and get it done.

Q: Is there a letdown when it's over?

A: Yes. It's like a wedding. You plan and plan, and then it goes on for five hours, and it's over.

To achieve sound unity in the one-on-one interviews, we want to maintain consistent mic distances at consistent volume levels. For this reason we decide to conduct all the interviews in a single location, and we plan to set up the camera, chairs, and microphones only once. We decide to use lavaliere mics (electrets) on each interviewee. We bring windscreens for the mics if we intend to record outdoors, and we also make sure to pack cables, extra batteries, mic clips, and gaffer's tape to secure the cables. For the panel discussion, we decide that three desk mics will be appropriate, and we pack the necessary desk stands to use with them.

We determine that the background noise of the site is something we wish to include. This natural sound, or **nat sound,** will add sonic texture, help identify the location for the audience, and help smooth out transitions when segments are shot out of sequence. To capture nat sound, we decide that the lavs we use in the interviews will be omnidirectional. (If we wanted to exclude nat sound, directional mics would be the better choice.) We also pack a separate tape recorder that we will use to record additional nat sound.

Production   On the day of the shoot, we arrive early enough to set up the equipment to execute the one-on-one interviews and the panel discussion. The lights, camera, tape recorder, VTR, audio mixer, mics, and cables are arranged. We secure loose cables with gaffer's tape to keep personnel in the area from tripping over them.

After positioning each interviewee and setting up his or her mic, we take sound levels to verify that everything is working. To record the nat sound, we start the tape recorder before each interview and let it run throughout the interview. At the same time, we record SMPTE time code on this tape for editing purposes.

For the panel discussion, we repeat the same procedures, this time setting up the desk mics. We make sure to space them properly to avoid interference problems, and we instruct panel members to keep incidental noise to a minimum. When the panel is assembled, we take levels. Then we roll tape and record sound before, during, and after the interviews. After the last field footage is shot, we thank all personnel for their participation, strike the equipment, pack, and head back to the editing suite to assemble the program.

Postproduction   At this stage, we check the nat sound on the videotape and supplement it, if necessary, with the sound we captured on the tape recorder. We can also add additional sound to the recorded material, such as background music, sound effects, or a voice-over with additional narration. All of these supplements may help eliminate dead air and improve transitions.

However, if we add music or sound effects, we should do so only after much consideration. For example, will music contribute to the program, or would silence contribute more? If we add music, how will it fit the mood of the show? We also question whether the music should contain lyrics. In most cases, lyrics under a voice-over are distracting and should be avoided. Although much of this thinking should occur in the preproduction stage, decisions about audio sweetening may not be finalized until postproduction.

At the postproduction stage, sound levels also become critical. If we have recorded principal audio and nat sound at full volume but now want to add narration or other sound, we may need to lower the volume of the field audio, or remove it completely, to make the new sound audible. We will return to these considerations in Chapter 11, where we discuss editing in more detail.

KEY TERMS

cueing *(140)*

open-reel tape *(142)*

signal-to-noise (S/N) ratio *(143)*

cartridge tape *(143)*

cassettes *(143)*

SMPTE time code *(145)*

digital audiotape *(146)*

nonlinear editing *(147)*

audio console *(149)*

VU (volume unit) meter *(151)*

feedback *(154)*

patch panel *(154)*

sound effects *(155)*

videotape recorder (VTR) *(157)*

voice-over (VO) *(158)*

backtiming *(158)*

audio sweetening *(159)*

nat sound *(162)*

QUESTIONS FOR REVIEW

1. How do CDs and DVDs differ from phonograph records in the way they store audio information? What are the implications of this difference for video production?

2. How does digital coding differ from analog coding, and what advantages does digital coding offer?

3. What advantages do tape-based media offer over CDs?

4. How does nonlinear editing differ from linear editing?

5. What are the functions of the audio console, and how can they help you in shaping the audio materials for a video production?

6. What is the difference between foldback and feedback?

7. What are some factors you need to consider in using nat sound or sound effects?

8 Graphic and Set Design

Several times in this book, we use the word *fabricate* to describe the creation of video content. The dictionary associates the word *fabricate* with terms such as *make, assemble, construct, manufacture,* and *shape.* However, the term *fabrication,* a close cousin, is also related to such terms as *fable, fakery, fiction, figment,* and *deceit.*

Just as MIDIs and digital audio workstations now enable producers to alter natural sound to fit the needs of a particular project, computers can be used to change visual images of events captured by the camera to fit the whim of the editor. For example, Figure 8.1 presents a wirephoto of President Clinton taken during a speech. The alterations show how readily computer graphics technology can change subject matter. Further, just as MIDIs can generate sounds without having to collect them from any natural source, computer graphics technology can similarly create images from scratch. Today video artists, using nothing but their imagination and a computer graphics program, can "paint" and store images (even animated ones) that can be fully integrated with regular broadcast signals captured from traditional video sources.

As technology continues to enable users to erase the line between fact and fiction, new questions emerge about what images presented on television are "real." Fakery is a particularly worrisome issue for journalists, whose job is to report news events faithfully. However, everyone involved in video needs to be aware of the ethical challenges today's technologies raise.

Keep the question of "real" versus "fake" in mind as you read this chapter and the next. This chapter begins with the topic of video graphics and moves on to set design. Basically, graphics include any two-dimensional visual design presented on video. Sets, of course, are three-dimensional, but many set elements that are visible on television are two-dimensional and therefore graphic in nature. In fact, new digital graphics systems are making it possible to substitute virtual sets for actual physical sets. Hence, more than ever before, design principles used in preparing graphics apply to sets as well.

The topics included in this chapter are:

GRAPHIC DESIGN

camera graphics • noncamera (computer-generated) graphics

TECHNICAL PRINCIPLES FOR CREATING AIR-QUALITY GRAPHICS

aspect ratio • scanning area and essential area • brightness and contrast issues • color context and compatibility • the role of graphic design in setting tone and style • preparing mechanical artwork

SET DESIGN

categories of scenery style

BASIC SET ELEMENTS

floor treatments • hanging units • standard set pieces • properties, furniture, and set dressings

PRODUCTION PHASES IN SET DESIGN

preproduction • production and postproduction

ELECTRONIC AND MECHANICAL EFFECTS IN SET DESIGN

electronic effects • mechanical effects • choosing between electronic and mechanical effects

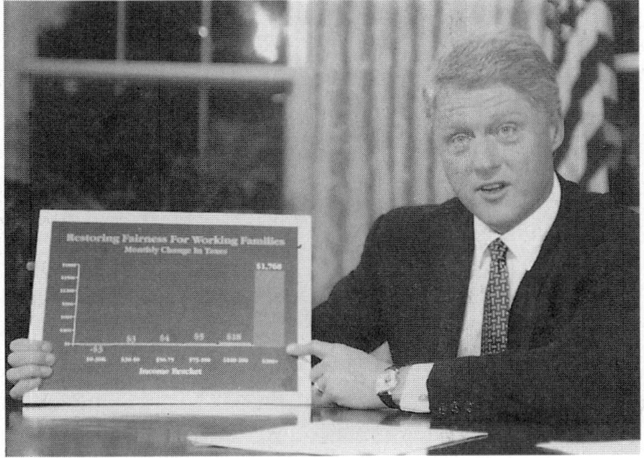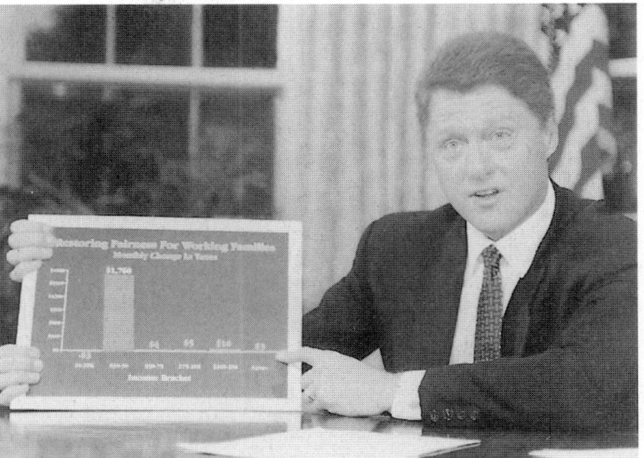

Figure 8.1   *A demonstration of how readily graphics can be altered with computer technology. (left) A photo of President Clinton taken during a speech. (right) The same photo altered: Clinton now has three hands and no watch, and his bar chart has been changed. What other differences can you see?*

GRAPHIC DESIGN

We may define *graphic design* as that part of video production concerned with the creation and use of two-dimensional visual images. Under this definition, any screen image of a two-dimensional display may be considered a graphic. Graphics include texts, pictures, and any variations, mixtures, or composites of the two.

The scope of graphic subject matter is infinite. Maps, charts, lists, graphs and other data displays, reproductions of signage and icons, pictures of people, places, and things, drawings, sketches, paintings, cartoons, and even animated presentations can all be rendered graphically for television.

Camera Graphics

As you might imagine, there are many ways to convey graphic information on video. For example, a camera aimed at a lighted card mounted with a photograph is one of the oldest methods of providing graphic material for television. Such cards (usually called *camera cards* or *studio cards*) have also been used to present titles and credits since the earliest days of television.

In news programs, camera shots of weather maps and other set pieces have long been a graphic staple of television. Cameras have also been used to capture visual information contained on slides. Of course, to avoid lawsuits, you should obtain written permission before using any copyrighted material.

Noncamera (Computer-Generated) Graphics

In recent years, noncamera video sources have taken the lead in video graphics. Most sources derive from computer-generated graphics machines. These devices operate at different levels of sophistication in terms of image-rendering capability, flexibility, speed, efficiency, and storage capacity. Some use a camera to capture graphic content that can then be manipulated by the graphic artist. Others internally generate and store material for later use. The following sections describe several such devices.

Character Generator   One of the most widely used electronic computer graphics devices is the **character generator (CG),** illustrated in Figure 8.2. The CG provides a keyboard that permits you to type letters and numbers on the screen in various fonts (type styles), colors, and sizes. Some CGs allow you to store hundreds of "pages" of text on disk and retrieve them as needed; others have more limited capacity. Some CGs let you enhance the lettering you select with drop shadow and border effects, as well as backgrounds, underlining, and other edging options. Another popular option available on some CGs is an **antialiasing** function, which softens the crude staircasing effect (*jaggies*) seen in diagonal lines and curved lettering. CGs equipped with antialiasing software create a blur around curves to reduce unwanted jaggies. Many CGs center text automatically when you display a page. Many offer the options to flash text on the screen, "crawl" messages horizontally on the screen, or vertically roll the messages up or down at variable speeds. A screen cursor indi-

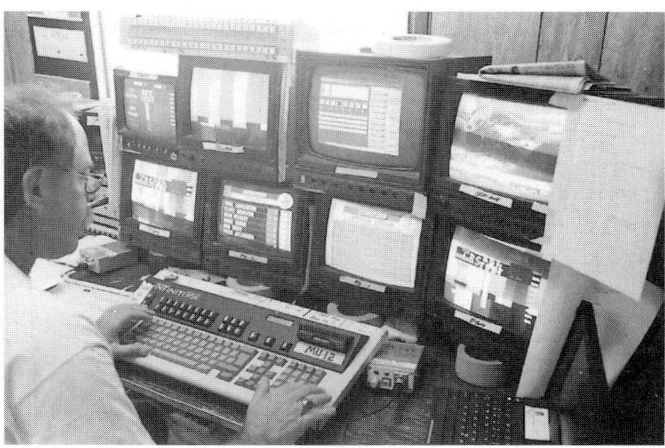

Figure 8.2   *A character generator.*

cates the location of the next character to be typed, just as in word processing programs.

Character-generated text can be combined with virtually any video source to create composite graphics. One such application is the placement of text at the bottom of the screen, over a camera shot, to identify on-air talent (Figure 8.3). Such text identifications (usually names, titles, and affiliations) are called **lower-thirds** because they typically occupy the lower third of the screen space. Character-generated lower-thirds are a good way to identify an on-air talent in a well-composed bust shot without blotting out the face. In contrast, cluttered or "busy" CG composite images, in which lettering blots out important picture information, can distract viewers and are generally considered not to be "air quality."

Other common uses of CG text include (1) centering a title over a program's opening shot, which can be provided by a camera, slide, tape, or any other source, and (2) rolling credits at the end of a show.

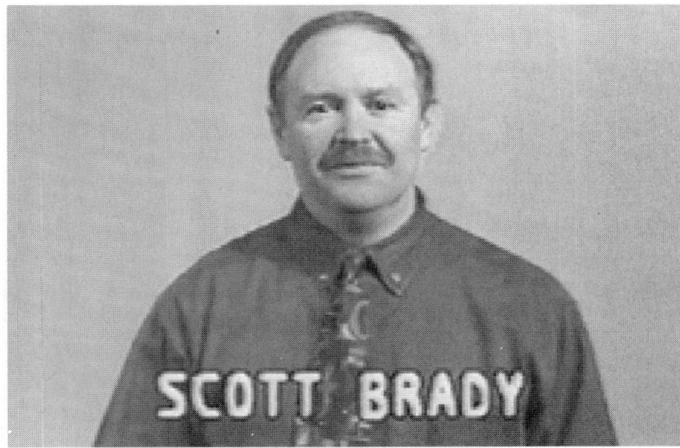

Figure 8.3   *A lower-third used to identify on-air talent.*

Digitizing Video Images   More advanced computer graphics devices can be used to digitize the video signal, allowing you to manipulate graphic content easily and flexibly within the screen space. Digitizing a video image provides the same advantages for picture manipulation that digital audio provides for sound. When a video frame is digitized from either a broadcast signal, analog videotape, or some other source, such as a slide or film, the continuous wave energy is sampled and discrete values are assigned to each sampled unit. In this way, each pixel is, in effect, assigned its own computer address, making it possible to change the color and brightness values assigned to it independently from any other pixel.

This system requires a massive amount of computer memory, but it enables virtually endless manipulation and control of video images. You can change their color, size, and shape, as well as the locations of image segments within the frame. Further, digital video can be reproduced repeatedly without degrading, and it can be transmitted over telephone lines to multiple locations. For example, Weather Service International (WSI) collects satellite images of weather patterns and distributes them to its subscribers by modem over telephone lines. TV stations that subscribe to the service can combine these images with camera shots of a studio weather map, text, and other pictures for local weather reports.

As with audio, the ability to disseminate video in this way opens the door to pirating copyrighted works. Sampling or altering copyrighted materials without permission can lead to costly legal litigation. For these reasons, as mentioned earlier, it is advisable to obtain written permission before using any copyrighted material in your productions.

Electronic Still Store   Another graphics device in widespread use is the **electronic still store (ESS)** system (Figure 8.4). The ESS enables you to save digitized single frames of video from any source (tape, slide, camera, and even CG composites already mixed with other video) and file them for later use on a high-capacity disk. Electronic still store machines are also called **frame grabbers**.

Some ESS systems can store only about a hundred "pages" of images, whereas others can store tens of thousands. Some offer immediate access to thousands of pictures, any one of which can be made available in a fraction of

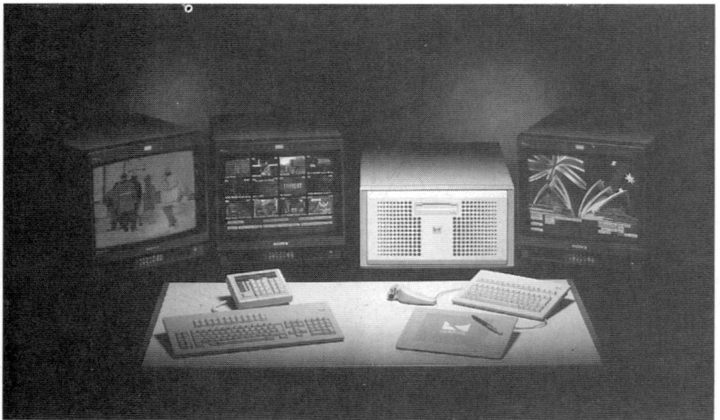

Figure 8.4   *An electronic still store system.*

a second. Some machines permit you to display image sequences in real time or in slow or fast motion. One common use of the ESS system is to display sports figures' pictures along with their vital statistics during sports programs. Different displays can be prepared for each player and filed by number for immediate retrieval at the appropriate time.

The captured video images can come from a television camera focused on a graphic or other object placed on a copy stand. Still store systems can also take images from other sources. For example, a machine resembling a Xerox machine, called a **flatbed scanner,** can digitize two-dimensional graphics, including slides and print photographs, which can then be captured and stored by ESS technology. News departments frequently use ESS with videotape, live camera shots, and CG text to create composite images, which can then be presented at the appropriate times during production. These composites may be further combined with other images when routed through a video switcher, as described in the next chapter.

Paintbox Programs   One way to eliminate the need for copyright permission is to create your own images using a video **paintbox** system (Figure 8.5). Digital paint systems permit you to create video graphics out of whole cloth, without going to any external source whatsoever, using a *stylus* and a *bit pad*.

The stylus functions as an electronic paintbrush, pen, or pencil. The bit pad is the computer's equivalent of the artist's canvas. Many programs feature an array of pull-down menus (the computer's counterpart to the artist's palette) from which you can select various options using a standard computer mouse or drawing stylus. By clicking on menu icons from the perimeter of the computer screen, you can choose various stroke widths, colors, and many other drawing and painting options. As you draw on the bit pad, the results appear on the computer screen. (Some machines allow you to draw directly on the screen.) When the work is finished, you can save it on a disk for later use or combine it with other graphics. (See Figure C5 in the color plates.)

Many systems go beyond providing a field on which to draw. They also offer **computer-aided design (CAD)** programs and more complicated graphics

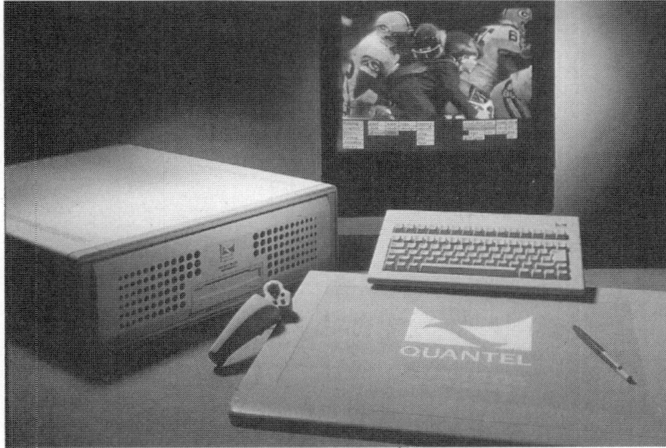

Figure 8.5   *A video paintbox system. Notice the bit pad at the lower right. The stylus, an instrument used for drawing on the bit pad, is not shown here.*

options. For example, some systems permit you to designate several points on the pad and then request the computer to connect them with lines. The computer can then be made to fill in the area defined by those lines with a selected color. The entire process and final results may be displayed on the screen. Other CAD and graphics options commonly include:

- *Sizing,* in which you instruct the computer to increase or decrease the overall size of an object or shape on the screen.
- *Rotation,* where the artwork appears to rotate on the x-, y-, or z-axis, either appearing to bring one edge closer to the viewer using perspective effects or merely canting the image from the horizontal.
- *Cutting and pasting,* in which you remove artwork intact to another portion of the screen. You may also make it occupy a smaller space, thus enabling you to work on other drawings while keeping earlier material in view.
- *Posterization,* in which the computer transforms the artwork in terms of only high and low luminance values. In other words, the middle values of light are eliminated, giving the artwork the look of a poster.
- *Cycling,* a simple form of animation in which an image (or a series of images) appears to be animated as its color or brightness is made to change in a repetitive sequence. Cycling has applications in weather graphics. For example, a simple cartoon image of the sun (often with a happy face and sunglasses) appears to shine at different intensities, thus increasing visual interest.

Software development is increasing the availability and complexity of specialized clip-art and storyboarding programs for television production, reducing the need for traditional mechanical work.

Three-Dimensional Graphics and Computer Animation As computer graphics programs become more sophisticated, their speed, versatility, and capability become more impressive. For example, three-dimensional modeling software now enables computers to present sequences of visual images that imitate what three-dimensional objects would look like if they were turning and moving in space. This is obvious to anyone who has seen the hit movie *Toy Story,* the first full-length feature animated film to use nothing but computer software to generate all of the images.

To do three-dimensional modeling, the computer first generates mathematical values that represent selected points on the object from a number of different perspectives. This information is then used to create images of the object from different viewpoints in sequence. Of course, this is not a simple task; it requires millions of calculations at extraordinary speed to present a convincing appearance of objects moving in real time. Such machines are therefore still at the high end of the computer graphics market.

In addition to providing color and the illusion of depth, computers enable you to program an endless variety of surface characteristics such as texture (smoothness or roughness), pattern, degrees of opacity-transparency, and reflection effects. Some programs known as *ray-tracing* software even provide for naturalistic light and shadow effects that vary the amount of illumination different parts of objects appear to receive, as well as varying the apparent angle, direction, and strength of the shadows the objects appear to cast.

When animated in real time, forms featured in computer-generated graphics can be made to look virtually real (thus the phrase *virtual reality*). Some are so convincing that it may be difficult, if not impossible, to determine whether or not their images were captured from preexisting things. Questions of authenticity become especially important here, as do the ever-important issues of journalistic and advertising ethics.

Some computer programs require a fair amount of time to render animated sequences of objects or scenes. In this case, the sequences are rendered one frame at a time off line during pre- or postproduction. Then they are edited into an animated sequence. In contrast, faster machines can render animated sequences on line. After a beginning image, a final image, and several interim images are identified, the computer fills in the rest. Regardless of the method by which the animations are prepared, when they are shown at thirty frames per second (the standard NTSC video format), they can be called *real-time animation.*

Some effects possible with more advanced computer animation programs include:

- *Morphing* (Figure 8.6a), in which one picture changes into another through a barely perceptible series of intermediate steps. An early use of morphing can be seen in Michael Jackson's *Black and White* music video, in which subjects of different genders and ethnicities are transformed into one another. A similar use of morphing is found in a well-known Schick razor advertisement when, in mid-stroke, one shaver's face changes into another's. Morphing is also used in the film *Terminator 2,* in which the title character seems to change from a cop into what appear to be pools of mercury.
- *Flips, pushes, and squeezes* (Figure 8.6b), in which images seem to go through various motions and distortions. These effects are especially popular in product advertising, because the motions lend visual interest and a sense of excitement to otherwise static objects.
- *Surface pinning* (Figure 8.6c), in which an image appears to conform to a preselected shape. Common uses of this effect can be seen in transitions between news stories when a news graphic or even a live location shot suddenly appears to change into a page in a book turning to reveal the opening visual of the next story.

Other computer graphics software permits you to overlay parts of images from different video sources (**compositing**) or to simulate fog and blur effects associated with motion. Some workstations combine video with audio capability. Companies specializing in computer graphics video technology and software include Chyron, Quantel, Alias Animator, NewTek, and Silicon Graphics. In Chapter 9, we will discuss further examples of special effects based on digital technology.

Digital Still Cameras Advances in still photography are also influencing video. For example, still cameras are now available for recording static images digitally on disk. These pictures can be transferred directly into video still storage without going through the step of digitizing an analog picture. If digital disk cameras continue to advance, we should see more and more use of them to produce still images for video applications.

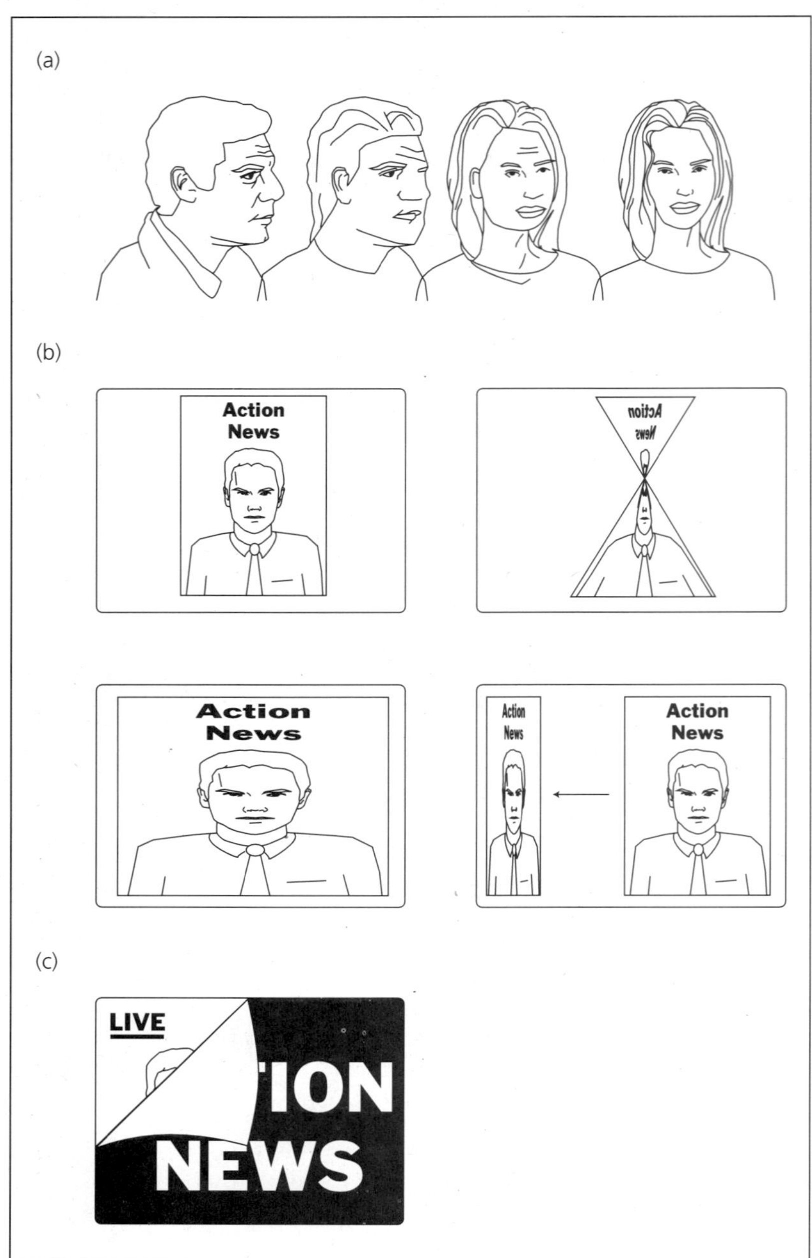

Figure 8.6   *Effects that can be created with computer animation programs. (a) Morphing. (b) Flips, pushes, and squeezes. (c) Surface pinning.*

TECHNICAL PRINCIPLES FOR CREATING AIR-QUALITY GRAPHICS

No matter what tools you use to create graphic images for video, it is important to design the images properly so they will be effective on a television screen. Successful graphic design requires understanding the video system's capabilities and limitations. What looks good to the naked eye may or may not succeed on television.

Remember that television as it currently exists is a low-resolution medium that, regardless of screen size, features a screen space with a fixed ratio of width to height, as well as a contrast range narrower than that available to your eye. The quality of the graphics you use is critically influenced by these and other technical factors.

Aspect Ratio

Successful graphics should conform to the rectangular television screen shape. Traditionally that shape, regardless of screen size, has been fixed at a ratio of width to height of four units to three units. As we saw in Chapter 2, this is designated as a 4:3 aspect ratio. This aspect ratio does not change from shot to shot or as program content changes. Therefore, regardless of the program material being shown, if it is to fit the screen without leaving dead space or unwanted border, it must be in a 4:3 aspect ratio.

With the recent adoption of a digital television (DTV) standard, as described in Chapter 2, the aspect ratio of television receivers will soon change from the traditional 4:3 to 16:9 (or slightly wider than 5:3) units of width to height. The new standard will also increase picture resolution significantly. Since DTV will produce higher-quality images on a wider screen, it will open up new graphic possibilities. For the next few years, however, the NTSC format will remain the industry standard.

In an ideal production world, all television graphics might be prepared in proper aspect ratio. But sometimes it is necessary or desirable to use graphics, particularly photographs, that do not conform to the correct aspect ratio. Such materials may have originated from news archives or other sources where video aspect ratio was not a concern.

To make use of such materials on television, you can show the entire image in a static shot with the dead space of the border visible (Figure 8.7a) and matte the border of the image with a color to make it more attractive. If you do this, try to fit the color you select to the tone of the production or at least make the border less obtrusive.

In severe cases where more border than image is displayed, this approach is not recommended. In such cases, try cropping the image so that you fill the screen space with only the most important information, perhaps eliminating the border entirely (Figure 8.7b).

Another approach is to feature the graphic with camera movement, either panning, tilting, or zooming on the image while on air (Figure 8.7c). This is more complicated than cropping to a static frame, and it may require practice with the camera to achieve smoothness of motion and proper pace. The

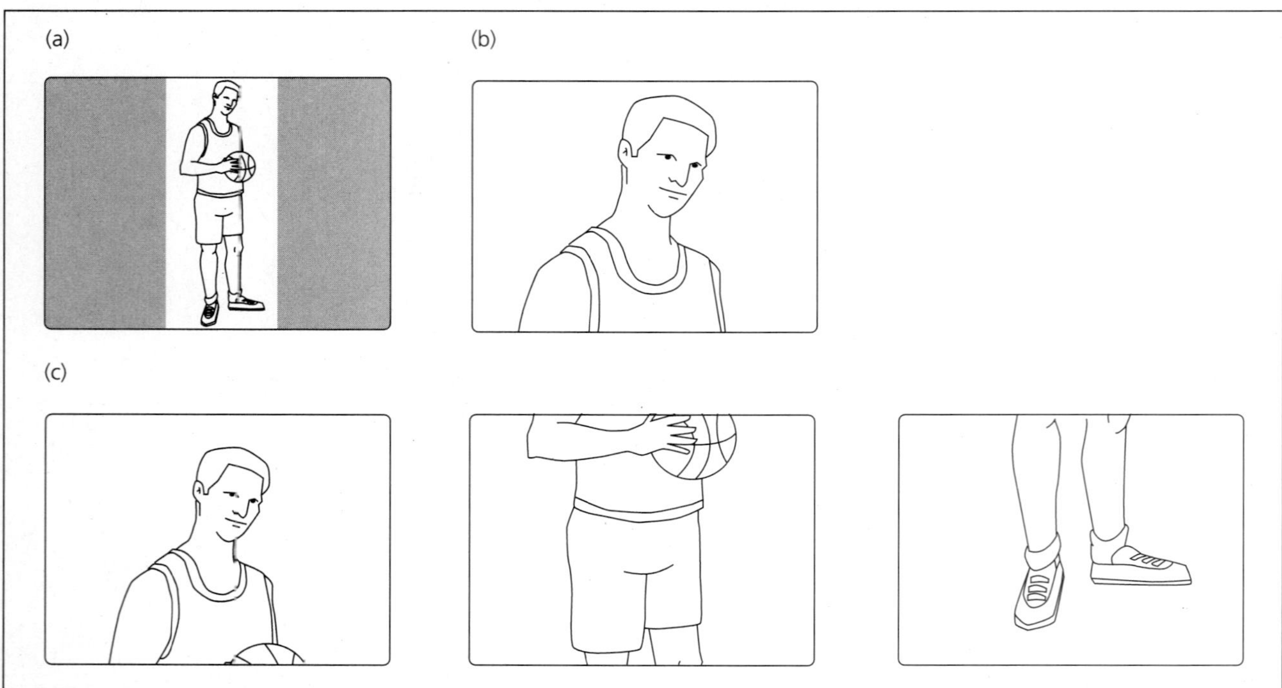

Figure 8.7   *Methods of treating an image that does not fit the TV aspect ratio. (a) Showing the entire image, but matting the dead spaces of the screen with a color. (b) Cropping the image to a shape more suitable to the screen. (c) Moving the camera to capture the image in successive frames—in this case, by beginning with a shot of the head and then tilting the camera gradually downward.*

biggest problems tend to be jerky camera movement, movement off the graphic entirely, and the tendency to use irrelevant or unnecessary movement that can be distracting.

On the other hand, effective movement can support the script's meaning. It may, for example, include reveals of important information. When done properly, such camera movement provides visual variety, heightens audience involvement, and emphasizes key aspects of the program. Keep in mind that executing this technique may require the cooperation of more than just the camera operator. You may need to pace the movements you have planned to the voice-over of an announcer. You may need card pullers or other personnel to integrate the moves into the rest of the program. This may mean quite a bit of rehearsal. However, the results can be extremely effective and engaging.

Scanning Area and Essential Area

Since there is always some signal loss during transmission due to masking of the perimeter of the home receiver and misalignment between the camera's *scanning area* (the total area scanned by the television camera) and that of the home receiver, it is important to confine essential information to the part of the screen that will reach the home receiver intact. To do this, video professionals designate an "essential" or safe title area. Roughly speaking, the

essential area of the television screen is the central rectangle of space comprising about two-thirds of the total area covered by the camera (Figure 8.8). Graphic information—particularly important facts such as phone numbers for viewers to call, titles, and credits—should be confined to the essential area.

Brightness and Contrast Issues

Regardless of subject matter, clear, discernible graphics require enough contrast among elements to enable the viewer to distinguish them from one another. This is true in any visual medium, but particularly in video with its relatively low resolution and limited range of brightness.

In terms of light reflectance, the brightness range of the television system varies from *TV white,* which is at best only about 70 percent reflectance value for monochrome receivers (with pure white set equal to 100 percent), to *TV black,* which is about 3 percent reflectance value (with pure black set at 0 percent reflectance). For color systems, the whitest TV white is only about 60 percent reflectance.

If we divide the brightness range, from 3 percent reflectance at one end to 60 to 70 percent reflectance at the other, into distinct steps, we get the **television gray scale** (see Figure 4.8b on page 62). The gray scale is often broken into nine steps; however, in the industry you may also encounter seven- or five-step scales.

One use of the gray scale is to determine differences in reflectance values between lettering and backgrounds. Most experts agree that to ensure legibility, *the minimum spread between text and background should be two gray-scale steps.* Obviously, this is only a rule of thumb, since two steps on a five-step scale may be much wider than two steps on a nine-step scale.

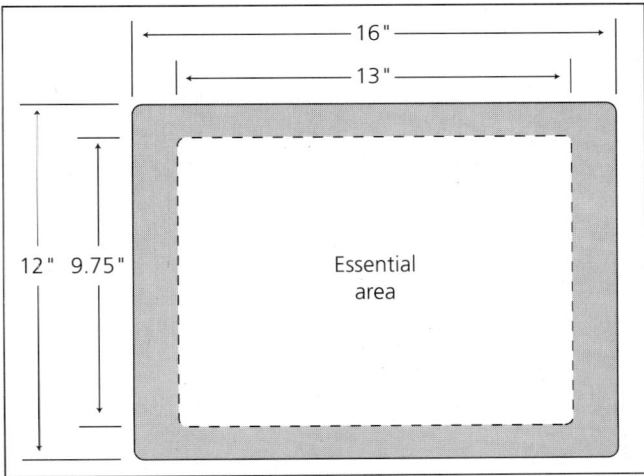

Figure 8.8 *The essential area of a TV screen. The measurements shown here apply to a screen with a 20 inch diagonal, but the proportions are the same for any screen size.*

PROFESSIONAL   POINTERS

Designing Text for a Video Screen

- To accommodate the relatively low resolution of the current television screen, keep typefaces simple. Avoid complicated serifs and overly thin and/or fancy letterforms (Figure 8.9).

- As a rule of thumb, make the text *legible at a distance sixteen times the height of the monitor.* For example, if the monitor is 12 inches high, type should be readable at a distance of up to 16 feet.

- When possible, enhance the clarity of letterforms by using *drop shadows* and/or *borders,* which help to distinguish text from its background.

- Organize the layouts of words into blocks of related material so that meanings are easy to grasp.

- When combining text with pictures, don't place the text over extremely intricate backgrounds. In short, avoid *busy* visuals.

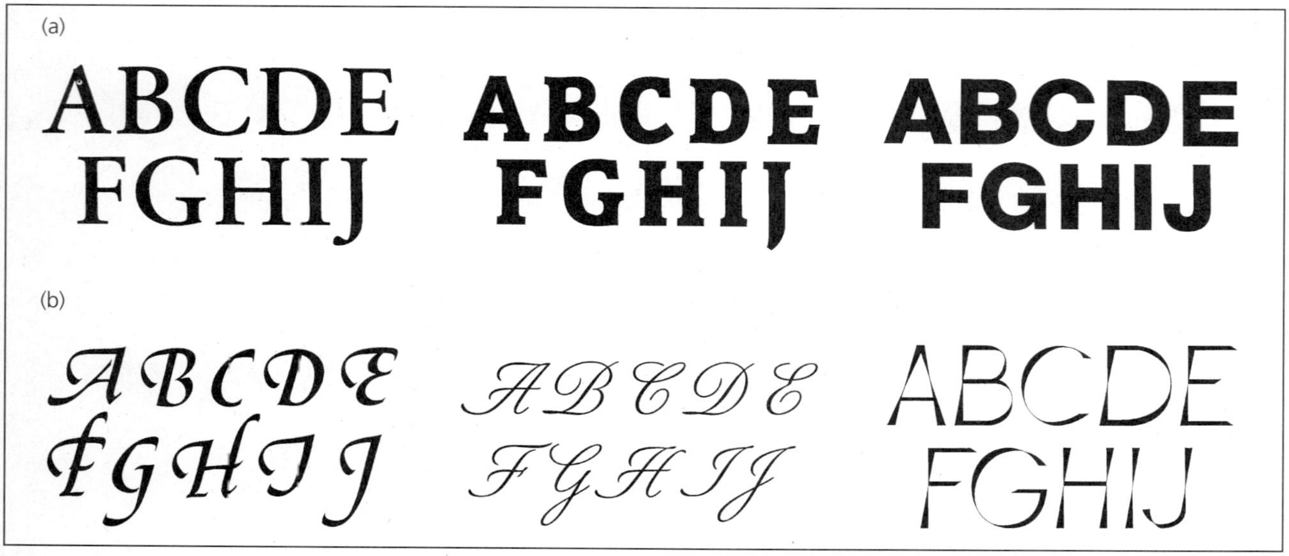

Figure 8.9 *Appropriate text styles for a TV screen. (a) Some fonts that are relatively easy to read on screen. (b) Some fonts that are much harder to read on screen.*

What if the text and background are different colors? In that case, can you ignore contrast? The answer is no, because different colors may have identical luminance (brightness) values. Therefore, to differentiate elements of the graphic image from one another, it is not enough to distinguish them by color; they must also differ in brightness. A waveform monitor, which provides a graphic display of light reflectance levels for different colors, can help you judge whether the elements of your graphics have enough variation in luminance values to make them air quality.

Color Context and Compatibility

Colors tend to change in appearance depending on their surroundings. A light color generally appears even lighter when viewed against a dark background. Further, light foreground elements appear to come forward more than dark ones, and they may appear larger merely as a result of their color (see Figure C.4 in the section of color plates).

In addition, some color combinations tend to create more eye fatigue than others, and some are harder to reproduce on television, even for the most expensive cameras. For example, the red-blue color combination is said to be particularly fatiguing, and red is one of the more difficult colors to render accurately.

Finally, different colors may have different cultural and psychological associations that may affect the tone of a program. For example, red is frequently associated with hot emotions such as anger and rage, as well as with signals of warning and danger, whereas blue is often associated with emotions such as sadness and cooler feelings. Of course, color associations are not forever fixed this way; they may change depending on program content and culture.

The Role of Graphic Design in Setting Tone and Style

Graphic design contributes significantly to a program's look and feel, tone, and style. For example, a program's title is often the first thing the audience sees. Therefore, the design of the title, in addition to simply stating it, offers the producer an invaluable opportunity to include enticing information about the show. Imagine the contribution made to a horror movie by a title that looks as if it were printed with candle wax drippings. Compare Nickelodeon's graphics with those of CNN.

If done tastefully, the entire graphic presentation—including the selection of pictures, illustrations, and other visuals, fonts, colors, sizes, layouts, and the pace—can convey a great deal of information to the audience about what lies ahead while adding thematic unity to the entire production. In all cases, of course, it is critical to preview all visuals to verify that they look the way you want them to look.

Preparing Mechanical Artwork

Even in a graphics world dominated by electronically generated computer images, it is worthwhile to know how to prepare paste-ups and mechanical artwork. Many television talent still use physical cards to convey key program content. For example, both Jay Leno and David Letterman use display cards to present comedy bits from their desks on their late-night television shows. The following sections briefly describe several methods for preparing studio graphics for your video productions.

Studio or Camera Cards   Studio or **camera cards** are sturdy sheets of card stock either held in the talent's hand or supported by an easel or a card stand. They can feature text and/or pictorial content. A manageable material for such cards is 11-by-14-inch rectangular illustration board. This size is large enough to support most artwork in the proper aspect ratio for video but not so large that it tips over card stands or overwhelms talents' ability to handle the cards comfortably. The space on this size of card can be easily captured by most video cameras without forcing the camera to either dolly out beyond the range of most studios or dolly in so tightly that transitions become difficult.

Illustration boards (or *art cards,* as they are sometimes called) can be purchased at reasonable cost in large sheets in a variety of colors from most art supply stores. They can be cut to size with a cutting machine or a matte knife.

Applying Lettering to Studio Cards   One of the oldest methods of getting lettering onto studio cards is to use *transfer letting,* either the plastic peel-off, stick-on kind or the rub-off type. This lettering, available from such companies as Letraset, Press-type, and Chartpak, is sold for several dollars per sheet at art supply stores. These products can be applied to flat studio cards, acetate, and even glass.

If you use transfer lettering to prepare text graphics, practice with a number of characters to master your technique so that you can control the spacing and placement of characters. Use light pencil guidelines that can be easily erased after guiding your work. Generally you want to arrange text so it is centered, even, and level. Be sure to contain all of the characters inside the essential area and to lay out the text in the proper aspect ratio.

Most production facilities have word processing and desktop publishing PCs that allow you to set type, do electronic paste-ups, and then either store the content electronically or print it out with a laser printer. Depending on the level of sophistication of the program and the printer you use, you can create text in a variety of point sizes in numerous fonts. For example, Illustrator, one of the most widely used graphic programs in video production facilities, lets you choose from an enormous variety of fonts, preset a page layout in the proper aspect ratio, alter many features of the lettering (such as its size, color, and shape), and add borders and shadows. Even with such a program, however, the paper stock your printer can handle may not be sturdy enough to enable you to skip the step of mounting the copy onto studio cards.

Images from Slides and Film   Although slide film is no longer used to display lettering on television, it is still useful for capturing pictorial images. Slides are 35 mm, static film images mounted in either 2-inch squares of cardboard or in plastic mounts of the same size enclosed with thin glass panels. Slides may be loaded into a slide carousel and projected into a fixed camera, usually in a film and slide facility in the control room called the **telecine** or **film island** (Figure 8.10). Kodak Ektachrome slide film is available with fast, convenient developing service at most photo shops, and for this reason it is

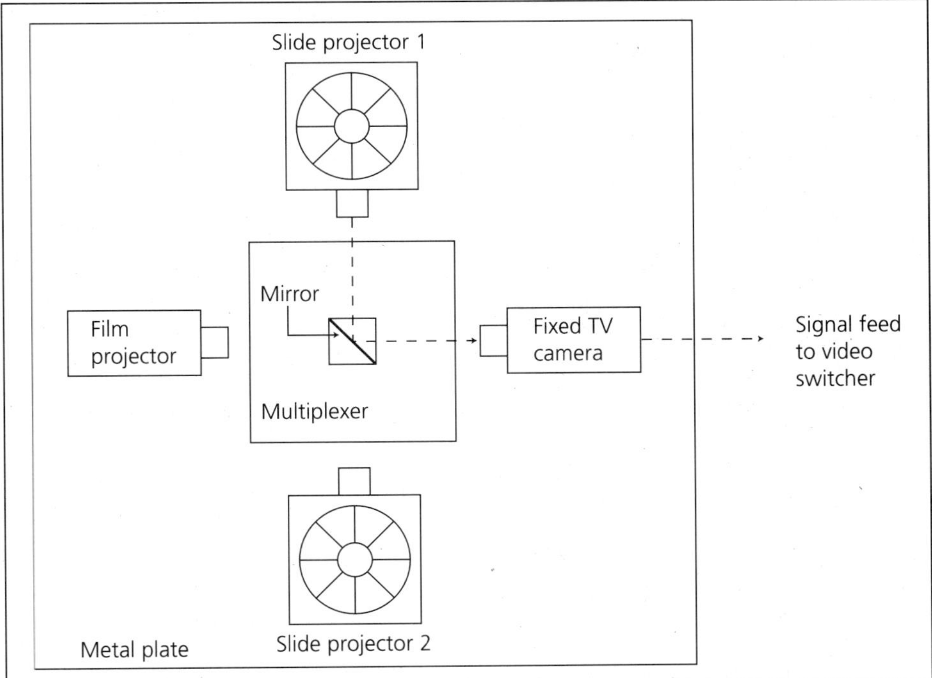

Figure 8.10   *An overhead diagram of a telecine or film island, used to feed slide and film images to a television camera. The multiplexer in the center contains a mirror that can flip to several positions to control the image the camera sees. In this diagram, the mirror is positioned so that the slide image from projector 1 reflects into the camera lens. The entire telecine is bolted to the floor so that all the image sources, as well as the camera, remain in fixed positions.*

I N D U S T R Y
voices

Cheryl Hurni
Graphic Designer, "The Late Show with David Letterman"

Q: How long have you been graphic designer of "The Late Show with David Letterman"?

A: A little over a year now.

Q: Tell me about your background in terms of schooling and so forth.

A: I majored in graphic design at the Moore College of Art and Design in Philadelphia. During my junior year, I freelanced in the summer at a design firm in New York City. I graduated in May of '95, and after freelancing that summer I got the job on "The Late Show."

Q: How did you get the job?

A: I was in the right place at the right time. A woman who graduated from my college three years prior to me had been working here about a year when the graphic designer left. She called my school and asked if anybody who was graduating was going to be in the city and, if so, what their background was. One of my professors gave my name and said I was really proficient in Quark and Photoshop. So I was asked to come in for a day. I did well, and she asked me to come back another day, and that turned into three weeks, and that turned into my job.

Q: How has the job developed over the last year?

A: This job is very different from a regular graphic design job. Graphic design that you learn in college is "nice design," very clean and simple, with not a lot of clutter. That's different from what we do here, which I suppose you can call "television design": very gaudy and eye-catching, not very clean. For example, in regular graphic design, there is a lot of white space. In television, there is no such thing as white space. Also, things that look good on your computer screen don't necessarily look good on the

television screen. So you have to be able to bridge the gap.

Q: How do you do that?

A: Well, we have a new piece of equipment called the Video Explorer, which lets me take what I'm doing and put it on a regular television monitor. That way I can test it and see how the colors look. This is necessary because the colors never look the same.

Q: You mean from computer RGB to the TV screen there is a difference?

A: Oh, definitely! The colors change even from program to program.

Q: If you were to break out the percentage of output devoted to physical art cards or camera cards as opposed to purely electronic graphics, what would it be?

A: I would say fifty-fifty. We make a lot of art cards during the day. There are also a lot of last-minute graphics for props, and many hand-held pieces.

Q: Talk a little about some of the ones you have here on display.

A: OK. Here's one we did a couple of weeks ago. This is a flip book we made, starting with a body image of President Clinton, with each succeeding page showing him getting heavier. To do it, we used the program Photoshop.

Q: Where did the original image come from?

A: We took a cartoon book of Clinton and scanned it on a flatbed scanner. Then we brought it in to Photoshop and altered it.

Q: Tell me about how you did this box over here.

A: This was a package we redesigned. We do our own packaging for parodies of various products. We rip the package open, lay it flat, and scan it in. We see how it's put together. Once it is scanned, we bring it into Photoshop, redo the outlines and the copy, make all our changes, and print it out.

Q: Did you print it out directly onto cardboard or paste it onto card stock after it came out of the printer?

A: After printing, I put it onto thicker stock. Basically, cardboard is not going to go through that printer.

Q: What about this prop that looks like a TV Guide*?*

A: Here we took a regular *TV Guide* and made a new cover. We scanned in the old one and fixed it up. We tried to make the color gold, but it came out less gold than we wanted. It's fortunate in cases like this that the television medium is so forgiving. We can get away with this on the screen, even though if you look at the graphic quality directly, you can see the color is not that good and the pixels are really noticeable.

Q: Do you have copyright problems?

A: It's a rarity. Most of the time we use stuff from photo labs, which is paid for. Often we use products, and the manufacturers like the free publicity, so they don't complain.

Q: Now let's talk about the other end of the spectrum. You also create things that are never seen by a camera or held in Dave's hand. For example, there are credits that roll at the end of the show. Do you do that kind of text?

A: No, that's not my department. However, I do the titles that introduce a skit. Also, if they need a picture for a background under a title for a skit, or if they want part of an image taken out, they send it to me.

Q: How do you make a title for a skit?

A: I start in a program called Illustrator. The writers tell me the title of the piece, what the piece is about, and what they might like. I pick a font that I think will enhance the piece or make it stand out. I type in the letters and design the type in Illustrator.

Q: If you were doing a Dracula *sketch and you needed dripping-wax letters, could you create them?*

A: Yes, in fact, let's do a title that way. Let's open up my font suitcase and pick out a font. [She clicks the mouse to start using the Illustrator program.] OK. Now I will open a preset page that has already been adapted to the aspect ratio of the TV screen. What should we type?

Q: How about Dracula's Graphic and Set Design*?*

A: OK. I can type it, then change the size, center it on the screen, color it, add drop shadow, stretch or condense it, add background, create outlines. [She does all this as she speaks.] We can make it bolder if we want.

Q: That's beginning to look scary, all right. After it's prepared on the screen, how does it get to air?

A: I can send it directly to video or print it out if I want. If I go straight to video, I do the type in Illustrator, then bring it into Photoshop, because that is the program that Video Explorer is compatible with. The Photoshop product is what gets sent to video. Then I look at it on the regular TV monitor and decide on colors because they are always so much darker on the computer screen than they are on regular video.

Q: What trends have you noticed in your field over the last five years?

A: Computers have totally taken over. In college, you still are taught manual cutting and pasting with X-Acto blades. They always say, "Well, what if the computer breaks down?" And that is excellent because you never know, the computer *might* break down.

Q: Would you still teach mechanical graphic design skills if you were teaching your field?

A: Yes, definitely, because it helps you understand design principles. By cutting and pasting, you get to move things around. With the computer, you tend to experiment less because the product looks so finished and clean on the screen.

The way I learned design, the process is to sketch first, get your ideas down, and don't use the computer until you have a complete idea. A lot of people go directly to the computer because they feel it looks so much nicer. But in my opinion, your best work is always done when you sketch first. Sketching puts you more in touch with ideas, whereas the computer doesn't help as much conceptually.

I would never have learned graphic design principles if all I did in college was point and click with a computer mouse. And even though graphic design and television graphic design are so different, I use the principles I learned in college every day here.

preferred over other slide film products. With all slide film intended for color photography, however, be sure to use the proper filters to avoid unwanted color shifts when integrating slides with videotape of the same subjects shot with artificial light.

A digital version of the telecine, called a **media record telecine (MRT),** enables you to digitize film images during their transfer to videotape. The MRT can also record bar code information on the vertical blanking interval of the tape, making accounting and housecleaning procedures easier. In addition, two other recent devices, the **flying spot scanner** and the **linear CCD array,** are being used to transfer film images to videotape. These devices electronically scan each frame of film, reversing the film image so that the image recorded on tape is a positive. To accommodate differences in frame rates between film and video (twenty-four frames per second for film versus thirty for video), odd-numbered frames of film are recorded on two video fields and even-numbered frames are recorded on three video fields. With this process, thirty frames of video are used for every twenty-four frames of film.

Mounting Artwork on Studio Cards   All mechanical artwork, including photographs, drawings, charts, and graphs, should be mounted on flat studio cards before being displayed on television. Do not rely on the ordinary paper stock when displaying images for the camera; in most cases, unmounted pictures will be too flimsy to stay flat, making it hard to achieve uniform lighting and focus. Mounting graphics on studio cards also provides additional border that helps to alert camera operators when they are shooting off the artwork.

How should artwork be mounted on studio cards? There are several methods to use, and some that should never be used. Recommended methods include using rubber cement, spray adhesive, and peel-and-stick adhesive sheets, all available at art supply stores. These choices result in flat, smooth artwork that is extremely long lasting if not permanent.

When using rubber cement, lay out the work to be mounted and outline the area the artwork will occupy on the studio card lightly with pencil. Place the artwork to be mounted face down on a sheet of newspaper or other expendable surface, and coat it entirely with a thin layer of cement. Then do the same to the surface of the studio card. *Go slightly beyond the area the artwork will occupy on the card.* When the cement no longer looks shiny (after a minute or so), apply one edge of the artwork to the place on the card where you want it. Then carefully lay down the rest of the artwork without trapping bubbles of air. (You may use a roller to make this part of the job easier.) Set the card aside for a few minutes until the cement is completely dry. Remove excess cement with an eraser or with the side of your finger. The card should now be ready for display and should last a long time without coming loose or curling up.

When using spray adhesive, after laying out the artwork and outlining the area as just described, place it face down on a clean, expendable surface such as newspaper and spray the back of the artwork with adhesive. *Do not spray the card.* Then carefully place the artwork on the card and use a roller to press it into place. Work from the edge or center of the artwork to avoid trapping air bubbles. Set the card aside for a few minutes until the adhesive is completely dry. The card should now be ready for display, and the spray adhesive is permanent.

Adhesive sheets are also easy to use. First, cut the adhesive sheet to the proper size. Then peel the sheet to reveal the adhesive material. Next, rub the

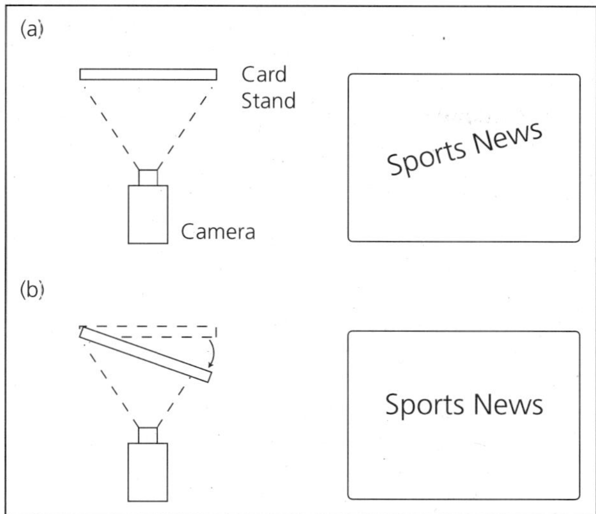

Figure 8.11   *Correcting a keystoned effect. (a)
The original position of camera and card stand
produces keystoning; on screen, the text appears to
run uphill. (b) The effect is corrected by rotating the
right edge of the stand closer to the camera.*

adhesive material onto the surface where you wish to fix the graphic. Finally,
put the graphic in place and rub it down to fix it.

Avoid using glues that cause the artwork to buckle and wrinkle. Also, do
not use tape ("invisible" or any other kind) to fix artwork to studio cards. The
tape shows, is unreliable, fails to eliminate buckling and warping, and, for all
these reasons, produces an unsightly effect. To ensure air-quality graphics,
use the recommended adhesives.

When using slides or studio cards, remember to number them (in pencil) so
they can be easily arranged if they get out of order. Studio cards may be identified with numbered tabs of masking tape, which makes them easy to
arrange and pull if you need a card puller during production.

Remember, also, to preview all visuals to confirm that they look acceptable
on air. One of the simplest problems to correct when previewing graphics is
the keystoned graphic. A *keystoned* graphic is one (usually text) in which the
material looks as if it is running uphill or downhill when you want it to appear perfectly horizontal. To fix this, while watching a shot of the card on the
line monitor, have the card puller pitch the left or right edge of the card closer
to or farther from the camera until it looks right (Figure 8.11). In most cases,
this will be all you need to do to solve the problem. May all of your graphic
problems be this simple to fix!

SET DESIGN

The term *sets* refers to all of the scenery, backgrounds, and furnishings intended to be seen by the television camera, including items associated with

interior locales, such as walls and floors, fireplaces, bookcases, doors, and staircases, and items associated with exterior locales, such as park benches, trees, storefronts, street lamps, and even mountain views. Furniture items and larger pieces brought in to fill out a scene are called set **properties,** or **props** for short. Smaller items, including embellishments such as lamps, books on bookshelves, and pictures hanging on walls, are called *set dressings*. There are also *hand props,* items held in the hand by on-air talent, such as telephones, guns, cooking and eating utensils, and so forth.

The function of a set is to provide scenery that supports, enhances, and evokes the tone and purpose of the video program. Needless to say, the set should permit ease of access, and it should be durable so that it doesn't come loose or fall apart during production.

Good set design must be shaped by what is *possible* as well as what is *desirable*. For example, when shooting a video about environmental issues, the producer may want the talent to stand in front of a mountain. In Montana, this may be easy to arrange. However, if the story is being shot in Indiana, the shot may require substantial costs for transporting the talent and crew to the desired location. Beyond such natural settings, time, money, labor, and carpentry talent, as well as access to materials, are often important constraints on set design.

It is the task of the producer-director to determine how the set should look. At the beginning of the set design process, the concept phase, it is best to proceed without regard for what is available. *The first concern should be to design the set on paper or in model form in line with the program's central concept or theme, without regard for what is possible.* Brainstorm. Ask yourself what the program is about and what its overall mood is. What degree of reality or unreality are you trying to capture? If you had unlimited resources, what would the ideal set look like? Only after deciding what the "best" set(s) might be should you allow limitations of time, money, materials, and effort to temper your plan.

Categories of Scenery Style

The main categories of scenery style include the following:

1. *Realistic.* Sets designed in a realistic style use authentic (or authentic-appearing) set pieces, such as sinks that have running water, refrigerators that actually keep food cold, doors with doorbells that ring, or pianos that are tuned and playable. Realism imitates nature so that the sets appear authentic from the camera's point of view. When a well-known site is reproduced realistically, the set is called a *replica*. Realism is often the style used in constructing sets for sitcoms, soap operas, and network series dramas (Figure 8.12).

2. *Representational.* Representational sets support and characterize the program material but are not authentic. For example, to convey a playroom setting, a children's show might paint pictures of windows and flowerpots on a set wall rather than use real window inserts (called *plugs*) and real flowerpots. Or a news special about the space program might hang several models of lunar modules, space shuttles, and space probes around the studio to indicate the subject matter of the program (Figure 8.13).

3. *Abstract-symbolic.* Abstract-symbolic sets merely suggest a setting with few elements actually present (Figure 8.14). For example, a shadow of re-

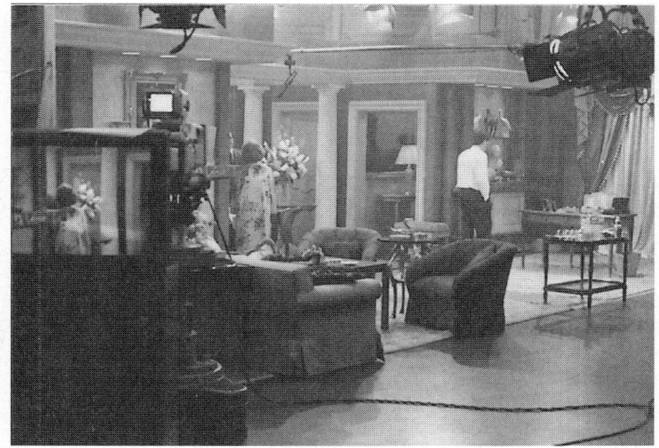

Figure 8.12 *A realistic set for a soap opera.*

peating vertical stripes projected on a neutral back wall might be used to indicate a jail cell; a desk and phone might be all that is used to indicate an office. With appropriate sound effects, such settings can be very suggestive. Obviously abstract-symbolic sets can save a great deal of time, effort, and money. They can also be quite open, giving cameras easy access to a number of different shooting angles. However, the sparseness of the set pieces can limit the number of desirable angles.

4. *Fantasy.* Sets that bear little resemblance to reality can be labeled fantasy sets. These can consist of bizarre, surreal, or unreal scenic elements, and they may include stylized renderings that merely suggest locations. For example, science fiction programs, dream sequences in dramas, and sets for music videos can all be appropriate program possibilities for fantasy sets.

5. *Neutral.* Neutral sets, which provide a blank backdrop or nondescript locale, are designed to maximize attention on the talent. For example, a blank

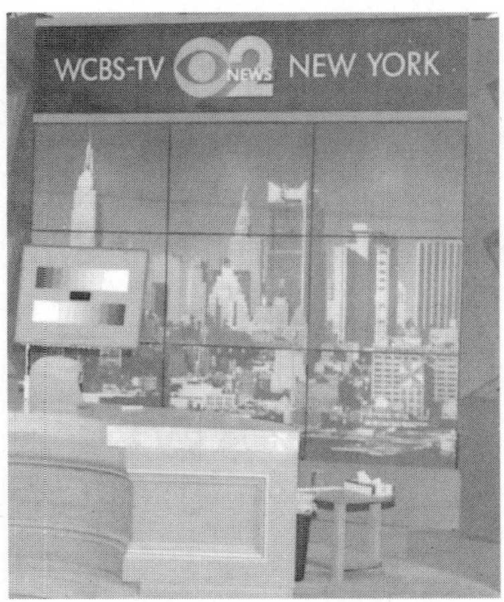

Figure 8.13 *A representational set for a news program.*

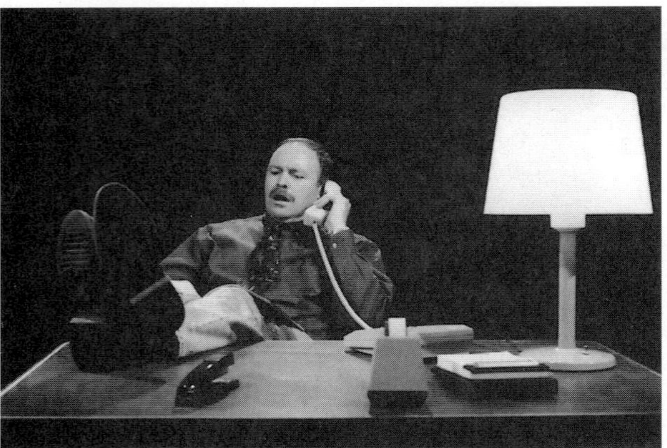

Figure 8.14   *A set of the abstract-symbolic type. This set is symbolic because it uses only a few real elements to suggest an entire office. There is no visible wall, floor, door, window, or interior decoration.*

wall or uniformly lighted cyc provides a neutral setting for a poetry reading or dramatic speech. **Cameo lighting,** in which the talent is presented in completely black surroundings (Figure 8.15), is an extreme example of a neutral setting. **Limbo lighting** presents the talent in a setting consisting of a single uniform color, giving the illusion of infinite distance (Figure 8.16). Neutral sets can be made more interesting with pattern projections and color variations at little cost in terms of time and money (Figure 8.17).

The scenery styles just outlined are only several points on a realism-unrealism continuum. In practice, the sets you develop for your programs will most likely combine stylistic elements from across the spectrum. Again, the main concerns should be what you want and need to support the central concept of the scene or program, tempered later by what is available, feasible, and affordable.

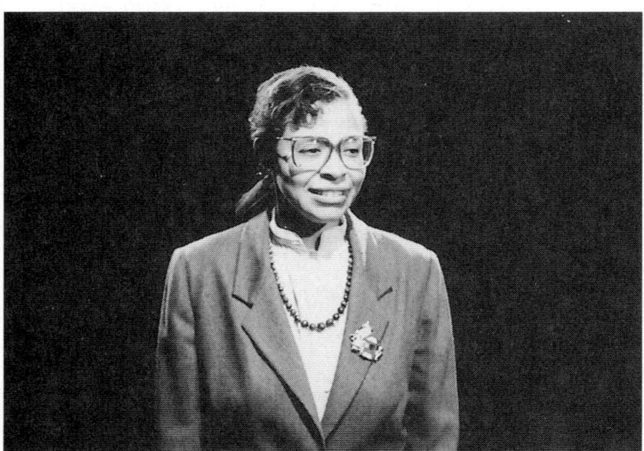

Figure 8.15   *Cameo lighting.*

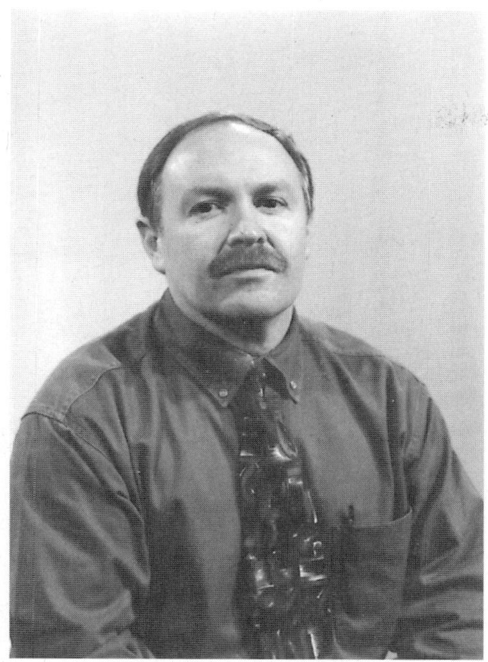

Figure 8.16   *Limbo lighting.*

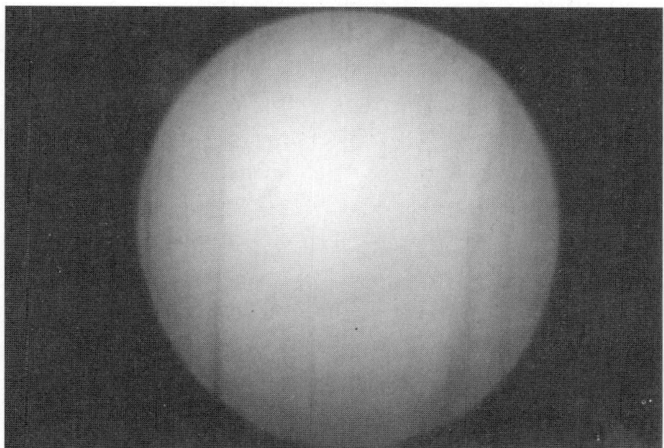
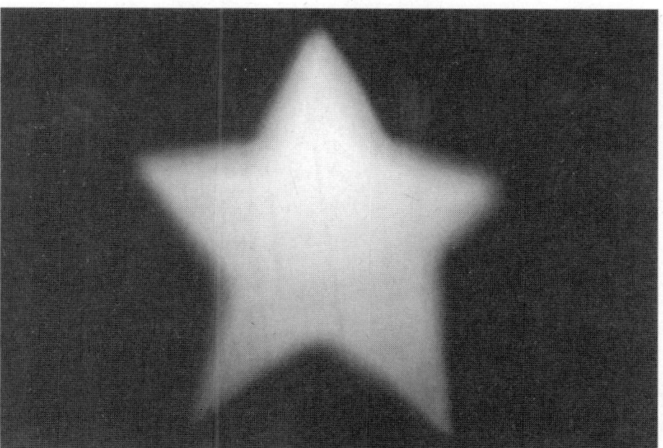

Figure 8.17   *The visual interest of a set can be increased by simple pattern projections. (left) An ellipsoidal spotlight on a neutral curtain. (right) The same spotlight shone through a star pattern cut out of tinfoil with scissors. A few devices of this sort can improve a set at very little expense.*

Since sets include all parts of the scene captured by the camera, there are many elements to deal with. Let's look at a few of them in more detail, working from the ground up.

Floor Treatments

The Floor Itself   Because the floor may be visible on camera, it must be thought of as a set element. It may be painted or covered with material such as carpeting or tile, depending on the program's needs. For example, a living room set in a soap opera or a sitcom may include carpets, rugs, or floor mats; a playroom setting for a children's show may use flamboyant painted patterns or drawings with bold, warm colors; an outdoor scene may require grass, bricks, or sand.

In preparing the floor for a video production, from the simplest to the most decorative treatments, ease of access of cameras, equipment, and personnel is a top priority. You should also be able to install and remove floor dressings quickly and easily.

If the plain floor is seen on camera, dress cables to hide them from view. For example, when possible, run mic cables behind talent, along the edge of the cyclorama, or along the edge of the floor near the wall to keep them out of view. This leaves the floor looking cleaner and less cluttered.

In general, when using rugs and carpets, tape down the edges to keep camera dollies from bunching them up if the camera rolls over them. Taping mats and rugs also keeps them from skidding. Sometimes rugs, mats, and carpets may be used to hide cables. (Still, electrical wires and cables hidden this way should first be secured with tunnel tape or protective plastic guard piping to protect them from damage in high-traffic areas.)

If you use tile or linoleum, make sure it can be easily installed and removed without leaving glue or cement behind. Tile can often be laid down without any adhesive at all. If you tape the outer edge in place, cameras and booms can have unimpeded access. After the production, removal is quick and easy.

Water-based paint permits floor decoration without interfering with camera movement. Cleanup, though, takes time and effort, and therefore the use of paint as a floor treatment may be more trouble than it is worth. If the floor must be quickly restored to its prior condition for an upcoming production, you may want to consider other alternatives.

If you use dirt or sand to simulate exterior locales, lay down a sheet of plastic first to make cleanup easier. Remember that flooring of this sort will limit the access of cameras and booms.

Platforms or Risers   Risers are platforms that permit you to elevate sets and talent from the floor. Among other things, they bring seated talent to eye level with the camera lens (Figure 8.18). This makes life easier for camera operators. Risers may also be used to distribute talent to different parts of the video space for visibility or dramatic effect—for instance, by creating tiers for singers in music groups.

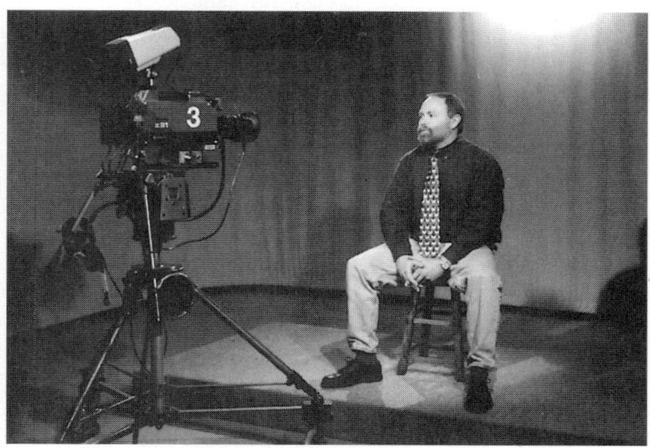

Figure 8.18 *A riser is often used to elevate talent to the level of the camera lens. Even a low riser, as shown here, can be a substantial help.*

Risers are usually constructed of either ½-inch or ¾-inch plywood on frames of 2-by-4-inch or 2-by-12-inch stock lumber, depending on the amount of weight the platform is intended to hold. Riser heights usually vary from 2 to 12 inches, but can go higher. In most cases, risers are covered with carpeting to reduce noise when talent walk on them. To reduce noise levels further, the hollow inside the riser may be stuffed with foam rubber.

Risers can also be constructed from polystyrene blocks covered with ¼-inch plywood. These may be painted black or carpeted. This type of construction results in much lighter set pieces, though they are less durable than the heavier plywood variety. When carpeted, these items are also quieter.

Most risers are made in 4-foot squares or in 4-by-8-foot pieces so they can be used in modular fashion. They can be configured in different layouts on the floor or stacked on one another. These modular units are also easier to remove than larger sections when it is time to tear down the set after production.

Risers may be placed on casters for easy movement. Sets may be placed on wheeled risers and then easily moved around. When in use, however, wheeled risers should be stabilized with wedges and sandbags.

Hanging Units

Cycloramas Cycloramas provide a seamless background around the perimeter of the studio. To keep the cyc taut, you may use small lead weights in the hem of the cyc near the studio floor or use ties to attach the bottom of the cyc to a ground pipe laid along the perimeter of the floor. To blend the bottom edge of the cyc with the floor, you can use a *ground row* (Figure 8.19). If the floor, cyc, and ground row are the same color (neutral gray is often preferred), you can achieve the illusion of an infinite backdrop.

Some cycs have a lightly woven curtain of gauzy material, called a *scrim,* hung in front of them to diffuse light hitting the cyc, further softening the

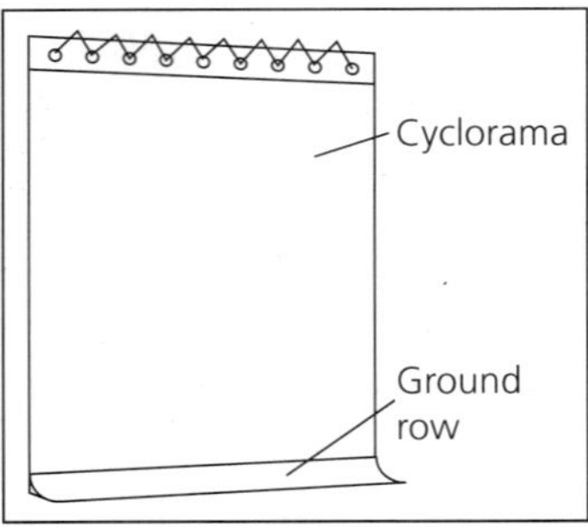

Figure 8.19   *A ground row helps to blend the bottom edge of a cyclorama into the floor.*

overall appearance of the backdrop. A gray cyc can be illuminated in a variety of colors, making it a flexible option for many production needs. In such cases, strip lights may be positioned between the ground row and the cyc itself to keep the lights out of the camera's view. Pattern projections can also be used to add detail.

A different kind of cyclorama, called a *hard cyc*, also offers an appearance of infinite space. A hard cyc is a smooth, plaster wall that curves into the studio floor, providing a seamless backdrop for the camera (Figure 8.20).

Black Velour Curtain   A black velour curtain may be hung like a cyclorama, but without lights or scrim. Such a curtain may be pulled along a metal track around the perimeter of the studio to provide a black background for talent. This arrangement is ideal when using limbo lighting, since the curtain ab-

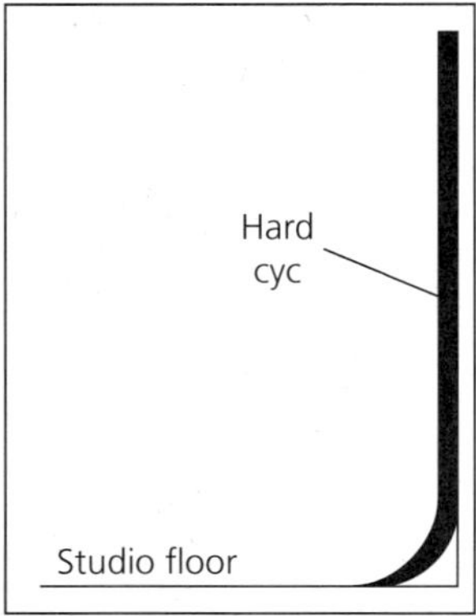

Figure 8.20   *A hard cyc, made of plaster, curves from the studio wall into the floor so that wall and floor merge seamlessly on screen.*

sorbs reflected light and directs all of the audience's attention to the talent. Black velour also makes it unnecessary to pull the curtain taut to obtain a smooth, uniform background. To achieve a total cameo lighting effect with black velour curtain, you can add a black floor covering to make the floor vanish.

Canvas Drops Canvas drops are wide rolls of painted canvas that can be hung from the studio grid on a roll and lowered into a set using fly lines (Figure 8.21). You can use a number of different drops during a production to provide various painted scenes. As a scene ends, you can use the fly lines to roll up the drop to reveal a new drop intended for the next scene. Drops can be repainted for use in other productions.

Seamless Paper A cheaper way to provide colored or painted backgrounds is with seamless paper, which comes in 9-by-36-foot rolls in many colors. By taping it to a studio wall or stapling it to a row of flats (described in the next section), you can provide attractive backgrounds quickly. With a small amount of paper, you can supply different backgrounds for stationary talent such as on-air reporters or announcers. When used appropriately, a variety of colored backgrounds can add visual interest to your productions while creating the illusion that talent are in different locales.

Standard Set Pieces

To form backgrounds such as interior or exterior walls, studio sets rely on standard pieces called **flats** (Figure 8.22). Flats may be either *softwall* or *hardwall*. Each has its advantages and disadvantages.

Softwall flats are generally composed of either 1-by-3- or 2-by-4-inch wood frames measuring 4 feet in width by 8, 10, or 12 feet in height, depending on production needs and studio ceiling height. The frames are covered with

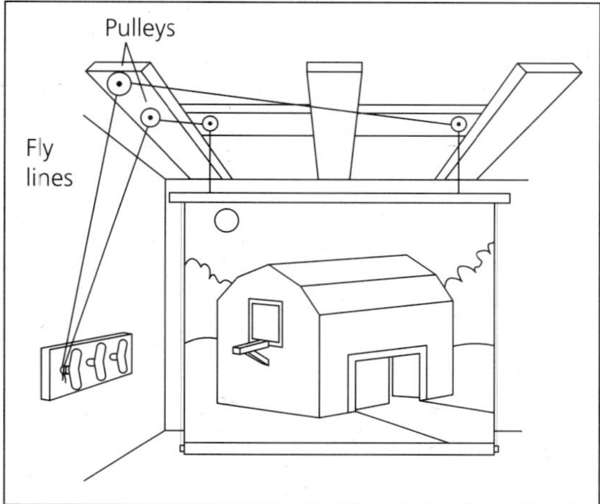

Figure 8.21 *A scene painted on a canvas drop can provide an appropriate background for the set.*

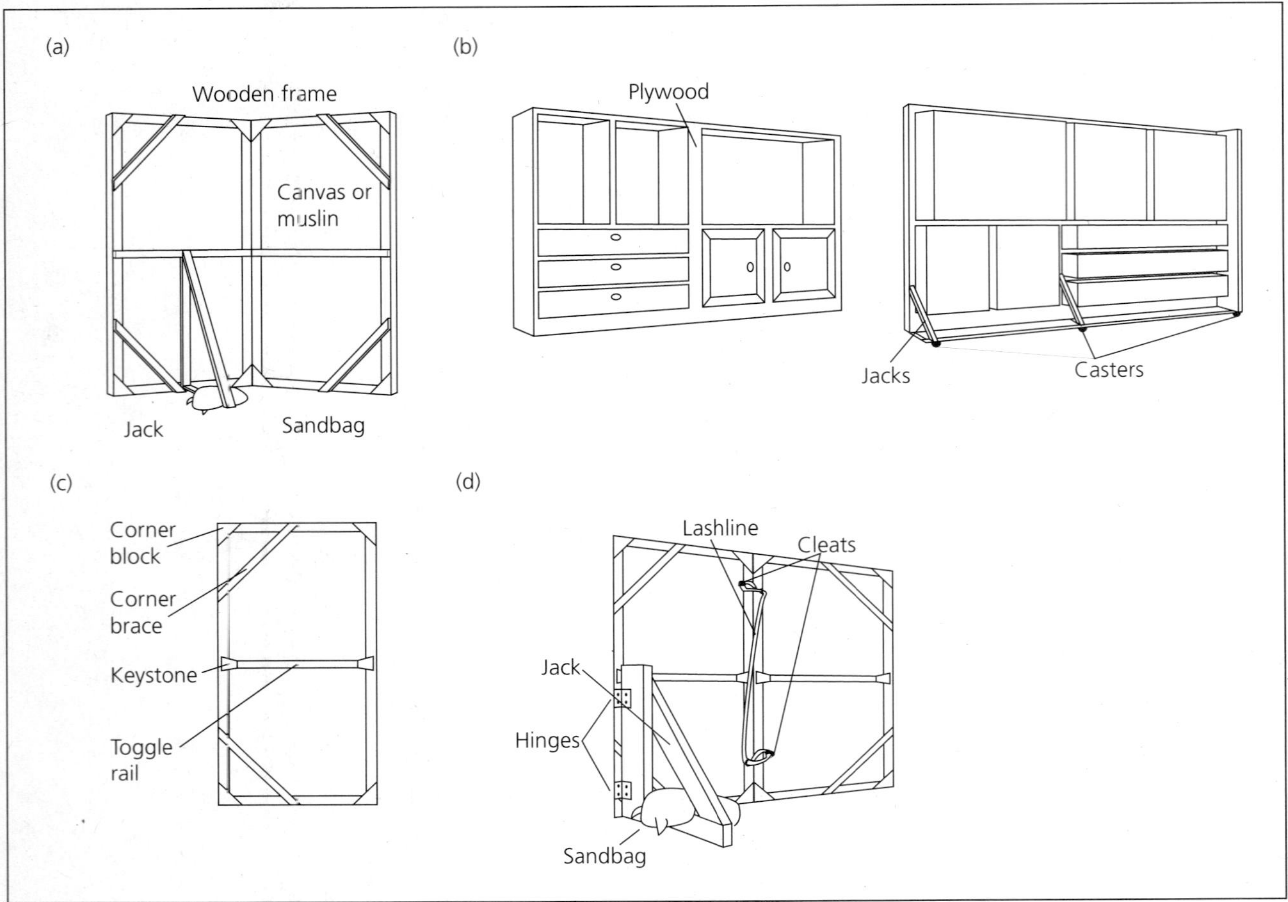

Figure 8.22   *Flats. (a) Softwall flat. (b) Hardwall flat. (c) Carpentry elements of a typical flat. (d) Items used to join and support flats.*

stretched canvas or muslin stapled into place. Flats are then attached to one another to make various backgrounds. The cloth face of a softwall flat can be painted (and repainted) to fit the tone of the production.

Hardwall flats use plywood, particle board, or wood paneling (usually ⅛ or ¼ inch thick) instead of cloth. This makes them more durable and less flimsy than softwall flats; they are also less likely to shake when doors are slammed or give way if a talent or piece of equipment moves into them. However, the wood face of a hardwall flat makes it heavy, often requiring more than one person to set it up.

Simple carpentry techniques keep flats from sagging, falling apart, or losing their right angles. *Corner blocks* and *braces,* as well as *keystones* and *toggle rails,* are used to make flats sturdy (see Figure 8.22c). Corner blocks are plywood triangles nailed to the frame joints running *with* the grain of the flat. Corner braces are strips of wood (1-by-3s) nailed across *adjacent* joint members. Keystones are small blocks of wood used to nail *parallel* frame members to a wood strip (the toggle rail) running between them to add stability.

In addition to a frame and a covering, flats need hardware to connect them together and a support element to stand them up perpendicular to the floor. Several methods are commonly used to connect flats to one another. One uses *lashlines* and *cleats* by running a rope back and forth between metal hooks fastened to the frames of adjacent flats from top to bottom, where the rope may be tied off (see Figure 8.22d). Another method uses *pin* and *hinge*–type hardware to fit adjacent flats together. Yet another fastens adjacent flats together with *C-clamps*. When joined, well-made flats should not show visible breaks between one another from the camera's perspective.

Support for flats to keep them standing perpendicular to the floor is provided by **jacks** or **braces** (see Figure 8.22d). The jack or brace is a triangular wood or metal unit joined at a right angle to the back of the flat with either a clamp or a pin and hinge. Then sandbags or stage weights are placed over the bottom section of the unit to secure it in place. Bracing is not required for every flat, but it should be used where the most weight or stress will be felt, such as at doorways and corners.

Hardwall flats may be permanently joined together with hinges (Figure 8.23). When two flats are joined in this manner, they are called *twofold flats;* when three flats are so joined, they are called *threefold flats*. The advantage of two- and threefold flats is that they are freestanding and therefore do not require bracing when partially opened. This makes them quick to set up. They can also serve different functions, including representing a corner or three walls of a room. However, these flats can be quite heavy and difficult to move. For this reason, they are sometimes placed on casters. They also take up a good deal of storage space. Some flats have sections cut out of them to accommodate door, window, and fireplace inserts (*plugs*). If not made properly, door and window plugs can lead to problems (or unintended comedic effects) if they malfunction. For this reason, it is best to leave the construction of such items to professional set designers.

Besides flats, other freestanding set pieces include *columns, screens, gobos,* and *polecats*. All of these set elements help break the studio space into areas designated for different activities.

- *Columns:* Columns come in four varieties: *pillars, pylons, periaktoi* (plural; singular is *periaktos*), and *sweeps* (Figure 8.24). All may be constructed from wooden forms and covered with painted cardboard. A pillar is a standard cylinder-shaped column. A pylon is a three-sided column that may be painted a different color on each side and turned to present dif-

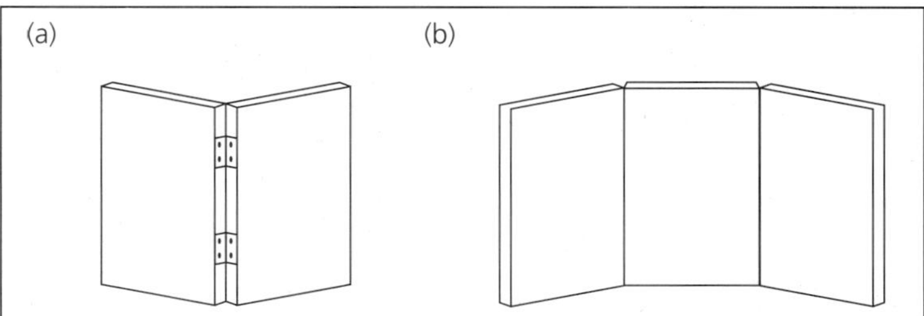

Figure 8.23   *Flats can be permanently joined to create free-standing set elements. (a) A twofold flat. (b) A threefold flat.*

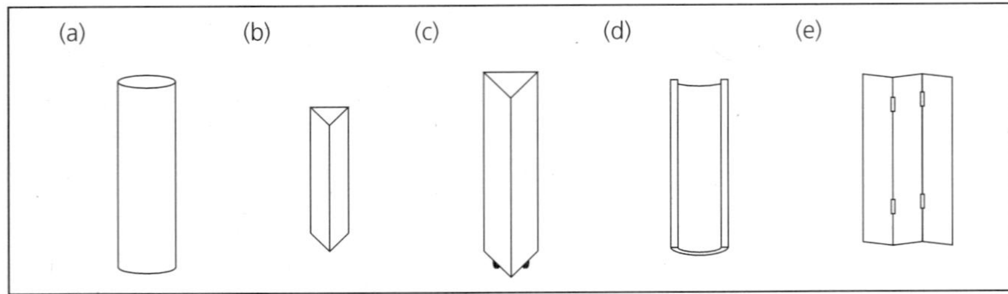

Figure 8.24   *Other freestanding set pieces: (a) Pillar. (b) Pylon. (c) Periaktos. (d) Sweep. (e) Screen.*

ferent painted surfaces to the camera, making it suitable for a variety of production needs. Periaktoi are pylons mounted on wheels to make scene changes even more convenient. Sweeps are columns shaped into partial hollow cylinders, making it possible to display either their inside or outside surfaces to the camera. Sweeps can be used to soften corners in flats or to add decorative touches. Columns can be used as door frames or to define the edge of a set in representational, abstract, or fantasy treatments.

- *Screens:* Screens (see Figure 8.24e) also divide the studio space into separate areas for different purposes. Screens may be made of wood or covered in fabric, but should not be too shiny or patterned too intricately.
- *Gobos:* Gobos are foreground set pieces through which the camera can shoot a scene (Figure 8.25). For example, a window frame (with or without the glass) might be set up in front of a camera to give the impression that the audience is peering into a dining room set from outside a house. Gobos emphasize depth by including foreground and background elements in the same shot.
- *Polecats:* Polecats are spring-loaded rods that extend from floor to ceiling. They can be used to support other set items.

Properties, Furniture, and Set Dressings

When selecting furniture, remember that talent may have to interact with it. Be sure it works! This means chairs should not creak or collapse, beds should not sag to the floor, tables should not wobble, and doors and drawers to cabinets should not stick. Get items that are truly serviceable, and test them. At the same time, keep in mind that heavy, bulky furniture can be both difficult to store and a real problem if it must be moved frequently. If you can get away with lighter and smaller rather than heavier and bigger, go for the easier alternative.

Further, establish size and style requirements for the furniture you wish to use. Keep in mind that there are hundreds, if not thousands, of types of chairs on the market. Some are huge, overstuffed lounge chairs, whereas others are stiff-backed skimpy wood things resembling little more than glorified stools. Be discerning about the needs of your program, and select items appropriate to it. For example, in interview shows, swivel chairs are often a distraction because they invite erratic movement from talent.

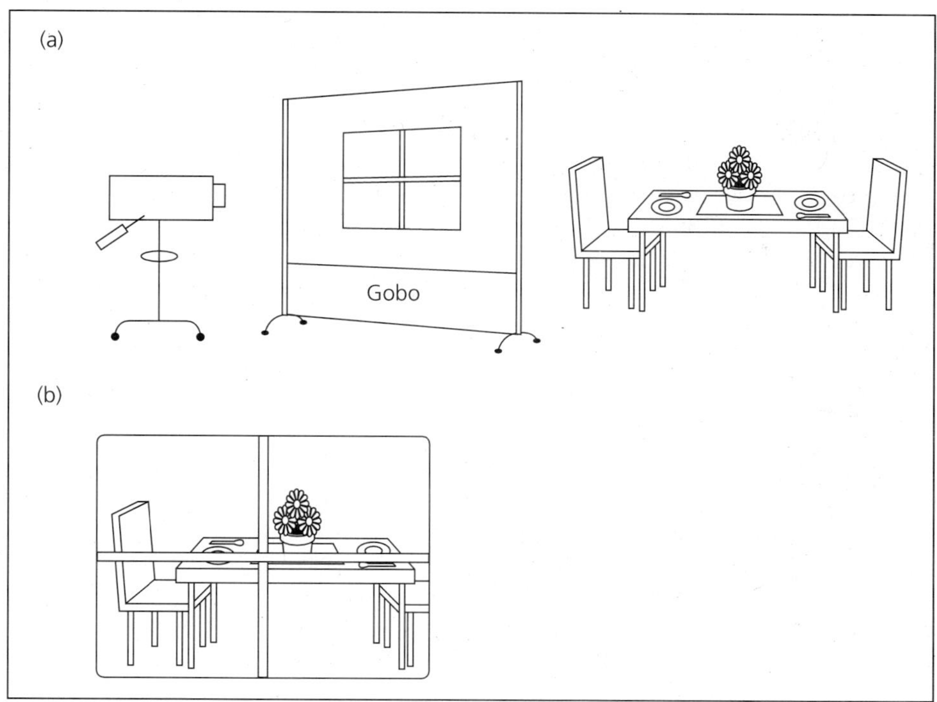

Figure 8.25 *Use of a gobo. (a) Shooting the scene through the gobo. (b) The scene as it appears on screen.*

Some items of furniture simply look better on camera than others. Intricate patterns on upholstery may create annoying moiré patterns when seen by the camera. Some pieces may be either too reflective or too dull.

Props, set dressings, and hand props should all contribute to the program's thematic unity. For example, in a televised production of the play *Death of a Salesman*, Willy Loman's sample case might be authentic and actually quite heavy. Of course, this does not mean you should use an authentic pistol for a shootout! Rather, the gun must appear authentic to the camera. Use what works. Look at the set dressings and props in close-up on the line monitor to ensure that they do the job.

PRODUCTION PHASES IN SET DESIGN

Preproduction

To develop successful set designs, preparation is critical. In preproduction, as with lighting design, the script is your road map. Read the script from start to finish, and list the things you will need item by item as they come up. Then construct a scale floor plan.

In some cases, you may move from the paper-and-pencil stage (lists and floor plan) to a miniature 3-D model of the set you are contemplating. Working small is a low-risk way to simulate the actual set. Like the carpenter's adage "Measure twice and cut once," using floor plan diagrams and small-scale models saves time and money. The following sections describe some advantages and details of this process.

The Scale Floor Plan   The floor plan is an aerial view of the studio (or production space if you are in the field) with the set in place (Figure 8.26). The advantage of a *scale* floor plan is that all items are shown in proportional size to one another, including the set pieces. This makes it possible to determine whether the desired layout is feasible. Floor plans should also include the locations of the studio doors, walls, lighting grid, and any other elements worth knowing about (location of the lighting board, electrical outlets, and control room entrance, clearance from flats to walls, and so forth).

The conventional scale of floor plans is ¼ inch = 1 foot. However, as long as you clearly label the floor plan, you can make it any scale you want. In metric units, floor plan scale is normally 1 cm = 1 meter.

The floor plan is an invaluable tool for crew members whose job is to set up scenery and props. It also helps in setting lights. The director uses it to block

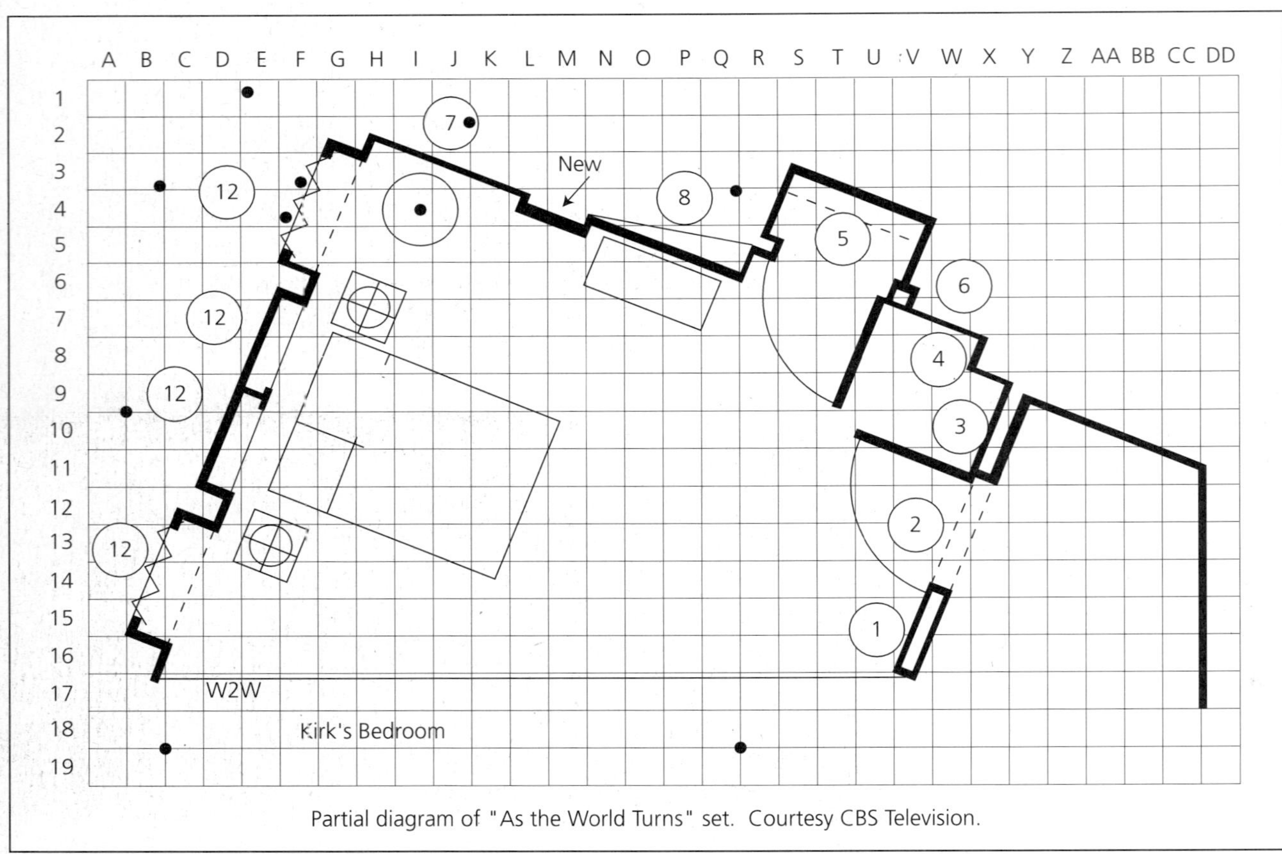

Partial diagram of "As the World Turns" set. Courtesy CBS Television.

Figure 8.26   *Part of the scale floor plan for the CBS soap opera "As the World Turns." This portion shows a bedroom layout.*

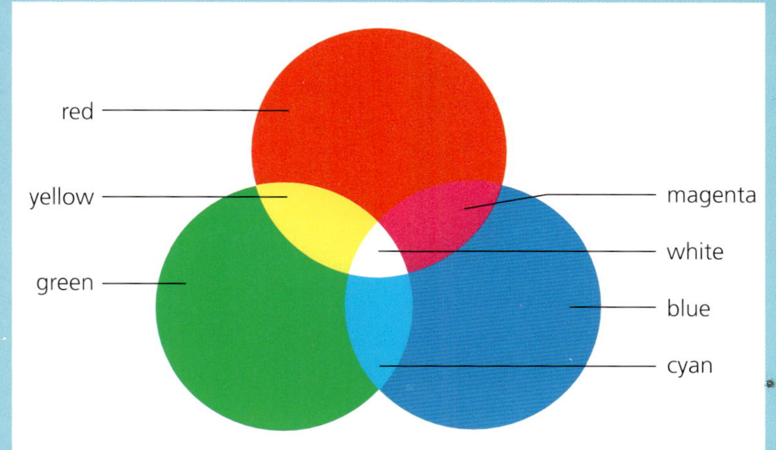

The additive color system in video. The three primary colors (red, blue, and green) combine to make the other colors seen on the screen. Notice how different this is from a subtractive color system; for instance, if you were painting on this page, you could not make yellow by mixing red and green.

Color temperatures expressed in degrees Kelvin (K).

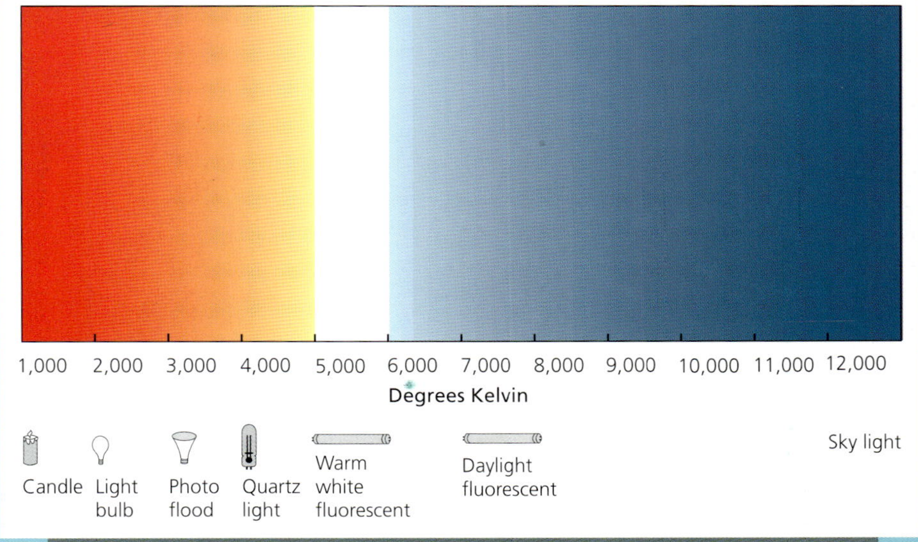

A

On a vectorscope, an electronically generated set of color bars can be used to color-balance a camera.

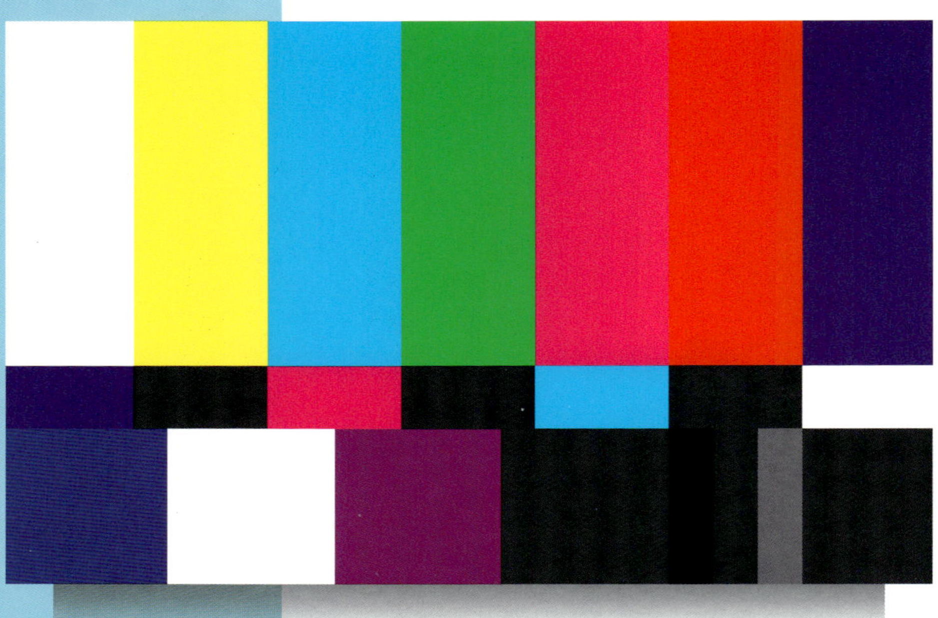

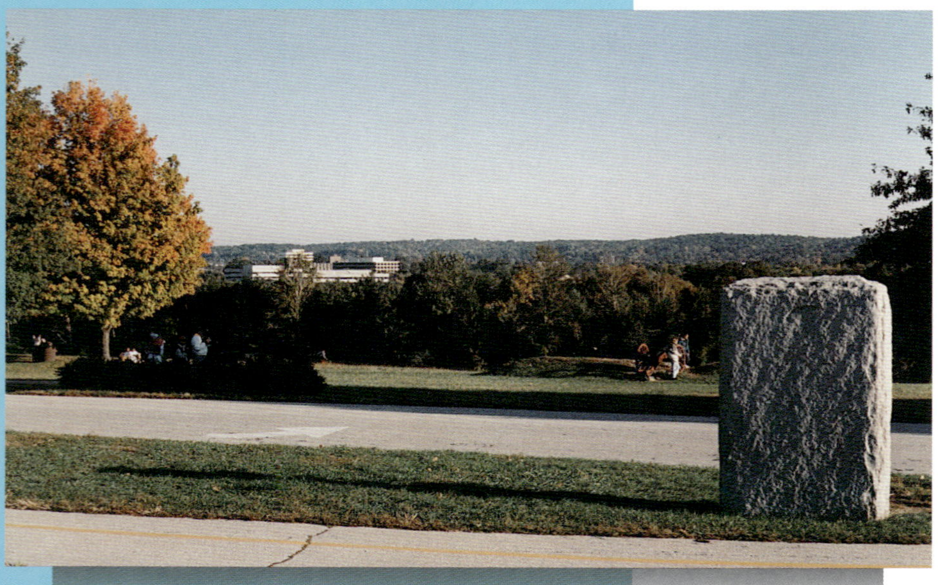

In a video image, brightly-colored, sharp, strongly-textured, high-contrast objects often appear to be closer than those that are duller or fuzzier. In this photo, notice how the stone's rough texture and the contrasting light and shadow strengthen its dominance of the foreground. Similarly, in the middle ground, the brightly colored trees on the left may seem closer than the dark tree on the far right. In the background, the bright building thrusts out of the darker, blurrier hills.

B

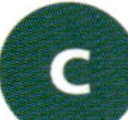

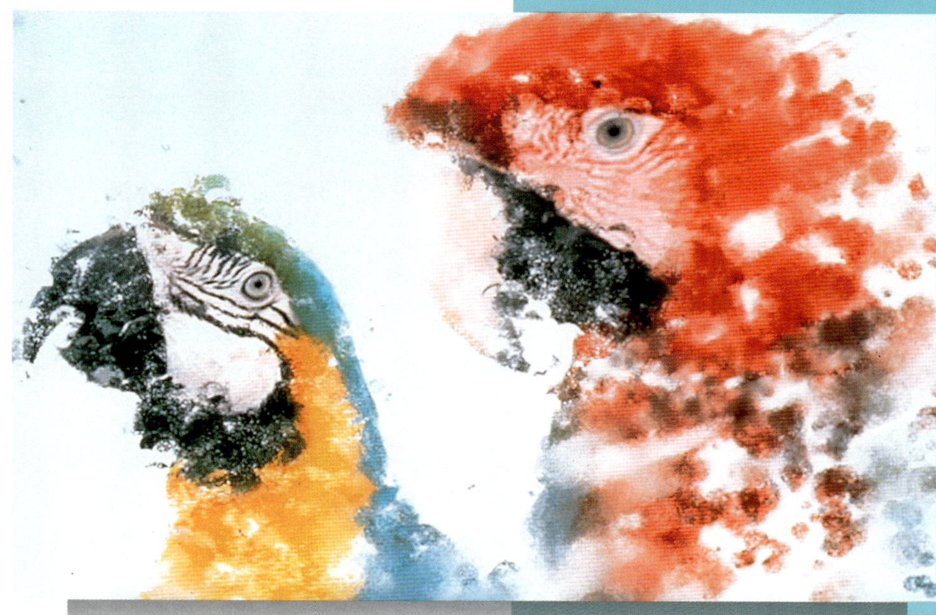

C.5

An example of the graphics that can be created with a video paintbox system.

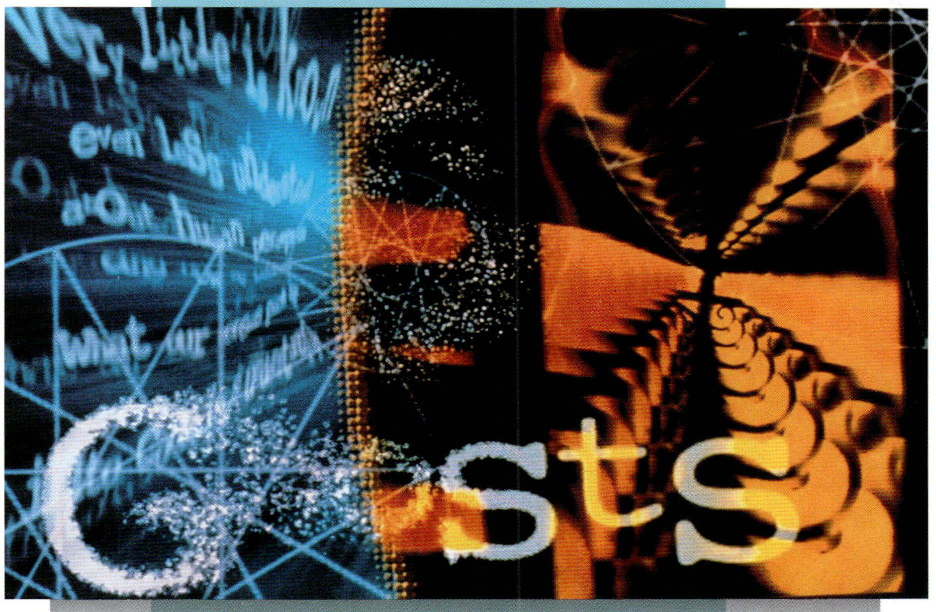

C.6

Effects like this—and many more—can be created using a digital video effects machine.

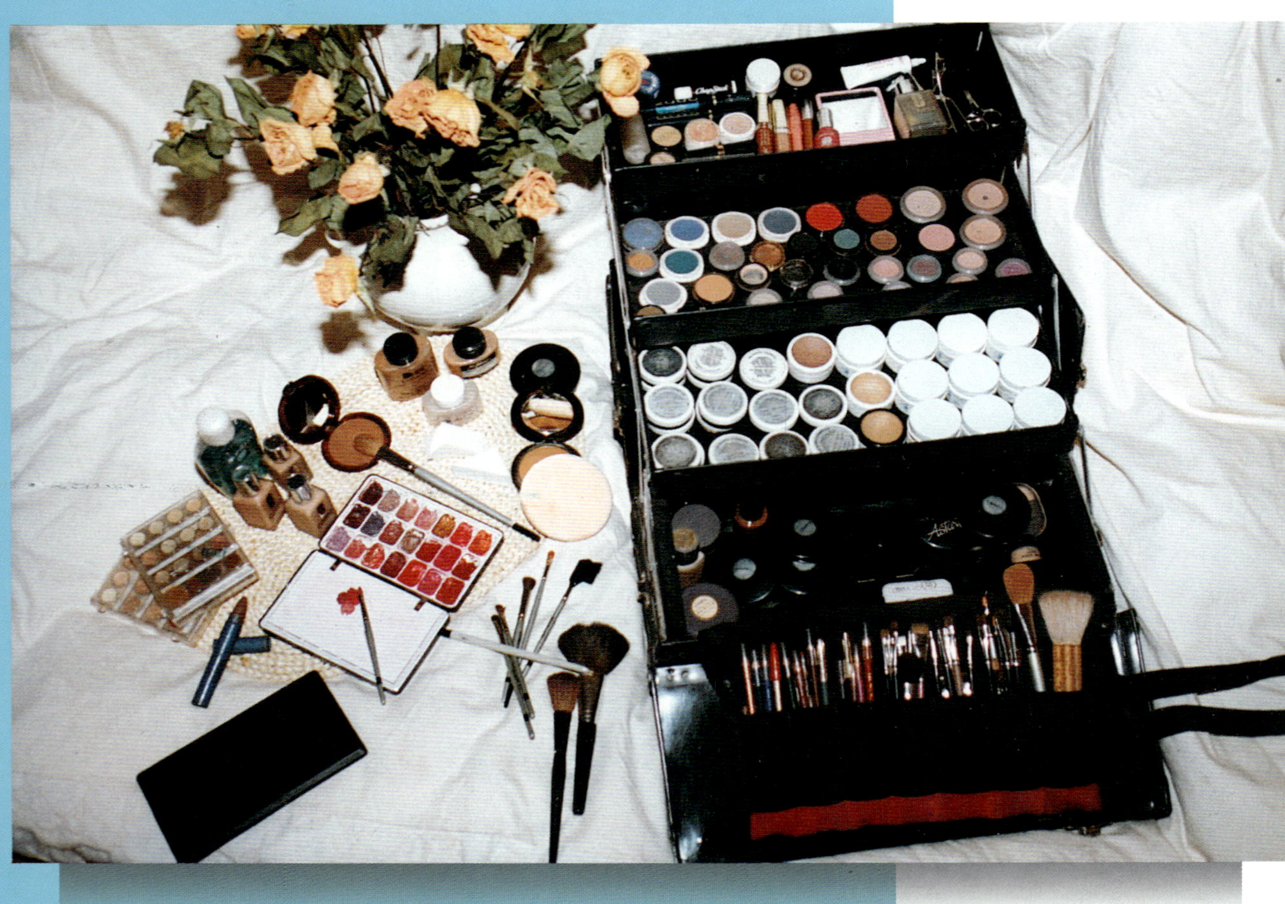

A basic beauty kit for video makeup, used for both male and female talent. To the left are lip colors along with liquid foundations, astringents, derma blends, and powders. In the box, from the top tray to the bottom, are lip balms, lip glosses, mascaras, and other eye makeup; eye shadows and rouges; foundations; liquid bases, loose powders, and powder compacts; and an assortment of brushes, lip pencils, eye pencils, and more.

talent and to confirm that cameras and other equipment can gain unimpeded access where needed. The audio operator uses it to plan microphone connections, locations, and cable needs.

The floor plan diagram may be drawn with the lighting grid imposed on it, or you may prefer to use an acetate overlay (a transparent sheet) that permits you to see the floor plan with or without the lighting grid. After the lighting design and floor plan are finalized, copies can be distributed to the director and crew for setup and rehearsal.

Of course, to be successful, the floor plan must have accurate scale representations of all set pieces as well as the major production items that will occupy space in the studio, including cameras and booms. This will require measuring every item or consulting an inventory list already on file. The list can then be used to draw items on the floor plan. For even cleaner results, you can create a stencil of each item and use that for drawing on the floor plan. An even better option is to use one of the CAD (computer-aided design) programs now available for such work.

Models   Miniature models of sets enable you to see and show others how a new set will look on the air before it is built. You can paint and repaint the model to modify the color scheme. Texture, fabrics, and other aesthetic factors can also be tried and approved or scrapped at this stage. Even camera angles and relationships among set elements can be envisioned.

Use balsa wood, construction paper, cardboard, and doll house furniture to create the model from sketches (Figure 8.27). All of this can be done with a minimal investment of time and money. Of course, if preproduction is hurried, you may have little time to spend in the sketch and model phase. In such cases, deciding how the preproduction time should best be spent is a judgment call. Picture the cost-benefit ratio of working with and without sketches and models. Imagine how the budget will be spent if errors are made with full-size sets, and weigh that against the costs of time spent constructing models.

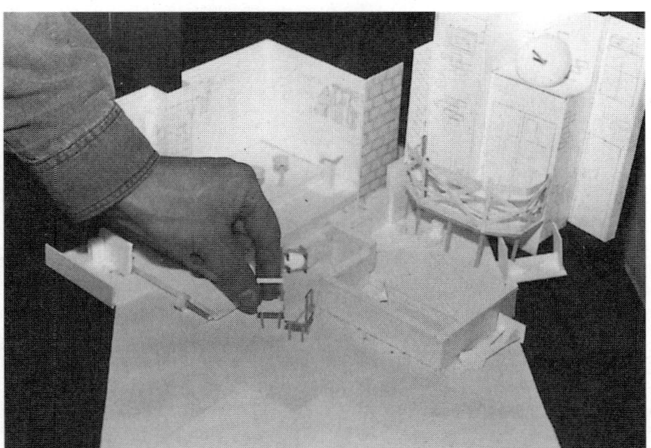

Figure 8.27   *A model of a set with doll house furniture.*

Sets and the Video Space

As with all other aspects of video production, it is relatively unimportant what the set looks like through direct observation. The key is how it looks on screen. Direct imitation of reality is often just a starting point. For set design, this means you will probably adjust your initial layout after seeing how your master shots and close-ups look on the line monitor as you block talent and cameras through the set.

Here are some common differences between the normal arrangements of daily life and those of video sets:

1. In real life, walls frequently meet at right angles, but video sets often feature walls that meet at much wider angles (such as 135 degrees) to allow easy camera access.

2. In real life, furniture is inches from walls, but in video the furniture and talent positions could be as much as 6 feet or more from any walls or flats. This distance helps keep talent shadows from falling on the set. Also, it keeps set lights separate from performance area lights—an important consideration because sets tend to be more reflective than talent and because you generally want the talent lit with higher intensities than sets.

3. In real life, interior locales are bounded on all sides by walls, but in video you can achieve realism even if you are shooting into an *open set*, that is, a set defined by a number of major set pieces but without a continuous background of walls or flats. Open sets allow you to position cameras to shoot from many different angles; talent are less restricted than in closed sets, lighting is easier, and the set itself is often simpler, making setup and strike less work. One disadvantage of open sets, however, is that fewer shots may be available featuring set pieces in the background, and this can reduce visual interest.

4. In real life, furniture and wall decorations are often placed where they are likely to promote the most utility, but in video they may be positioned to maximize a key shot. For example, it might be foolish to hang a painting at the talent's true eye height if doing so drops it from the frame of a key shot. Similarly, whereas the space between a dresser and a bed might be several feet in the real world, on television it might be better to push them closer together to avoid dead space when the talent reaches from the bed to get something.

5. In real life, the eye can adjust for great differences in brightness between people and the backgrounds in which they are seen, but in video, if brightness values on backgrounds exceed those on talent by more than several magnitudes, the iris of the camera lens can close down so far that the talent's face will look too dark. In general, brightness differences between talent and backgrounds should not exceed three or four gray-scale steps.

Production and Postproduction

After being designed and approved, sets are erected for rehearsal and production. Test the set during rehearsal. Block talent through each shot sequence, and make final adjustments by watching the rehearsal on the line monitor. At this stage, all aspects of the set's contribution to the program should be assessed, including color, layout, and ease of access by crew and talent. Dead spots can be eliminated by moving talent and set pieces closer together. Glare can be reduced by changing the lighting, turning the offending surface slightly away from the camera, or treating it with dulling spray. In some cases, rubbing the surface with a bar of soap can eliminate the problem.

If the set must be struck for another production and then reset the following day, mark the position of each set piece with masking tape, if possible. You might also shoot videotape or still photos of the set to help in its reconstruction. After the production is completed, label all set pieces and props before storing them so they can be retrieved easily if needed.

ELECTRONIC AND MECHANICAL EFFECTS IN SET DESIGN

Electronic Effects

Electronic overlay technology has made it possible to create electronic "sets" or set elements that can be combined with live studio action on a prepared background. One of the most common methods is a technology that has been used for many years: chroma key.

Chroma key[1] technology exploits the fact that the camera's color signal can be used to select the point where part of a picture may be replaced with an external video source. That is, if the camera picture shows a blue background and nothing else in the picture is blue, an external video source can be electronically substituted for the blue portion each time the camera scans in blue. For example, if a talent is shot standing in front of a blue screen and a second camera shoots a forest scene featuring a cascading waterfall and a river bed with trees and rocks, the forest scene can be substituted for the blue background, making it appear as if the talent is standing in the forest. The same technique is used to show a weather map behind a TV weatherperson. Each time the electron gun in the camera scans a blue signal, that portion of the picture is replaced by the external video source.

The reason blue is one of the popular industry standards for chroma key effects is that blue is generally not present in flesh tones, making it easier for the technology to switch between video signals when talent is on camera. Other colors can also be used.

---

[1]The term *key* in this context should not be confused with a key light. In the current context, keying is an electronic effect that inserts an image or part of an image from one video source into another. In addition to chroma key, there are matte keying and insert keying; all of these are discussed in greater detail in the next chapter.

TABLE 8.1

Some Common Mechanical Effects in Video Production

Effect	Method	Notes and Cautions
Fire and smoke	To convey a working fireplace, industrial setting, or cooking stove, use a few drops of oil on a burning charcoal tablet.	Use extreme caution. Have a fire extinguisher on site. Protect surfaces with metal trays.
Flickering flame	Cut slits in a strip of black paper. Rotate the paper in front of a studio light aimed at a fireplace plug. Inside the fireplace, hide a flashlight for additional illumination. Hide short pieces of colored silk in the fireplace and blow them up from below with a small fan.	
Fog	(1) Rent a smoke machine to simulate fog in the studio. Use patchy lighting and mild air movement.	Option 1 can cause breathing problems for the crew and talent.
	(2) Use fog filters or a stretched nylon stocking over the camera lens. Or, for static shots, place a white card a foot in front of the lens. Cut a circular hole in the card just large enough to leave a slight border around the shot. Then light the card from behind the camera to cause it to flare slightly.	
	(3) Tape a sheet of glass or plastic wrap, smeared with petroleum jelly, over the lens.	
Lightning	Discharge flash units from behind the set. For a more convincing illusion, add thunder sound effects.	
Mist	Drop pieces of dry ice in hot water. This will create heavy white clouds that hug the floor and hide it from view. Keeping the area cool and damp makes the effect last longer. Dry ice machines are also available.	The mist is composed mostly of water. Dry ice should be handled with gloves.
Rain	(1) You can use real water to simulate rain trickling on a window. Use a trough to catch the water.	Supply and cleanup of real water in the studio can be more trouble than the effect is worth. It can also be dangerous because of the electrical equipment. Consider shooting rain footage elsewhere against a black background and superimposing it electronically.
	(2) A more convenient variation is to use a *rain drum*, a cylinder about a foot in diameter covered in black paper with white "raindrops" painted on it. While someone turns the drum, shoot footage of it slightly out of focus. Then superimpose this footage over the scene.	

continued on next page

TABLE 8.1

Some Common Mechanical Effects in Video Production (cont.)

Effect	Method	Notes and Cautions
Snow	(1) Buy plastic granular material or paper confetti at a theatrical supply house. Have crew members throw it up so that it falls down in front of the camera. Augment its motion with a wind machine.	Remember to light the area through which your "snow" is falling so that it can be seen on camera.
	(2) As another option, cut a canvas cloth with slits and have crew members hold it above the scene. Place the plastic grains or confetti on the canvas and shake it back and forth to make the "snow" fall through.	
	(3) For snow on the ground, use fine sawdust from white soft woods. For extra whiteness, sprinkle it with talcum powder, plaster, or flour. Add some glitter for sparkle.	Be careful that an accumulation on the floor doesn't impede the camera wheels. If you use flour in badly ventilated areas, you may cause a dust explosion.
Steam	Cold-water humidifiers create water vapor that looks like a forceful flow of steam. For a gentle flow, use dry ice.	
Wind	Wind machines are fans made to operate quietly. They come in various sizes and operate at different speeds. For a light breeze, use small, conventional fans and aim the fan at the talent's hair to create just the effect you need.	With conventional fans, watch mic placement to avoid noise problems.

With chroma key, a virtually infinite variety of backgrounds can be made to appear as electronic overlays. In recent years, moreover, computer graphics technologies have made it possible to layer many more images more intricately into video signals. We deal with this topic in greater detail in the next chapter; for now, note that multilayered digital images can be combined with other video signals, making the final product quite complex. Some music videos are produced as virtual "cybersets" that never existed in physical space. A set of this type appears in Michael Jackson's *Scream* video, which features Jackson in a number of disorienting, futuristic, unreal, computer-generated environments.

Mechanical Effects

Though multilayered electronic key effects are increasingly being used to provide sets for video, we should not overlook the numerous mechanical means for embellishing sets. Table 8.1 lists a number of such effects and describes how they are created. A thorough treatment of mechanical special effects in video, including pyrotechnics and breakaway furniture, is available in Wilkie (1979).

Choosing Between Electronic and Mechanical Effects

The choice of electronic or mechanical effects to create or modify sets should be based on the same considerations involved in all production issues: the ability to meet production goals, feasibility, and cost. For example, it might be easy to create blur and fog effects or simulate snowfall with the computer equipment you have available. In other situations, though, computer effects could be expensive, could take a long time to create, and might even be disappointing in their final results. Base your choice on a cost-benefit analysis—whether you are getting what you want for a price you can afford—not simply on a desire to use the hottest tools.

In some cases, old-fashioned mechanical effects may be the best resource. On the other hand, computer effects are sometimes easier and cheaper than their mechanical alternatives. Kevin Costner might have saved a lot of money making the film *Waterworld* if he had used state-of-the-art overlay and matte techniques to simulate underwater environments rather than building expensive, life-size sets on location in the ocean, where some sank during filming. In short, no one route is best all the time.

KEY TERMS

character generator (CG) *(166)*	telecine (film island) *(179)*
antialiasing *(166)*	media record telecine (MRT) *(182)*
lower-third *(167)*	flying spot scanner *(182)*
electronic still store (ESS) *(168)*	linear CCD array *(182)*
frame grabber *(168)*	properties (props) *(184)*
flatbed scanner *(169)*	cameo lighting *(186)*
paintbox *(169)*	limbo lighting *(186)*
computer-aided design (CAD) *(169)*	riser *(188)*
compositing *(171)*	flat *(191)*
essential area *(175)*	jack (brace) *(193)*
television gray scale *(175)*	chroma key *(199)*
studio (camera) card *(178)*	

QUESTIONS FOR REVIEW

1. Describe the types of visual material included in the term *graphics*.

2. Besides the character generator, what other digital graphics devices are available?

3. How do aspect ratio, essential area, color, and contrast issues influence the way graphics are prepared for video?

4. Give some examples of how graphics can influence the tone and style of the programs in which they appear.

5. What kinds of elements and treatments are available for use on the studio floor?

6. What are some standard backdrops and set pieces?

7. What factors should you consider in choosing props, set furniture, and dressings?

8. Describe the typical stages involved in designing a set.

9. What is the chroma key effect?

9 Video Processing

hapter 8 made it clear that videographers currently have a great (and growing) variety of sources for video images—not only the camera but also sources such as character generators, still store systems, and digital effects generators. All of these devices can help you create dazzling visual displays.

But how are the outputs from this rich variety of video sources made ready for use precisely when needed? For programs produced in real time, including both live broadcasts and programs taped live for later use, the device that routes, controls, and shapes video images during production is the **video switcher**. The switcher does with video images what the audio console does with sound. The crew member who operates the video switcher is called the *technical director (TD)*.

Video switchers are also used for programs assembled in postproduction, such as in fieldwork using a single-camera format. Because such programs enjoy freedom from the instant time demands of live television, postproduction switchers may differ from production switchers, in terms of both how they are built (their internal architecture) and the way they are configured with other video sources. Nevertheless, both production and postproduction switchers perform essentially the same functions: to provide special effects and smooth transitions among video sources used in a production and to output the finished product. In this way, the video switcher transforms a grab bag of video snippets into a cohesive program.

At first glance, the video switcher looks quite intimidating. Most have rows of buttons (called *buses*) accompanied by several levers or toggles. There may also be clusters of other buttons, switches, and knobs and even digital readouts. With all these items available to shape incoming video signals, it is little wonder the novice might get confused. However, understanding this piece of equipment, even high-end versions, is really not difficult.

This chapter explains how the TD operates the video switcher, beginning with the most basic transitions and progressing to more exotic special effects. In addition, it discusses several peripheral devices needed for the proper functioning of the switcher. To illustrate how switchers work, we will "build" one from scratch. The switcher we construct will be just one among several common types of compact switchers. The goal here is to learn general principles, not to memorize specific buttons to push, since different switchers delegate functions in different ways. The switchers you encounter on the job may look very similar to the one described in this chapter, or they may look different, but in either case the basic principles of operation will be the same. The topics covered in this chapter include:

THE VIDEO SWITCHER

simple transitions: cuts or takes • mixing video signals • from mixes to special effects • key effects • wipe effects

MORE COMPLEX EFFECTS AND TRANSITIONS

mix/effects banks • cascading: multiple M/E systems • computer-assisted switchers

DIGITAL VIDEO EFFECTS GENERATORS

I N D U S T R Y
voices

Jay Dorfman
Principal, Unherd of Productions, New York

Q: How long has Unherd of Productions been in business, and what do you do?

A: We've been here since 1985. Our approach starts from the philosophy that all visual communication is driven by a unique design for every type of content. Each concept is developed from the point of view of finding a unique presentation for each job we do. We also believe that "new" technologies should be used not because they are new but to propel new ideas.

We have been involved in many business start-ups and new marketing applications. We begin each new job with the view that the idea is the driver of the product we design.

Q: Tell me about some of your latest work.

A: We recently worked on contracts with three companies: one to produce a documentary on the making of a new album by a major record company, another to develop a video for a company's new business plan and a design for an interactive television network, and a third for a documentary shot on location in Malaysia on how to advance computer manufacturing in the Asian Pacific.

Q: Sounds like a diverse body of work. Who are some of your clients?

A: A & E Cable Network, American Movie Classics, MTV, Nickelodeon, Showtime, National Geographic, GTE, US West, Johnson and Johnson, Hearst New Media, MCI, QVC, SONY, Polygram, BMG, EMI, and Virgin Records.

Q: What changes in the industry have you seen over the last ten years that have impressed you and influenced your work?

A: The entire business has undergone revolutionary changes. Most important, desktop systems have brought costs down on broadcast-quality production and postproduction. As a result, we have seen a great proliferation of content developers, producers, and designers entering the business. They bring with them an assortment of new disciplines and talents. Graphic designers are now animators, photographers are becoming directors, musicians are becoming composers, all with the help of new hardware and software. Creatives are able to extend their reach and expand their creative vernacular. The spectrum of techniques has grown, and the ability to communicate an individual point of view has grown along with the technology.

With all that expansion of technology, the production activity has become a lot more intuitive. The end effect is a sort of "Kamikaze approach" to production that puts a greater premium on technique and the ability to improvise with the new tools. The result is we are beginning to see a lot more experimentation and personal preference being expressed, more personal freedom in broadcast production. There is a lot less rigidity and adherence to the rules of the game than ever before.

Q: Are there any techniques you have used that are now outdated?

A: I would not say that we would never use some of the older technologies again. I can tell you that the design elements for an animated piece we did for MCI were produced with a thirty-year-old slide projector, some miniature wind-up toys, a set of multicolored gels, and a Hi-8 camera. The important point is that we did it that way because we decided that would be the right way to bring the desired look to that project. We won a Tellie award with the piece. It was also nominated for a Monitor award. So we let the idea drive the production plan. We don't let the new technological toys at our disposal dictate what to do.

Q: What do you look for in an intern?

A: Creative ability is the most important driver in our business. It's relatively easy to pick up skills in how to operate the latest tools. That's not a problem. But creativity is less easy to teach. The individual who is capable of articulating ideas, who is familiar with human nature and understands the motivation behind why an audience attends to a message and why the producer is making it for a particular audience—that person will represent the greatest asset to any creative organization.

Simple Transitions: Cuts or Takes

The simplest transition between any two video sources—for instance, between two cameras or between a camera and a videotape—is the cut or take. A **cut** or **take** (synonyms in this context) is an instant transition from one source to another. The hardware used to accomplish a take is a *program button* on the switcher. If takes were the only change ever used, the switcher would need just one row of buttons, called a **program bus** (Figure 9.1), on which each button represents a particular video source.

Let's examine how to use the program bus by *taking* the image seen by camera 1 and using it for an instant replacement of *video black* on the **program** or **line monitor** (Figure 9.2a), the monitor that shows the actual output the audience will see. The director accomplishes a take by giving a ready cue and then a command to the TD as follows: "Ready one. Take one." Upon hearing the ready cue, the TD poises a finger on the button labeled CAM 1. Then, at the "take" command, the TD pushes the button. It's that simple. When the button is pushed, the scene being pictured by camera 1 is instantly displayed on the line monitor, as in Figure 9.2b. Note that video black is an internal video source generated by the switcher itself.

Usually the buttons on the program bus are uniquely colored red (or some other bright color) to indicate that when they are pushed, they send video sources to your actual program. In short, these buttons are "hot," the "real deal," and when one of them is pressed, the source wired to it becomes part of your program.

Also notice that each button is labeled to indicate the video source it represents. For example, the first button on the left is labeled BLACK. The next button is labeled cam 1. Then we see buttons marked CAM 2 and CAM 3. Additional sources include FILM, VTR for a videotape recorder, CG for a character generator, and COLOR BARS, TEST, and COLOR BKGD (color background) for additional internal signals generated by the switcher itself. Of course, the bus can be expanded to include even more sources, such as other cameras, satellite feeds, and digital video effects.

It is helpful for the TD and the director to have the output of each video source displayed on a separate, labeled monitor in front of the switcher console (Figure 9.3). Seeing a bank of source monitors enables the director to call the next shot simply by looking up and selecting a source. Usually the individual sources are displayed on smaller black-and-white monitors, whereas the program monitor is somewhat larger and in color.

The bank of source monitors may be likened to an artist's palette from which the artist selects paints to use on a canvas. Just as the artist dabs a brush into various paints and transfers them to the canvas, the TD presses buttons

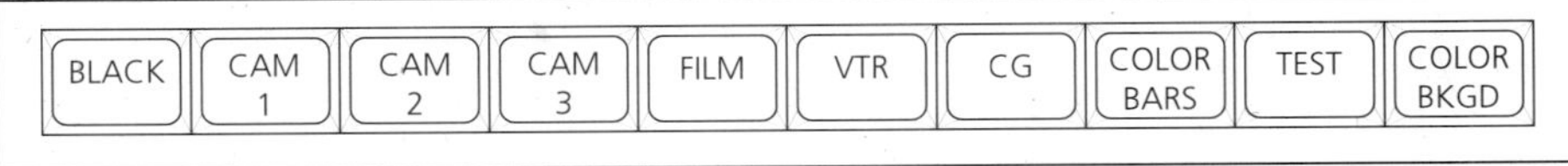

Figure 9.1   *A program bus.*

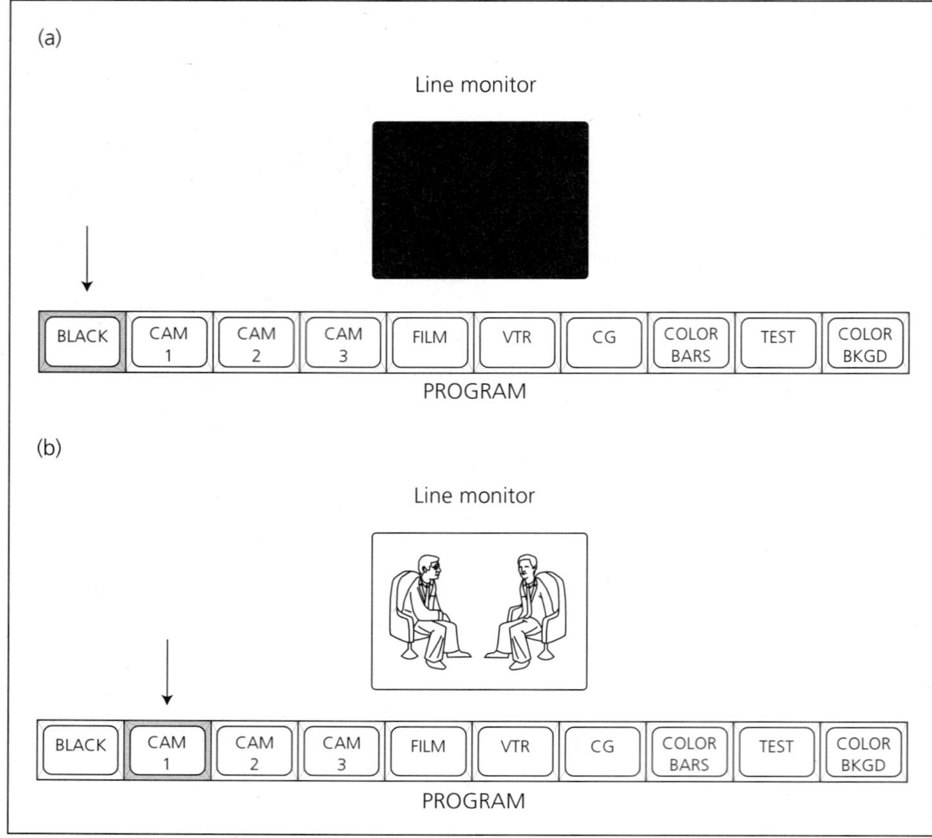

Figure 9.2   *Using the program bus for a simple take. (a) The* BLACK *button on the program bus is pushed in, and the line monitor shows black. (b) When the* CAM 1 *button is pushed, the line monitor shows the scene being shot by camera 1.*

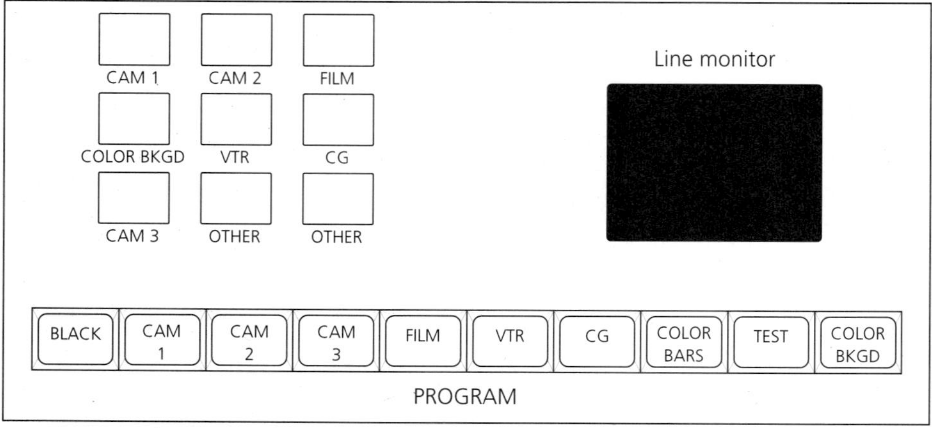

Figure 9.3   *A bank of source monitors added to the program bus and the line monitor.*

on the video switcher and transfers images from source monitors to the line monitor. In this sense, source monitors are an electronic palette, the video switcher is an electronic brush, and the line monitor is an electronic canvas.

Synchronous and Nonsynchronous Sources   Actually, takes are not exactly instant because full frames of video are not available continuously; remember, they begin only every thirtieth of a second. Although this is a very short time in the context of a video presentation, it is still a critical factor because video signals coming from different sources generally are not precisely in step (*synchronized*) with one another.

To put signals from different video sources in phase with one another, in-house sources are fed through a synchronizer. This permits transitions among signals from purely electronic sources such as cameras, switchers, and CGs to be glitch free and stable. Sources such as videotape machines, which are less stable because of their mechanical components, generally use a machine called a **time base corrector (TBC)** to smooth out jitters, thereby creating an output as stable as purely electronic sources are. For sources external to the studio facility, such as a network feed, a **frame synchronizer** is used. The frame synchronizer delays incoming video until it can be output in phase with the in-house sources. The term *genlock* is used to refer to the ability of a device to synchronize various video signals to one another. Genlocking means adjusting various outputs to synchronize their signals.

When signals from various sources are fed to the video switcher, the actual moment of switching is delayed electronically so that it occurs only during the vertical interval of the scanning process. Hence you will sometimes hear the term *vertical interval switching*. This additional function helps stabilize signals so they may be switched cleanly.

Mixing Video Signals

Fades   Instead of using instant cuts between sources, you may want to make a gradual change from one source to another. For instance, when you are starting a program, you may wish to go gradually from black to a nonblack visual. Such slow transitions to or from black are called **fades**. Instead of pushing a button, fades require the gradual moving of a *lever* or *toggle*. Also, as Figure 9.4 illustrates, we need a second bus, called a **preview** or **preset bus**. We also add a color preview monitor to display video sources punched up on the preview bus.

The preview or preset bus contains exactly the same sources the program bus does in exactly the same order. The main difference between them is that sources punched on the program bus are *always on air*, whereas sources punched on preview are *always off air*. A source displayed on preview is like the on-deck batter in a baseball game: ready to participate but not yet involved in the action.

Figure 9.4 shows what would happen in a fade from black to camera 2. The process begins with *black on all lines,* that is, video black on both the program and preview buses. On hearing the cue "Ready fade two" from the director, the TD punches the button labeled CAM 2 on the preview bus, causing the image seen by camera 2 to appear on the preview monitor. Then, on hearing the

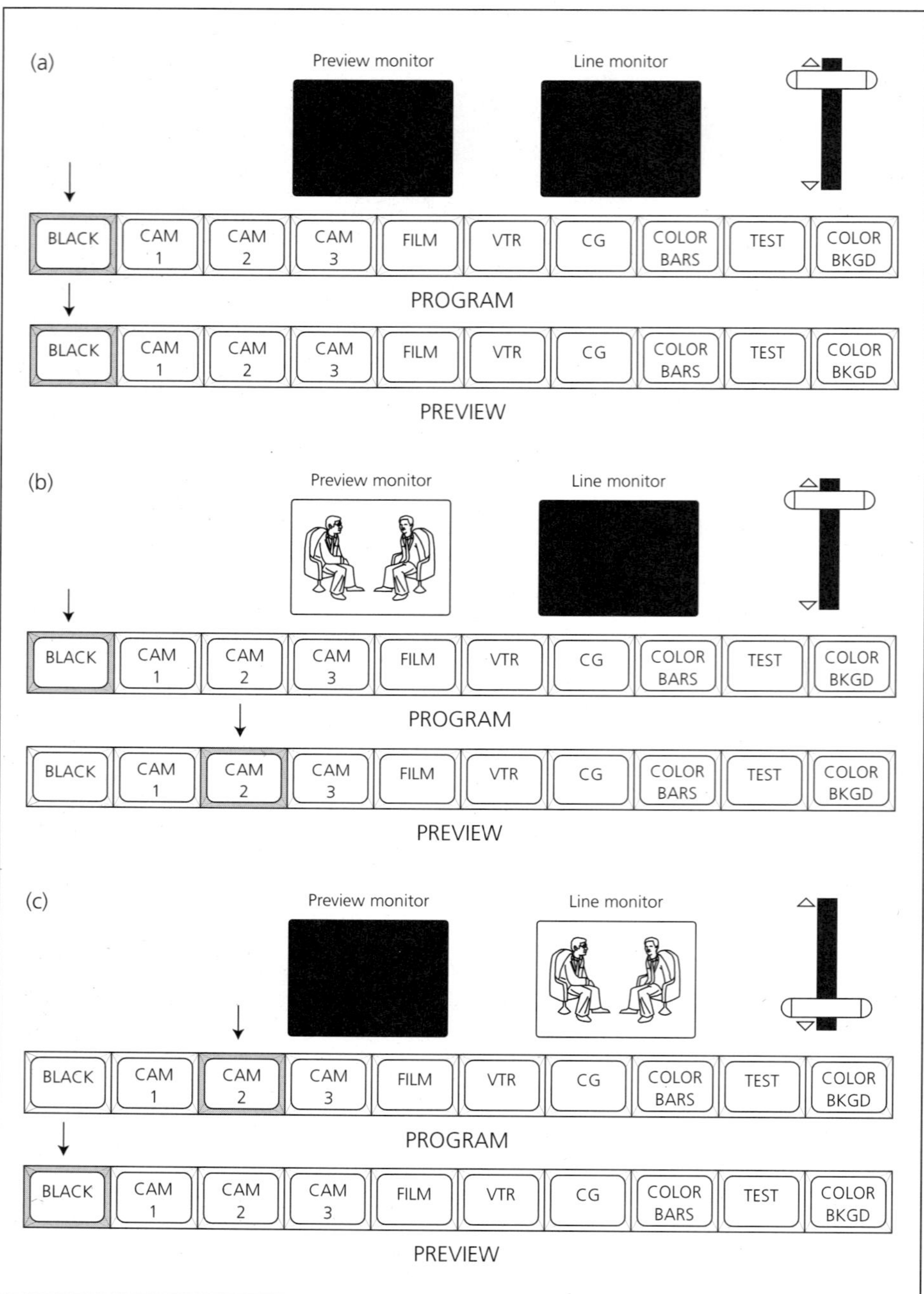

Figure 9.4 *A simple fade using the program bus, preview monitor, and line monitor. (a) Black on all lines. (b) When CAM 2 is punched on the preview bus, the image from camera 2 appears on the preview monitor. (c) As the toggle lever is moved down, the scene from camera 2 gradually appears on the line monitor, and the preview monitor fades to black.*

command "Fade two," the TD moves the lever through its entire range of motion. As the lever moves, the scene being pictured by camera 2 gradually transfers to the line monitor, while video black gradually disappears. The speed of movement involves an aesthetic judgment, which may be worked out with the director during rehearsal.

Most switchers are made so that when the toggle completes its full range of motion, the two sources being manipulated (in this case, video black and the feed from camera 2) switch places or *flip-flop;* that is, video black now flips to the preview bus and monitor, and the shot on camera 2 goes to the line bus and monitor. This makes it easy to fade back to black simply by moving the toggle through its full range of motion once again with no additional set-up.

One advantage of a separate preview monitor is that it enables the director, TD, and assistant or associate director (AD) to see and refine upcoming shots before they are put on air. Such refinements might include instructing the camera operator to frame or focus the shot more effectively. Making final adjustments in a shot while it is still on preview is called **shot sweetening**.

Dissolves Exactly the same mechanics are involved when making gradual transitions from nonblack sources to other nonblack sources, except we do not call such transitions *fades;* instead, we call them **dissolves**. As an example of a dissolve, imagine that you have camera 1, providing a close-up of a singer's face in the left two-thirds of the screen, punched up on the program bus. Camera 2, punched up on the preview bus, shows a long shot of the singer's whole body in the right third of the screen. At the appropriate moment, you can move the toggle to make a slow dissolve from one source to the other, giving the viewer a nicely composed transition from the tight image of the singer's face to the long shot of the same subject. As the change is taking place, you can instruct the camera operator with the long shot to begin zooming in slowly on the subject while centering up. This could be done in tempo with the music to create a smooth, fluid effect.

Superimpositions You can create still another mix of images, called a **superimposition** (or *super*), by leaving the toggle in the halfway position. That is, after the move of the toggle begins, do not go all the way through. Instead, leave it about half the distance through the entire range of motion. This will make the images from both sources remain on the line monitor at about one-half the picture strength for each. In our singer example, a super would result in a close-up of the singer's face in the left two-thirds of the screen and a complete body shot of the singer in the right third.

When using supers or slow dissolves to create video compositions and transitions on screen, you need a clear idea of what you want your cameras to do. Otherwise, you may end up with mixes that are cluttered or unbalanced. Or your supers and dissolves may lack visual interest because they feature essentially the same visual content from only slightly different angles. To get the most out of your video mixes, plan shots that use the screen space to its fullest potential. You can improve the camera work by having camera operators watch the line monitor in the studio while building shots and executing movements. Also, some camera viewfinders have an *external* feature that can carry the line-out signal so that the camera operators can see composite images in their viewfinders. In either case, rehearse camera moves to familiarize the crew with what you are trying to achieve.

From Mixes to Special Effects

Thus far, we have designed our video switcher to execute transitions and mixes among several video sources. However, the pictures we have described have not exceeded the signal strength of a single source. In other words, even when we used composite images from more than one source, the total combined signal strength of the signals we mixed on the line monitor never exceeded 100 percent.

In the early days of television, such mixes were the only means available for combining video sources, and some drawbacks were obvious. For instance, when supers were used to place titles and credits over program video, the composite images left a ghostly and transparent appearance for the lettering as well as for the talent and set shots. The lettering was sometimes difficult to read, and the overall shot tended to be dimmed. By today's standards, such image quality is unacceptable.

To provide a more robust signal featuring full picture strength for different sources in a composite image, a **special-effects generator (SEG)** has been added to the video switcher. The SEG enables the switcher to insert images electronically from one source over another while maintaining full signal strength for all remaining portions. Replacing part of one signal with part of another is called **keying**.

Figure 9.5 shows additions to our switcher to allow keying. Notice the groups of *effects* and *key control* buttons in a part of the SEG called the *effects keyer*. Figure 9.5 also shows a *key bus* above the program bus and an *effects transition control group* to the right of the program and preview buses. A digital timer, called *auto transition rate,* has also been added at the top right of the switcher console to control the time it takes for selected transitions to occur. Of course, the locations of these new components may differ from one switcher to another.

Key Effects

With a setup such as the one in Figure 9.5, pushing KEY in the effects transition control group tells the SEG to activate the key mode. Then several different types of keying become possible.

Luminance or Insert Keying   In **luminance keying** (also called **insert keying**), background video is electronically replaced at full picture strength by a second video source. The area replaced by the insert is defined by brightness or luminance differences of the key source signal.

Luminance keying is frequently used to insert titles from cards, slides, or a CG. Such sources usually feature white letters on a black field. When the SEG scans the image, it finds the bright portions (the letters) and uses them to replace the corresponding parts of the background.

As part of the effects keyer, Figure 9.5 shows two buttons for controlling the key level. The **clip** button determines what level of source video brightness is used to cut the key image onto the background. If lettering looks too thin and wavering, or overly thick and bleeding around the edges, the clip level may be adjusted. The **gain** button, on the other hand, controls the transparency or solidness of the key image. When using key effects, it is wise to readjust the

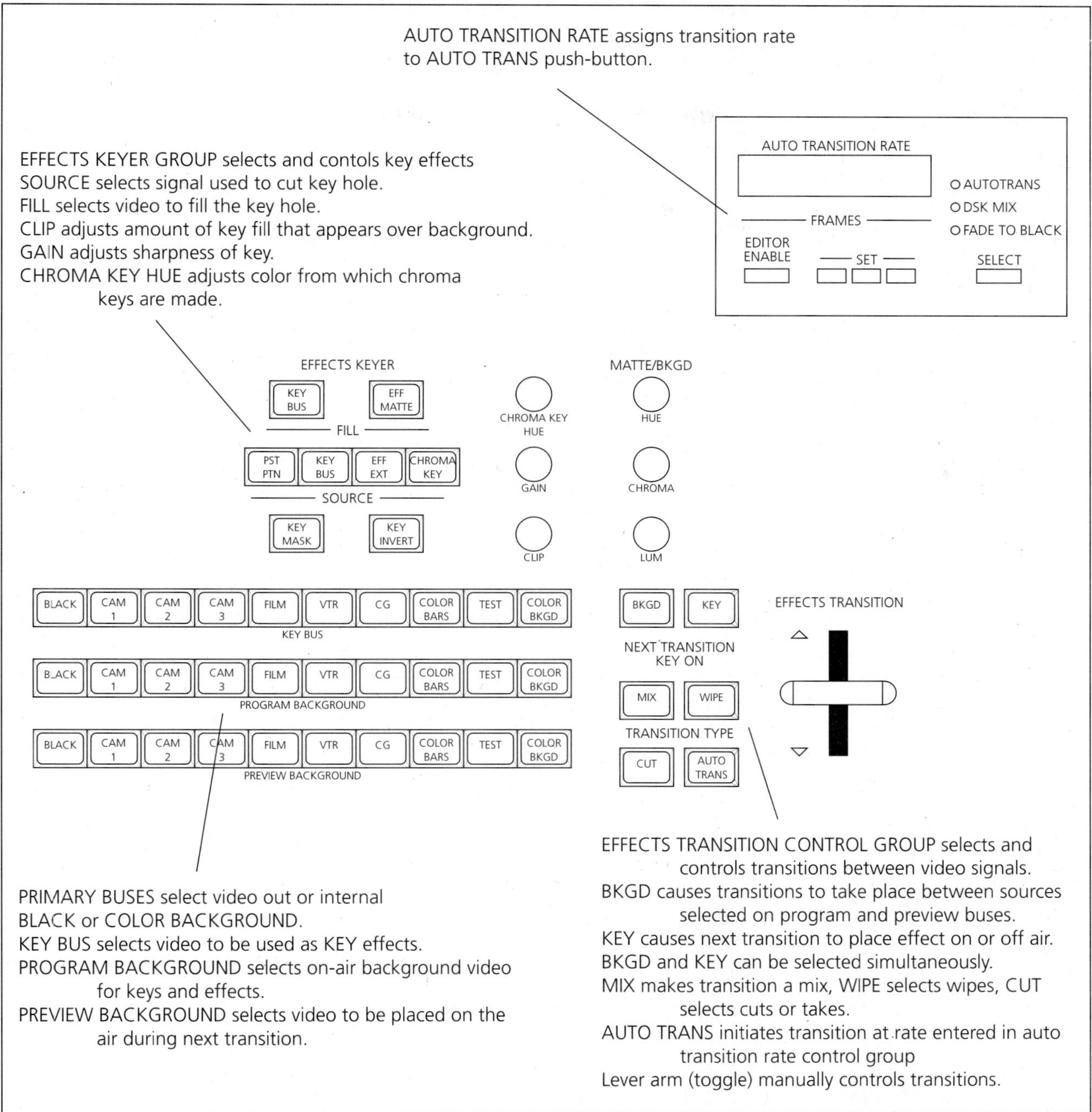

Figure 9.5 *A further development of the video switcher from Figure 9.4. Four special-effects groups have now been added.*

clip and gain controls as lighting conditions change. Some switchers are equipped with a memory to make such adjustments automatically.

Let's set up a luminance key of a lower-third ID from the character generator over a bust shot of an interview guest on camera 1. Imagine that camera 1's bust shot is already on the line (the button for camera 1, marked CAM 1, is punched up on the program bus). To prepare the key, we punch the source to be keyed (in this case, the CG text) on the key bus above the program bus and the source it is to be keyed over (in this case, camera 1) on the program bus (delegated to backgrounds). We also press KEY BUS on the effects keyer and BKGD on the effects transition control group.

The TD now waits for a ready cue and a command from the director. On hearing the cue "Ready key CG over one," the TD poises a finger on the CUT button from the effects transition control group. Then, on hearing the command "Key CG over one," the TD presses the CUT button. This takes the camera shot of talent, now with the lower-third keyed over it, on the air (Figure 9.6).

Of course, before taking the insert to air, it is wise to preview it to determine whether it is air quality. Previewing allows you to sweeten the insert by adjusting clip and gain control. Once a key is on preview, you also have the option to gradually insert the key with the toggle instead of "popping" it on air instantly. To add the effect gradually, we would select the MIX button in the effects transition control group instead of the CUT button. Then we would use the toggle to add the key. As an alternative to the toggle, the AUTO TRANS button would allow us to execute gradual transitions with the push of a button. In such cases, the duration of the transition is controlled by the auto transition rate, displayed in frames per second (see the top right portion of Figure 9.5).

The specific procedure we just followed to set up a luminance key is not identical across all brands or models of switchers, but the general process is the same. Most switchers also allow you to add shot variety and visual interest by undercutting or overcutting inserts. *Undercutting* is done by cutting among various sources on the program bus, delegated to backgrounds, while keeping the same key source (in our example, the CG) on the air. Conversely, *overcutting* is done by cutting among various sources on the key bus, delegated to key sources, while keeping the same background image on the air. For example, while credits are rolling from a CG at the end of a show, we

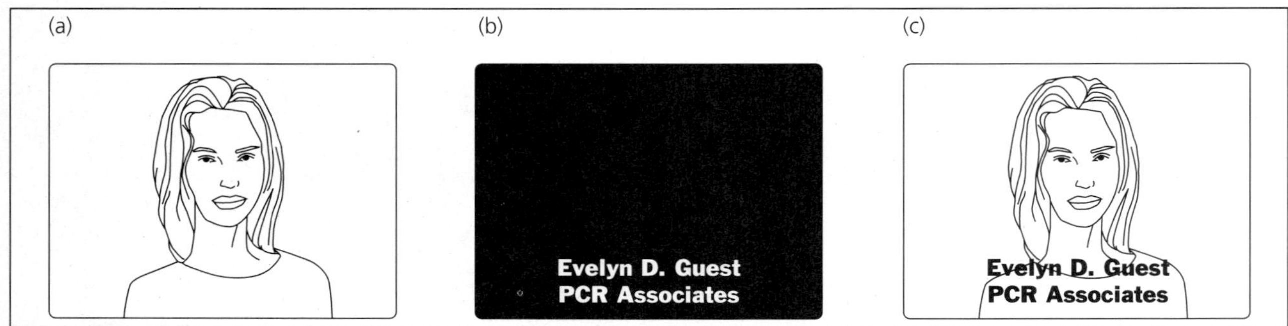

Figure 9.6   *Luminance (insert) keying of a lower-third ID. (a) The talent's bust shot on the line monitor. (b) The lower-third ID prepared on the source (CG) monitor. (c) The key executed, combining the two images.*

might use an undercut to cut from a shot of the host and guest chatting on one camera to a close-up of the guest reacting on another camera. Similarly, just before fading to black as the show ends, an overcut might replace the keyed credits with the show's title.

Most switchers feature some other luminance or insert keying options as well. For example, a KEY INVERT button (shown in Figure 9.5) reverses the luminance values of the key source so that dark areas become the portion of the source keyed rather than the light portions. To become familiar with the options available on any switcher you may use, it is helpful to experiment with the key source, background, and fill. To learn even faster, study the manual that comes with the switcher. Finally, keep in mind that some options on the switcher you are using may not be installed exactly as described in the manual. If your switcher seems to have a mind of its own at times, you may need to talk with the engineering department.

Chroma Key Effects   As mentioned in Chapter 8, chroma key effects replace a camera signal of a preselected hue with a signal from an external source. For example, if a weatherperson stands in front of a blue background, the appropriate weather graphics feed can be electronically inserted behind the talent in place of the background (Figure 9.7).

Setting up a chroma key effect takes several steps. Usually the source to be substituted (most often a background scene) must be punched up on either the preview or program bus, and the source carried by the chroma key camera must also be selected. On our switcher shown in Figure 9.5, we would select the BKGD and KEY buttons from the effects transition control group. Next, we would select CHROMA KEY as the key source and key bus as the fill on the effects keyer. Most switchers also allow you to adjust the chroma key hue, gain, and clip controls to clean up the effect before it goes to air.

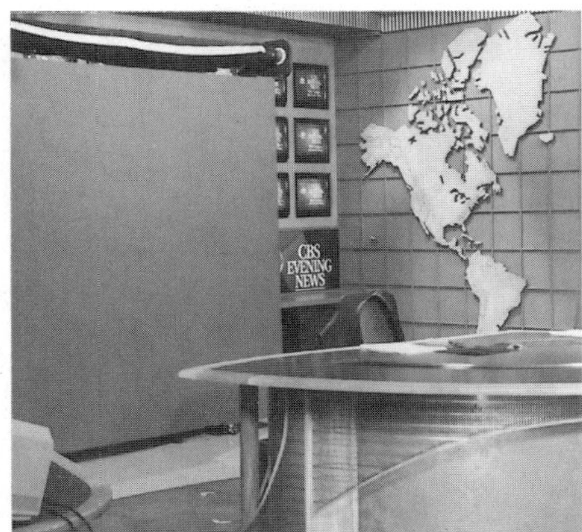

Figure 9.7   *On this news set, the blank screen at the left is the chroma key screen. On the air, the blank space can be filled with any appropriate graphics feed. This kind of screen is typically used by weatherpersons, among others.*

Ensuring Good Chroma Key Effects

- Consider the talent's wardrobe carefully. Avoid using clothing with colors that might confuse the signal. For instance, if the chroma key color you have selected is blue, the talent should not wear blue.

- Keep the lighting on the chroma key background relatively bright and uniform. Keep shadows off of it, if possible.

- Keep the talent an appropriate distance from the background. In addition to the problem of shadows, if talent stands too close to the background, light reflected from the backdrop can cast the chroma key color onto the back of the talent's head. This may confuse the signal and cause *tearing,* a wavering of the signal between sources.

- As another way to offset tearing, backlight the talent with pale yellow or amber gels. These tend to create better color separation, especially in the case of dark-haired talent.

To take the effect to air, we would use the same procedure as for luminance keying. Of course, we could dissolve or fade the chroma key picture either by punching the mix button on the effects transition control group and then using the toggle or by using the AUTO TRANS button.

Some recent digital innovations in chroma key technology have improved the appearance of the chroma key signal. One such innovation is a *shadow key,* which uses digital technology to sense shadow detail in a foreground object, casting it into the background portion of the overall picture. Another innovation involves the change from a *hard key,* where objects appear with a hard edge against the chroma key background, to a *soft key,* where foreground objects appear to have a softer, more natural outline in the overall picture. Soft chroma key eliminates the impression that foreground objects have been pasted onto the background. The softer appearance is made possible by a digital gain control that permits the scanning signal to be sampled in proportion to the color and luminance values presented. This move to digital sampling also permits more realistic rendering of translucent and transparent objects such as glass, water, and smoke, making them look more fully integrated into the overall picture.

Matte Keying   Color can be added to key sources by using a *matte key* feature. Whereas in a luminance key the area of screen space cut by the key source is filled with the key source's own content, a matte key adds an extra effect by filling the space with a selected color. Matte colors are often used to fill characters from a CG or a graphics card.

Internal Versus External Keying   When a portion of a background is cut out and replaced by video defining the key area, we call that type of key insert a *normal, self,* or *internal* key. However, it is possible to have a key source define an area to be keyed over but then provide a third source to fill in the space. Such a key insert is called an *external* key. To create external keys, the switcher needs an external key button, shown in Figure 9.5 with the label *eff ext.*

For external keys, three signals must be selected: (1) a background scene into which the key effect is to be inserted, (2) a key source that cuts a hole into the background (called the **keyhole**), and (3) a fill video signal to fill in the keyhole. Matte color signals may be considered a form of external keying, since the color signal comes from a source other than that used to provide the shape of the key that is inserted. However, more elaborate external keys may be provided by cameras or tape machines designated as the external source. For example, if the switcher is set up to accept a videotape source as an external key, footage of a fire can be used to fill lettering keyed from a CG over a background shot, creating the effect that the lettering itself is made of fire (Figure 9.8).

Downstream Key   Most switchers include a special keying function called a **downstream key,** often abbreviated DSK (Figure 9.9). Downstream keying enables you to insert titles and other text immediately before the composed signal leaves the switcher. The advantage of the downstream keyer is that it permits a final manipulation of the video signal—for instance, adding a title, lower-third, or closing credits—without occupying a bus. This frees a bus to

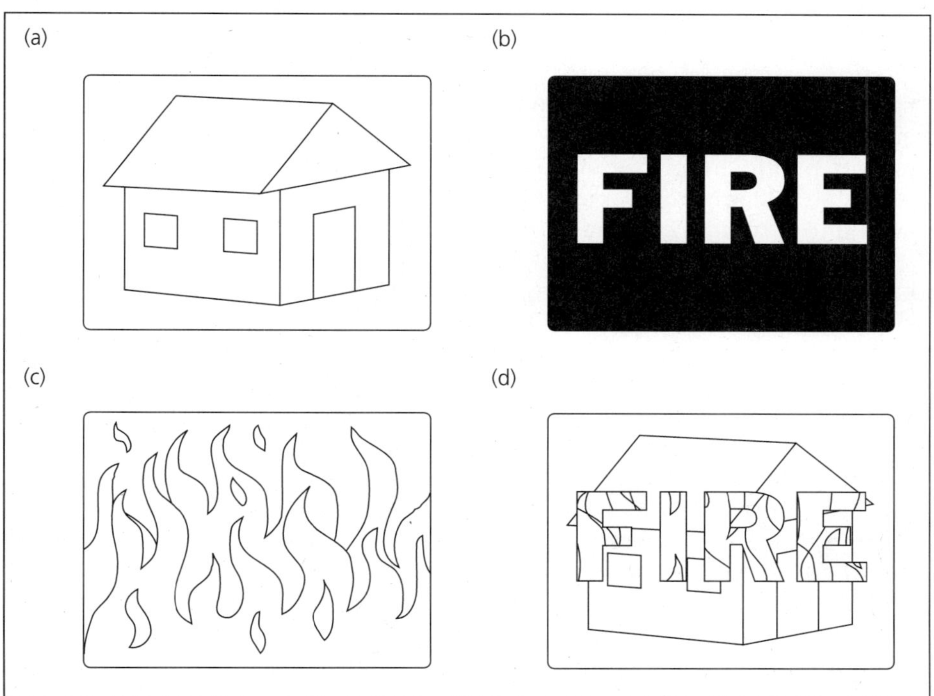

Figure 9.8   *External keying. (a) A background shot of a house. (b) The "keyhole," in this case the word* FIRE *prepared on a character generator. (c) Videotape of an actual fire. (d) The key executed. Now the letters, filled with flames, are superimposed on the house.*

serve other functions, thus adding greater flexibility to the switcher. In news and sports productions, DSK is used for lower-thirds, where text must be inserted over pictures already composed from several sources. *Border, outline*, and *drop shadow* buttons enable DSK lettering to be edged with selected outlines to distinguish it from its background. Figure 9.9 also shows a DSK MATTE button that would allow the user to matte-color DSK text. Switchers with DSK are often equipped with a master fader (labeled *fade to black* in the figure) to lose the DSK graphic along with the entire video signal when, for example, the program ends.

Wipe Effects

A **wipe** is a transition between sources in which one image seems to move across another and "wipe" it off the screen. The effects transition control group usually features a *wipe* button, which, when activated, permits you to create such transitions. In a *background wipe,* for example, one image appears to obliterate the image beneath it. Wipes can be made in a variety of patterns, including circles, diamonds, stripes, rectangles, and squares. Figure 9.9 shows two *pattern control groups* of buttons, an upper one that indicates the available patterns and a lower one for additional shaping and control of the selected patterns.

Wipes may be used to indicate a change of time or place. Wipe transitions were used frequently in the 1970s hit series "Happy Days" to signal scene changes from Arnold's malt shop to Fonzie's place to the Cunningham residence. Intricate comic checkerboard wipes were often used to punctuate the transitions.

Using the toggle, a horizontal wipe can be stopped halfway to set up a split screen between two sources. Figure 9.10 shows two common uses of the split screen: (1) to carry a phone conversation between two talent portrayed as being in two different places and (2) to set up a corner wipe of a talent gesturing in deaf sign language during a televised speech. The same corner wipe may be used to carry news graphics behind newscasters.

Several features of the wipe pattern may be adjusted. For example, the BORDER knob shown in Figure 9.9 would permit us to add a border and adjust its width around the wipe pattern. The edge of the border could be made soft or hard with the SOFT control. In the MATTE/BKGD control group, the HUE, CHROMA, and LUM controls would allow us to adjust, respectively, the color, intensity, and brightness of the border. Using the lower pattern control group, we could adjust the height-to-width ratio of selected patterns, such as rectangles and diamonds, by selecting the ASPECT ON button and then adjusting the ASPECT control knob. The white areas on the pattern symbols of the console indicate where the wipe will overcut the screen image. However, selecting REVERSE from the upper pattern control group would reverse this process so that the black areas on the pattern symbols would indicate where the overcut would be. Selecting REVERSE would also change the direction of the wipe. Finally, selecting the POSITIONER button in the upper pattern control group would activate the *joystick,* permitting us to move the pattern anywhere on the screen.

Most switchers provide these options for controlling and shaping the wipe patterns you select, though the knobs and buttons may not be located in exactly the same places shown in our sample switcher. Actually, the switcher we have "built" in this chapter is based on the GVG 100, a compact model manu-

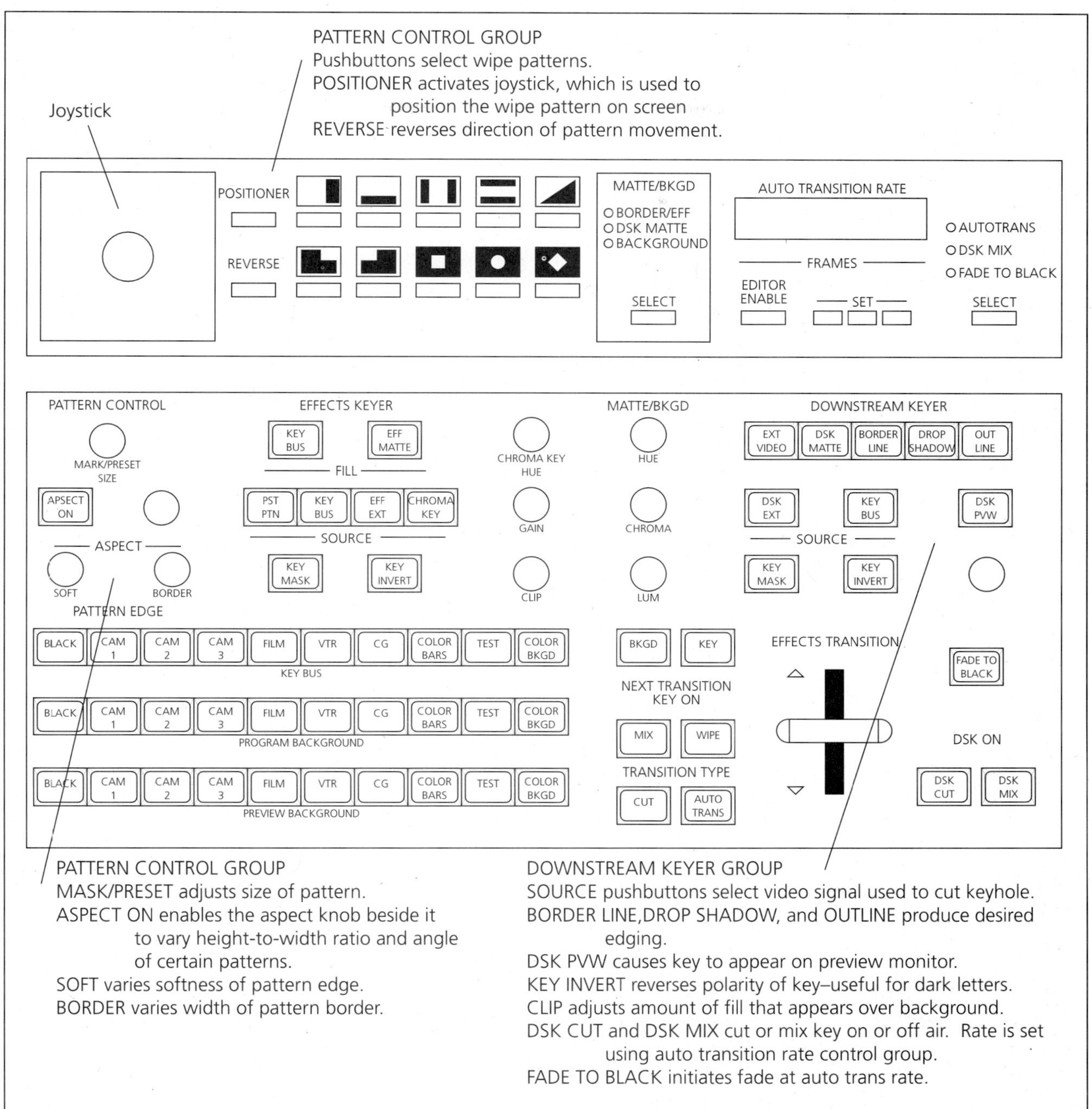

Figure 9.9   *The video switcher from earlier figures, with the addition of a downstream keyer group and two pattern control groups. The downstream keyer selects and controls downstream key effects. The pattern control groups select and adjust wipe patterns.*

factured by the Grass Valley Group in Grass Valley, California. See the box "Additional Switcher Options" for a list of some other features available on the GVG 100 and a number of other switchers.

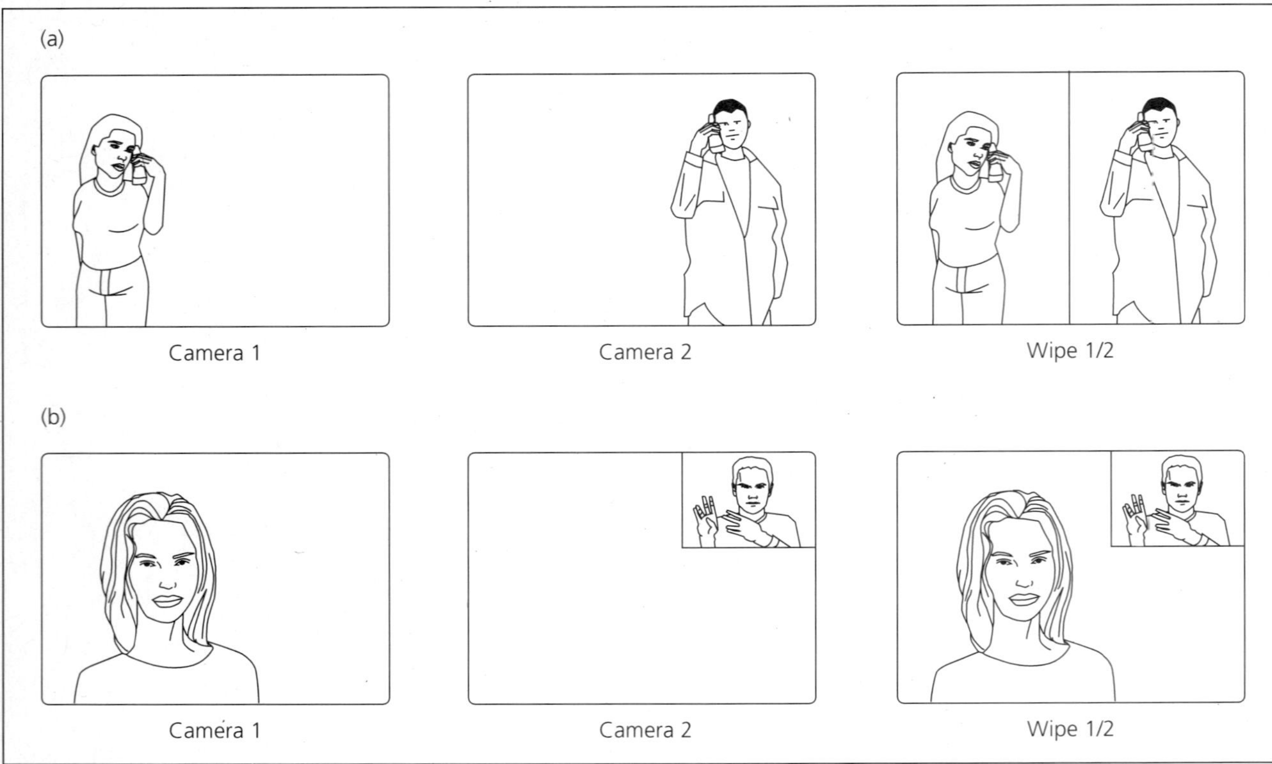

Figure 9.10   *Typical uses of a wipe effect with a split screen. (a) Two people speaking to each other on the phone, shown first in separate shots and then together on a split screen. (b) A speaker captured with camera 1 while camera 2 provides a shot of the sign language interpreter; a corner wipe then puts the two together on screen.*

Additional Switcher Options

Besides the effects described in the text, many switchers offer the following features:

- *Spotlight.* A highlight effect produced by supering full-strength video shaped by a wipe pattern over an attenuated signal from the same source.
- *Key mask.* A key function that enables you to use a wipe pattern to prevent undesirable parts of a key source from cutting a shape into the background video.
- *Editor.* A control system (often computerized) that permits you to control videotape machines, the switcher, and other devices from a single control panel. Editors make it easier to produce ready-to-air programs from numerous sources.

MORE COMPLEX EFFECTS AND TRANSITIONS

In the switcher we have been describing, the program, preview and key buses, plus a single toggle, are adequate to handle changes between one video source and another, including takes, fades, dissolves, and supers, as well as limited transitions between sources involving special-effects compositions. But what happens if we want to take or dissolve from a super directly to another preset super or between wipes featuring multiple sources? The above system, though quite impressive, is simply not enough for such transitions. Let's look, then, at features we need for more complex effects.

Mix/Effects Banks

To handle complex transitions involving multiple sources, more advanced switchers add a set of buses above the program bus called a **mix/effects bank** (or *M/E bank*), with its own toggle. Figure 9.11 presents a simplified diagram of a switcher with two M/E banks.

Generally, mix/effects toggles differ from program/preview toggles in that direction is important. That is, on M/E banks, if the toggle is in the "up" position, it routes the video source punched on the top mix bus, whereas if it is in the "down" position, it routes the signal punched on the bottom. In short, the mix/effects toggle does not provide flip-flop capability.

Notice that the buses in the M/E banks repeat the same set of video sources in the same order in which they appear in the program and preview buses. However, Figure 9.11 adds M/E buttons beside the program and preview

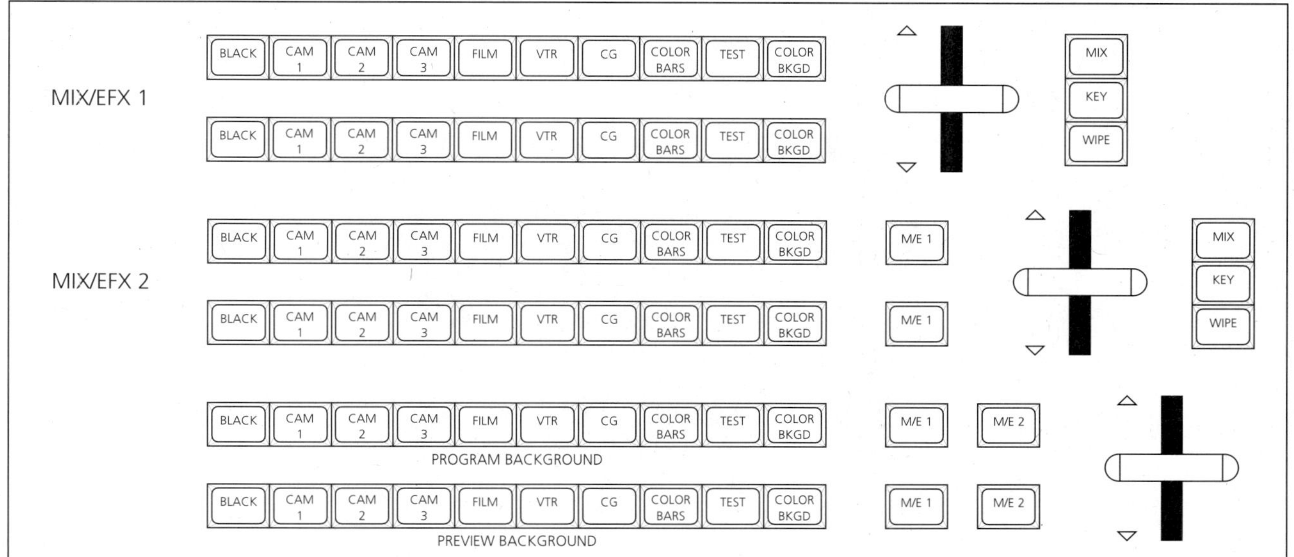

Figure 9.11   *Diagram of two mix/effects banks above the program and preview buses on the video switcher.*

buses so that we can feed whatever signals we mix or compose on the mix/effects banks to either the program or preview bus or both. We call the mix/effects buttons *delegation buttons,* because they route video sources from one part of the switcher to another but do not carry unique video sources of their own.

Using a mix/effects bank, we can set up a super, for example, along with other effects such as wipes. By punching appropriate buttons on the preview bus, we can sweeten these effects with the mix/effects toggle. (When the mix/effects toggle moves through roughly half its range of motion, both buses on the M/E bank are routed to the preview bus at about 50 percent picture strength.) When we like what we see and want to cut to the super we have created, we can simply press the M/E button on the program bus.

Cascading: Multiple M/E Systems

In high-end switchers, there may be two or even three M/E banks. In such models, additional mix/effects delegation buttons are added to the ends of some M/E banks, as well as to the program and preview buses, to route composite video signals in an orderly manner. Other toggles are also provided. Switchers with more than one M/E bank are said to have *multiple M/E systems.* Figure 9.12 shows a photo of one such switcher.

Switchers with multiple M/E systems can produce highly complex video images. A video effect composed from two or three sources on one M/E bank (for example, a camera shot of a studio talent standing in front of a chroma key background with a lower-third CG text) can serve as a single source feed to another M/E bank. Combining this output from the M/E banks with still other sources at the program/preview part of the switcher can create an extremely complicated output.

Switchers with multiple M/E systems work according to a **cascading** principle, whereby the composite video signal flows from the top M/E bank downward, much like a stream of water flowing downstream. Switchers that route the output of one M/E bank to another for further processing before it is finally

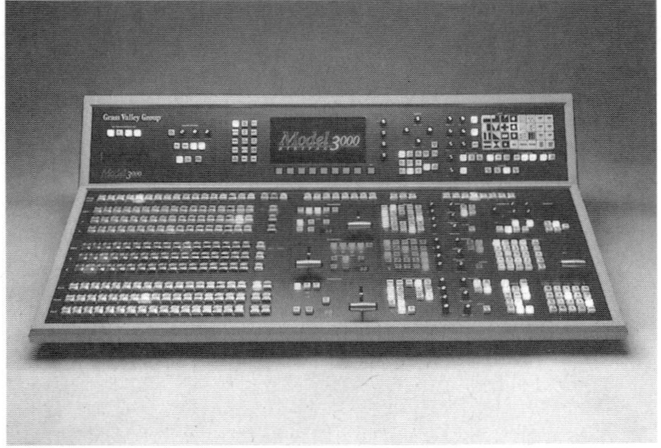

Figure 9.12   *The Grass Valley Group Model 3000 video switcher.*

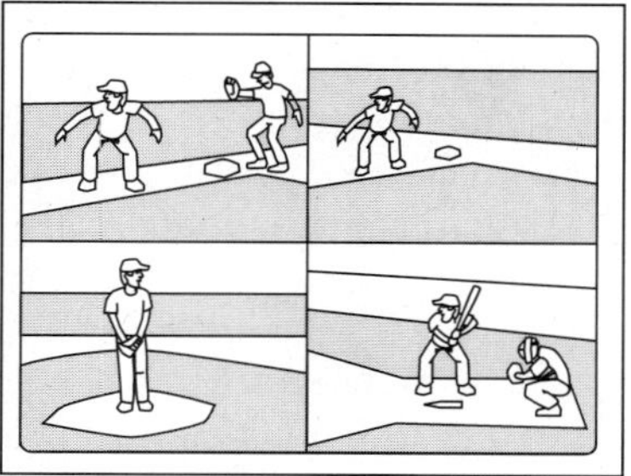

Figure 9.13   *A quad-split.*

sent to the program bus are said to be capable of *double re-entry.* The cascading architecture of multiple M/E system switchers explains how the downstream keyer got its name: some signals are made to flow downstream but not up.

One effect possible with multiple M/E system switchers is the *quad-split,* in which the screen is split into four quadrants, each carrying a different video source (Figure 9.13). The quad-split is often used in news and public affairs programs to carry video of several individuals located in different cities discussing a debate topic. The four sources are often separated by a thin wipe border. Another effect is *layering,* which combines multiple key and wipe effects to build a complex image. For example, a quad-split image of four speakers can form the background picture for a chroma key insert in an over-the-shoulder shot of a studio news anchor (Figure 9.14). As the speakers answer questions, lower-third graphics may be keyed into the appropriate quadrants to identify each speaker.

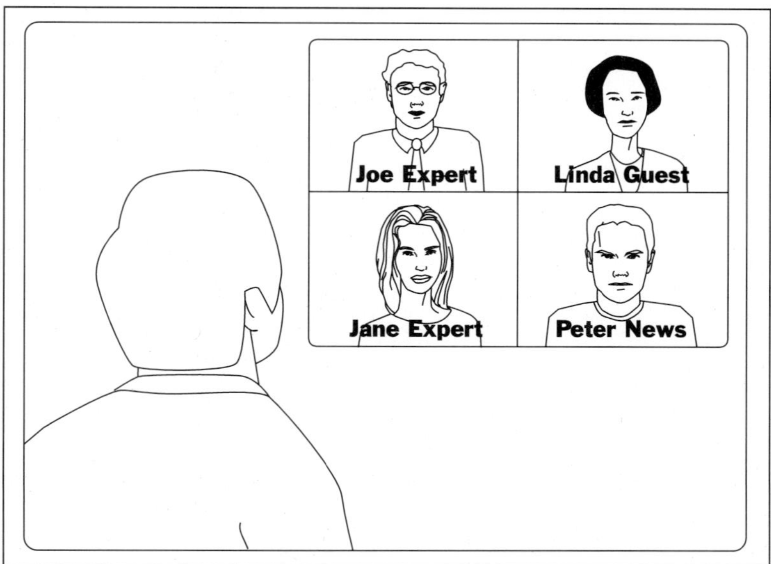

Figure 9.14   *A layered use of the quad-split: separate shots of four interviewees are combined into a single chroma key insert behind the news anchor.*

I N D U S T R Y
voices

Ethan Becker

Digital Authoring Specialist, Data Translation Multi-Media Group, Marlboro, MA

Q: How do computers influence television production these days?

A: Once a video image has been digitized into a computer, it can be manipulated into many different effects. Most of the effects I create I make with our Media 100 digital authoring system. I use it to acquire and edit material, and then I use Macintosh software programs to manipulate the image. This way I benefit from the best of both worlds. I get creativity, power, and the ease of use of the Mac and the professional image quality of Media 100.

Files digitized with Media 100 can be stored using Apple Computer's QuickTime format and can be taken into any QuickTime application, making it possible to use hundreds of digital effects and have a very high-quality final result.

Q: What other tools have you used?

A: By far one of the most powerful tools for digital video postproduction is Adobe's After Effects. Besides allowing you to composite up to 65,000 layers of video, it has a feature called Motion Tracking that, in effect, lets you select a few pixels in a frame and follow those wherever they go throughout a video shot. For example, there's a commercial for Coca-Cola featuring penguins skating around on ice. Each penguin has a Coke in its flipper. The designer used Motion Tracking software to select pixels on the flipper to make the Coke can track and stay in the penguins' flippers no matter where they went.

Q: Obviously we've come a long way with digital graphics since the first CG and chroma key machines, haven't we?

A: The kind of control you have nowadays with the image, once it has been digitized, is useful in newer types of chroma key. In the past, if you wanted to chroma key an image into a shot, you had to shoot with a blue or green screen in the background and use a keying or matting system. Now there is no longer a color requirement. With After Effects, you can select out any color for keying and key out more than one color at a time, if you want.

Although you don't need a specific color anymore, it still makes things easier to use a single color in certain applications, because you tend to light more evenly when a specific color is chosen. In some cases, using a single color still makes for smaller headaches in postproduction.

Q: All the hot new tools must present a challenge for the editor. Is it tough to wrap your mind around the endless possibilities in postproduction?

A: The conceptual planning for editing has become a huge part of the production process. Because you are now editing in a nonlinear manner, there is often a period of time near the end of the production process when there is a new indecisiveness that never existed before. In the past, you knew you were finished with a show when you laid down the last shot. You never thought too much about changing something in the beginning of the show because it was too expensive or too much work to re-edit everything after that change. Now things are totally different. Using a Media 100 system, you as the editor can organize a bin-window that will display thumbnails of all the shots you have digitized.

This function of the system is like the old-fashioned practice of creating a paper edit, where you write a description of each scene or shot on an index card and then lay them all out on a big table or the floor to decide how you want your show to look. You keep changing the order and layout of the cards while you conceptualize the program until it makes sense. With the bin-window feature of the current system, after all your shots have been digitized, you can sort and organize them in the order you want. Then you select the shots and drag them to a time line, which is equivalent to your master tape. The new method takes you from initial concept to rough cut much faster. It is at this point that the producers, directors, and editors become digital authors. They can tweak the show, select special effects, animations, cutaways, titles, and all the other goodies.

The impact of this capability is that there's never an "OK, we are laying down the last shot." You can continue tweaking and editing. The trouble comes if you fail to recognize when you have finished your show. I have fallen into this trap with producers who are paying by the hour but keep making changes. With producers who are new to digital editing, it often becomes necessary to explain these pitfalls so that we can speed up the editing process.

Computer-Assisted Switchers

Computers are commonly used to set up and execute complicated transitions. Programmable switchers can memorize long sequences of transitions and execute them according to a strict time schedule precise to the thirtieth of a second (that is, *frame-accurate*). When a sequence is executed on the air using a computer-assisted switcher, it is first entered into the computer's memory during preproduction. Then it can be set in motion by the TD during showtime with the push of a single button.

Smart switchers, as these machines are called, also find application in postproduction work where the instant time demands of live production are not a factor. Smart switchers can be easily integrated with digital editing systems for precise postproduction tasks, where the emphasis is on creative transitions and effects. In this role, computer-controlled switchers handle transitions among many digital video sources, including digital video effects generators, edit units, and digital audio machines.

When used this way, the computer can be programmed to pre-roll tape machines before taking material from them, as well as load and use material from still store machines and graphic paintboxes. Because computers can be reliably programmed to coordinate switching among many sources, programmable switchers are becoming more common in master control rooms throughout the nation's production facilities, adding a robotic quality to the production environment. Of course, the computers that control complex switching functions are themselves incapable of making decisions. They must be programmed to do their work, and this task still depends on people.

Postproduction switchers often differ from their production counterparts in their internal design or architecture. The production switchers we have described use a *linear architecture*, which employs the cascading principle in which signals move downstream from one bank to another for successive manipulations. In contrast, many postproduction switchers work according to a *parallel architecture*, whereby the manipulation of the video signal occurs, roughly speaking, all in one place. This difference makes it possible for postproduction switchers to program complex transitions and special effects with only one M/E bank and to execute them all with the push of one button.

DIGITAL VIDEO EFFECTS GENERATORS

Digital video effects (DVE) machines are computers that turn video signals (both analog and digital) into digital graphics. Once captured, the images are subject to virtually infinite manipulation without losing signal quality. When used as a video source and routed to the video switcher, the effects produced with a DVE can be integrated into those available from the switcher itself, resulting in a dazzling tour de force of image fireworks. (See Figure C6 in the color plates.) Switchers built to accept only analog signals must be outfitted with *black-box interfaces* to be able to accept digital signals, and vice versa. Some examples of the most common DVE image manipulations include:

- *Slide effects.* One video source appears to slide to one side, revealing another source that appears to be hiding underneath.
- *Freeze frame effects.* The screen displays a series of static frames of an original live image.
- *Zoom, bounce, spin, squeeze, and other motion effects.* An image appears to change from full size down to zero size or to expand from zero to full size (*zoom*); or behaves like a ball, bouncing from one screen edge to another while being compressed by the bounce (*bounce*); or tumbles and rotates in selected directions (*spin*); or expands or compresses as it moves through different size changes and screen locations (*squeeze*).
- *Position and perspective changes.* An image appears to pan and tilt from its starting position, or to stretch either vertically or horizontally, or to twist, bend, or lean in different directions.

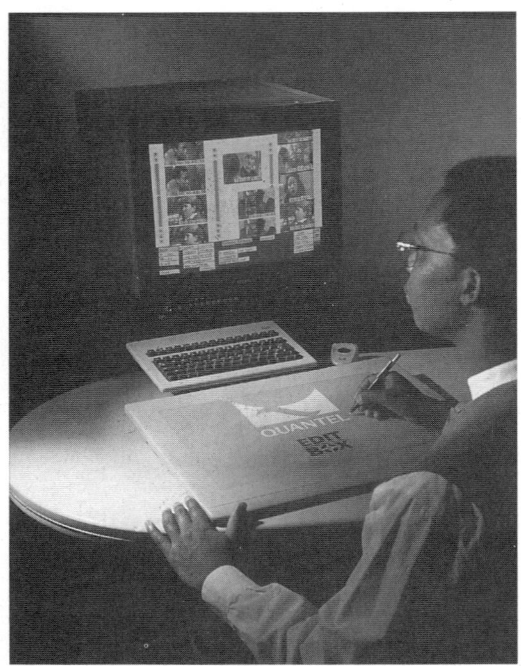

Figure 9.15   *Numerous image manipulations can be done with a digital video effects machine.*

- *Posterization or solarization.* The luminance values of an image are reduced to a limited range of values, giving the image the high-contrast look of some poster art.
- *Mosaic and tiling effects.* An image is broken into a number of squares resembling tiles, reducing picture clarity while creating a stylized pattern of the original image.

A DVE machine and some of the effects it can create are pictured in Figure 9.15. Many of these effects are bound by time constraints. For example, if a speech is to be punctuated by a digital effect used as a transition to the next speech, and the time between utterances is three seconds, it is critical to tell all of the machines involved when to execute each of their functions so that everything works as planned. Some systems alert the TD with a beep when a sequence of commands does not add up. Other systems are not that smart. Depending on the system you use, you will need to be more or less vigilant about time factors when programming sequences.

Though DVE effects are tempting, programs featuring dazzling sequences of effects for their own sake will probably not sustain an audience's interest very long. Avoid the temptation to use digital effects just to use them. When considering a fancy special effect or graphic display, ask yourself, "How would this enhance my program or move it forward?" "Is this effect motivated by my program purpose and content?" If the answer to these questions is no, it is probably wise to exclude the effect. When in doubt, leave it out.

KEY TERMS

video switcher *(204)*	superimposition *(211)*
cut (take) *(207)*	special-effects generator (SEG) *(212)*
program bus *(207)*	keying *(212)*
program (line) monitor *(207)*	luminance (insert) keying *(212)*
synchronous and nonsynchronous sources *(209)*	clip *(212)*
time base corrector (TBC) *(209)*	gain *(212)*
frame synchronizer *(209)*	keyhole *(217)*
fade *(209)*	downstream key *(217)*
preview (preset) bus *(209)*	wipe *(218)*
shot sweetening *(211)*	mix/effects bank *(221)*
dissolve *(211)*	cascading *(222)*
	digital video effects (DVE) *(225)*

QUESTIONS FOR REVIEW

1. Why do different video sources have to be made synchronous with one another? What machines accomplish the synchronizing process?

2. What are the differences between fading and dissolving video sources?

3. What are some effects possible with the special-effects generator (SEG) on the basic video switcher?

4. What is the difference between internal and external keying?

5. What are some common uses for the downstream key (DSK) feature of the video switcher?

6. What special production considerations regarding lighting and set design would guide you in planning for a successful use of the chroma key effect? What would you tell the talent?

7. What is the cascading feature of multiple mix/effects bank switchers? How can digital video effects (DVE) machines enhance your video presentations? What are some of the effects you might use to shoot a music video?

10 Field Production

An increasing amount of daily television fare currently originates from nonstudio locations, including news and sports programs as well as segments for sitcoms, dramas, commercials, and variety programs. Increases in *field production,* as this work is called, are due largely to breakthroughs in technology in two areas. First, because of the smaller size and weight of television equipment and improvements in technical performance, you can now collect video content under conditions that were once inaccessible to television. Second, once you have collected it, you can instantly transmit programs globally from any location through a variety of means. In short, recent advances have increased television's accessibility and transmission capability.

State-of-the-art field equipment is more portable, reliable, durable, efficient, and higher quality than ever before. Newer dockable camera/recorder and *camcorder* designs combine into one unit the once separate camera and videotape recorder, eliminating the need for a cable connection that could pull loose during a shoot (Figure 10.1). Single-unit design also dispenses with the need for a recording assistant to run alongside the camera operator to capture key shots, and the operator can work alone without a cumbersome recorder dangling from the shoulder.

The move to charge-coupled device (CCD) technology, as described in Chapter 2, also makes field cameras lighter and smaller than their tube predecessors, which makes them easier to pack and transport. In addition, solid-state technology makes field cameras more durable; they are less subject to registration problems if they get knocked around.

Field cameras are also more light sensitive than ever, so they can work in lower light than was once possible. This means they can use denser lens systems, affording them greater focal range under more varied conditions. They also operate with lower energy demands—on batteries—for longer periods of time. Moreover, because of the greater light sensitivity of cameras, lights can now be used at lower power levels, making light kits smaller and lighter. The move toward lower power also means less intrusion and distraction for on-camera personnel; thus, a more naturalistic atmosphere can be achieved.

Microphones have also become smaller while increasing in sensitivity and directionality. In addition, wireless systems have largely been perfected, and now are quite reliable over longer distances in both studio and field settings.

Once cameras, mics, and lights have been deployed in the field, the producer has the option to either transmit to a home base for immediate live broadcast to virtually anywhere on earth or record material on tape for later editing and/or broadcast. Methods of delivering a signal to a home base include traditional coaxial cable and terrestrial (land-based) microwave relay links, newer fiber optic lines, and even satellite *transponders* (radio devices aboard satellites that receive and then retransmit video signals originating on the ground).

This chapter explains how field production works and offers guidelines for preparing for and executing successful field shoots. All the aesthetic and technical principles covered in earlier chapters remain important in field production, and you should review them when necessary. The topics covered in this chapter include:

THE GULF WAR: A CASE STUDY IN FIELD PRODUCTION

THREE BASIC CONCEPTS

reach • range • interactivity

ELECTRONIC NEWS GATHERING

the ENG mobile unit and equipment • signal transmission and relay facilities • the ENG preproduction stage • the ENG production stage • the ENG postproduction stage

ELECTRONIC FIELD PRODUCTION

the EFP mobile unit and equipment • the EFP preproduction stage • the EFP production stage • the EFP postproduction stage

MULTICAMERA REMOTE PRODUCTION

the MCR mobile unit and equipment • the MCR preproduction stage • the MCR production stage • the MCR postproduction stage

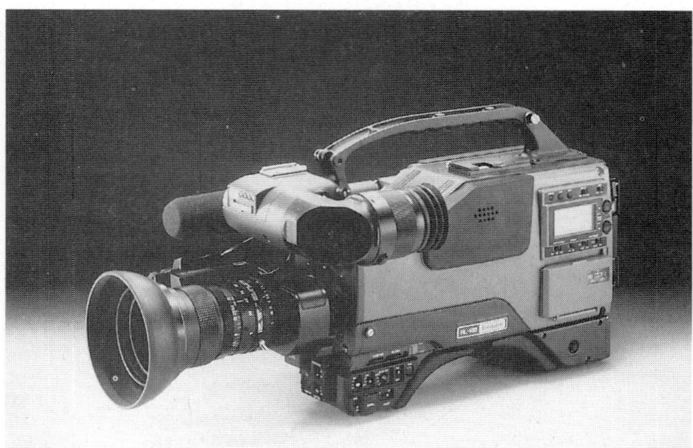

Figure 10.1   *A dockable camcorder.*

THE GULF WAR: A CASE STUDY IN FIELD PRODUCTION

Perhaps no event in recent history more dramatically illustrates television's current robust field capability in terms of both immediacy and pervasiveness than the Persian Gulf War of 1991. In addition to conventional broadcasting technology, television news services relied on cellular telephones, satellites, computers, microwave relays, and fiber optic technologies to cover that conflict.

Through various configurations of these technologies, television news (however sanitized by the sources) was broadcast worldwide. Throughout the war, satellite television, most notably CNN and other news services, fed both sides of the conflict, as well as neutral nations, information as events unfolded. Television news coverage was viewed not just by citizens but by intelligence officers and military personnel, who in some cases used such information to make real-time strategy decisions. Some events were televised *as they happened:* SCUD missiles were seen exploding both on the battlefield and on American and other nations' TV screens virtually simultaneously. This unique circumstance enabled policymakers in the audience to order responses that affected the prosecution of the war itself, giving rise to the term *telediplomacy.*

What made this possible was not just the technical capability of a single device or several devices but the *convergence,* or linking, of media technologies with one another. In fact, some of the communication technologies used to cover the Gulf War were neither new nor technically advanced. For example, CNN used a four-wire phone to set up a telephone line from its Baghdad location to a base in Jordan. The four-wire is a plain telephone with a direct link to just one destination, exactly like the old army field telephones. Further, for battle shots, all the networks used Sony "Hi-8" 8 mm camcorders available off the shelf at most retail camera stores. These cameras were cheap enough to abandon on the field, if necessary, and they made smaller targets than conventional field cameras.

Among the more advanced devices used was the **flyaway video satellite uplink,** a satellite earth station that can be packed into suitcases and then flown anywhere on a small, commercial airplane. Upon arrival at a field location, flyaways can be set up on a truck and moved to any place the truck can go to provide global, live television coverage.

The laptop computer was another relatively new item used by journalists. Linked to a modem, laptops provided e-mail capabilities for reporters in the field, as well as access to the Internet and various databases, enabling reporters to file and read stories from around the world within moments of their creation. News agencies could also download files of historical and archival data for use by field reporters. In addition, home-base television stations and news agencies frequently forwarded news items from competitors to reporters for use in editing their own stories. Hence, the convergence of new media technologies with older ones not only has made global journalism more pervasive and immediate but has also contributed significantly to its content.

It is unlikely that you will be thrust into a field setting such as the Persian Gulf War. More likely, you will have to cover news or a live sporting event for a local television station. Or you may work a live, multicamera remote of some special event, such as a parade or a government official's inauguration. However, regardless of what you cover, the field production principles you use will be the same as those used to cover the Gulf War, and you may confront a similar variety of conditions. Some field reports may be broadcast live, and some may be taped, edited, and broadcast later. Some may be shot at night and some in the daytime. Some may be shot outdoors and some indoors. Some may be done under clear skies and bright sun and some under heavy clouds or in stormy weather. Some may be done in the heat and some in the cold. Some may feature professional talent and some the greenest amateurs. Some may have huge budgets, and some may be done on a shoestring. Many productions will have tight deadlines, and a few will have the luxury of time. Some will use radio energy (such as microwave links and satellites) to deliver live feeds from the field to the home base, whereas others will use phone lines or some other physical connection. Some will permit favorable access to key locations for cameras and equipment; others will not. Finally, some hosts, subjects, or jurisdictions will impose few restrictions (legal or otherwise) on the use of the materials you shoot, whereas others will impose such severe limits that you may wonder whether it was worth shooting anything at all. *In all cases, you will be expected to function professionally with a finite supply of space, time, materials, and personnel, and within ethical guidelines of acceptable journalistic practice.*

THREE BASIC CONCEPTS

To understand the media infrastructures relevant to field production, three general concepts are useful: reach, range,[1] and interactivity. The following sections describe these concepts and give examples of their application.

[1]The terms *reach* and *range,* as they are used here, were first introduced to the author by Calloway in her essay in McCain and Shyles (1994), pp. 56–59.

Reach

The term **reach** refers to the proportion of all relevant parties that can be connected to one another quickly and automatically by a communication technology. Ideally, maximum reach is attained when any relevant party anywhere can communicate with any other party, as might be possible in a perfectly operating worldwide telephone or postal system. In our Gulf War example, the worldwide reach of the broadcasts was impressive. But the concept of reach also applies to more common situations, and it relates to communications among the production team as well as communication with the audience.

For example, in producing a multicamera remote of a golf tournament, one group of relevant parties is the field crew, including the commentators, who may be deployed at some distance from one another and from the greens they must cover but must be instantly informed about which player's shot will be taken next and when others must be taped for later broadcast. To keep informed, the camera operators and other production personnel, many beyond earshot and out of sight of one another, need access to the director's voice. If voice contact is lost, the operation can grind to a halt. Hence, reach, at least in terms of voice contact, is critical.

Range

The concept of **range** refers to the different *types* of information (data, voice, live-action video, taped replays, and so on) that are handled by a given system. For example, the telegraph has great reach; however, since it uses Morse code but no voice or video, it has limited range. Likewise, the conventional telephone system has great reach (now made even greater by the advent of the cellular telephone), and it also enjoys greater range than the telegraph, since it can handle data transmissions such as e-mail and faxes in addition to voice communication. However, if the phone system suddenly became capable of transmitting live, real-time video—say, through the addition of handsets with video screens—we would say that its range had increased significantly.

The golf tournament example shows how the concept of range helps in understanding field production. Imagine a commentator having to describe different golfers' shots from distant locales. The commentator may be cued by a director's audio feed that it is time to describe actions at the ninth tee or the twelfth green, but audio alone would not be enough to show the commentator the action; that is, the audio feed can *reach* the relevant party, but it lacks the *range* to provide the necessary information, namely, the video feed the commentator needs to see to be able to report intelligently. By adding a line monitor with the video feed, we increase the range of this communication system enough to make it fully functional.

Future growth in the range aspect of communication systems lies in adopting systems with digital platforms. By reducing all forms of messages (data, voice, and video) to series of 0s and 1s, digitalization makes them potentially universally compatible with one another. In theory, digitalization of information may eventually enable messages to be integrated and shared in all formats. The rapid rise of multimedia technology is only a hint of what the future will bring.

Interactivity

Finally, **interactivity** refers to the ability of a communication system to permit users to encode and decode messages simultaneously in real time. The telephone is an obvious example of interactivity, but how does the concept apply to video? Consider "The Larry King Show" on CNN, an interview show in which the host and guest take phone calls from viewers. This program has impressive reach (it is broadcast to dozens of countries) and some range (because it includes the viewers' telephone calls, though it does not carry video of the callers). It also has a limited amount of interactivity: it broadcasts calls in real time, providing live communication between host and caller, but only for one caller at a time. In the golf tournament example, interactivity in voice communication is provided by headset or phone connections for all relevant parties who must speak to and hear one another during a broadcast, including the field crew, commentators, the director, and other personnel at the home base.

Keeping these three basic concepts in mind can help you plan and execute both live and taped field shoots. The proper configuration of communication systems linking relevant parties to one another in the field is essential for producing coherent field productions. In the remainder of this chapter, we will examine the major categories of field productions: the multicamera remote type of field production alluded to in the golf tournament example; electronic news gathering, as in the Gulf War example; and electronic field production, such as might be used for a corporate video or an interview of a government official. These categories differ in emphasis, but they should not be viewed as entirely distinct from one another. Many of the same principles apply to all of them, though each approach has unique qualities that make it appropriate for different production situations and program goals.

ELECTRONIC NEWS GATHERING

Roughly speaking, **electronic news gathering (ENG)** refers to that type of field production used for on-the-spot daily news coverage. Since news events happen in different locations without prior warning, ENG production is often marked by rapid response to fluid situations and by tight deadlines. In the news business, it is essential to be poised for mobility, to get the scoop, and to be first with the late-breaking story. As a result of severe time constraints, ENG production often has relatively rough (though still air-quality) production values, including hand-held camera shots; imperfect lighting; simple, often unplanned blocking; and less than optimal audio. These imperfections are overlooked when the story is dramatic enough, however. For instance, when a Gulf War reporter described bombs exploding outside his window, no one expected the audio to be flawless. On the other hand, difficult conditions are no excuse for sloppy work, and the quality expected in today's news operations is generally very high.

The ENG Mobile Unit and Equipment

The purpose of the ENG mobile unit is to enable a crew to move and deploy video equipment quickly and efficiently to the site of a fast-breaking news story. At the simplest level, the ENG mobile vehicle may be nothing more than a car into which a camcorder and microphone have been loaded, with a reporter and an assistant who doubles (or triples) as driver and audio/video operator. Under more ideal conditions, the ENG crew will have three members (including an engineer who can be stationed inside the vehicle for operation functions and security needs), and the vehicle will be a van or minibus equipped with all of the equipment needed to cover a story, either live or on tape, from the field under varied conditions (Figure 10.2). The van or minibus is a good choice because it has enough room to carry what almost any situation requires while remaining small enough to park and maneuver with relative ease. Table 10.1 lists the typical contents of a fully loaded mobile unit.

TABLE 10.1

Typical Contents of an ENG Mobile Unit

Cameras: at least one, preferably two

Camera tripods, body mounts, and/or shoulder braces

Lenses and lens shades

Flags and reflector boards

White balance cards and test equipment to set up and adjust transmission signals

Complete light kit (two-instrument minimum; three are better) with extra bulbs

Microphones: several wire and wireless lavs and hand-helds, as well as a shotgun mic for more distant pickup, all with windscreens

Portable audio mixer and headset for monitoring audio levels

Batteries

An additional power source, such as a generator powered by the vehicle's engine

Two-way radio for communication with the home base

Walkie-talkies for on-site crew members

Microwave transmitter

Satellite dish for live transmissions (optional)

In a well-outfitted vehicle, all the items are arranged in an orderly manner to make the setups and strikes efficient. Panels of peg board may be fixed to the inside wall of the vehicle for securing cables and miscellaneous items (gaffer's tape, clamps, extension cords, fishing line, and so forth). Custom-made containers with safety straps for securing cameras and other major items may also be attached to specific locations in the van. In the glove compartment, it is a good idea to keep a flashlight and road map, as well as any police permits the crew may need to gain access to restricted areas.

Signal Transmission and Relay Facilities

A common method of delivering video signals (either live or on videotape) from a field location to a home base is through the use of microwave radio energy, which provides line-of-sight transmission from an antenna mounted atop the ENG van (see Figure 10.2). The effective distance for microwave relay signals is roughly from thirty to a hundred miles, depending on signal power and terrain conditions. Once this distance is exceeded, a **repeater station** is used to boost the signal for another trip. With a series of about thirty repeater stations, television signals may be transmitted from coast to coast.

To link an ENG camera with a home base, two connections are needed: (1) a connection between the camera and the antenna atop the mobile van and (2) a link from the mobile van to the home base (Figure 10.3). To link the camera with the van, you can use a physical cable. In cases where this is not possible, you can connect to the camera a battery-powered microwave transmitter with

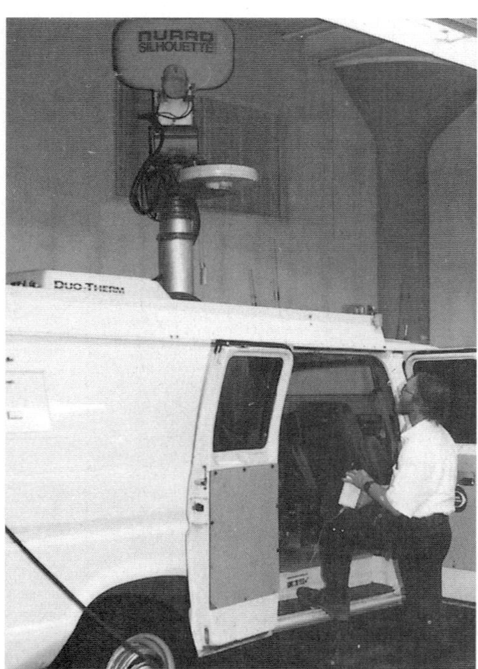
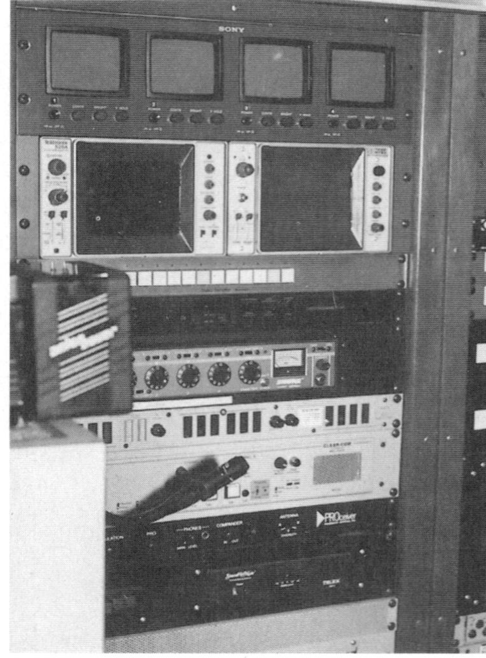

Figure 10.2 *An ENG van, with a microwave transmitter on top and ample video equipment stored safely inside.*

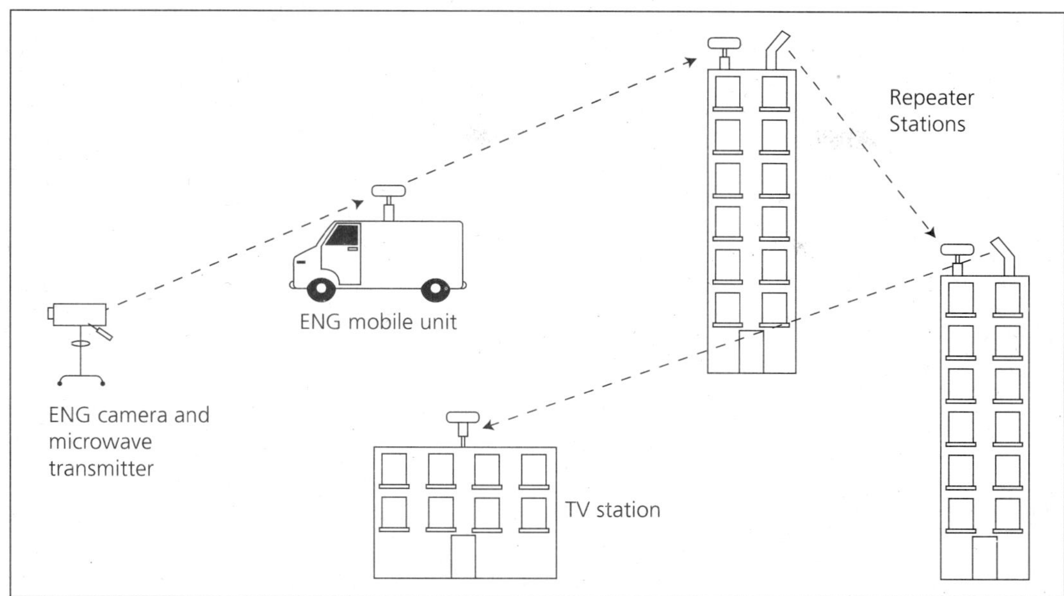

Figure 10.3 *Typical microwave links in an ENG operation.*

about a one-mile range. For reliable operation of this microwave transmitter, a clear path must be maintained from the camera to the van's antenna.

The link from van to home base is provided by the van's microwave transmitter. Sometimes either the distance is too great or the line-of-sight is too obstructed to permit successful transmission from the van directly to the home base. In such cases, repeaters are used to get the signal back to the home base for tape or broadcast. In most cities, repeaters are conveniently located at several geographical high points, atop tall buildings or hilltop towers. To get the best transmission, experienced crews carry a list of locations that have worked well in the past. Sometimes helicopters or tethered blimps, equipped with portable towers, can hover over news or sports locations to serve as repeater stations. A cheaper, though perhaps less reliable way to get signals to receiving antennas is to simply bounce them off a nearby building.

When microwave delivery is not feasible because of distance, power, terrain, or other limitations, ENG transmissions can be sent using a satellite uplink (Figure 10.4). Since the mid-1980s, **satellite news gathering (SNG),** as it is called, has extended the reach of ENG operations by using satellite **transponders,** which are orbiting microwave receiving/transmitting stations. A satellite uplink aboard an ENG van is aimed at a preassigned transponder aboard a satellite traveling in a geosynchronous orbit 22,300 miles above the earth. (A satellite's orbit is geosynchronous when the satellite stays above the same spot on the earth's surface throughout its orbit.) Aboard the satellite, microwave radio signals received by the transponder are converted to another frequency to avoid interference or jamming problems and then sent back to Earth. Since signals from orbiting satellites come from more than 22,000 miles away, the coverage pattern, or **footprint,** blankets about a third of the earth's surface. Because the coverage pattern from a satellite is so great, satellite communication is termed *distance insensitive.*

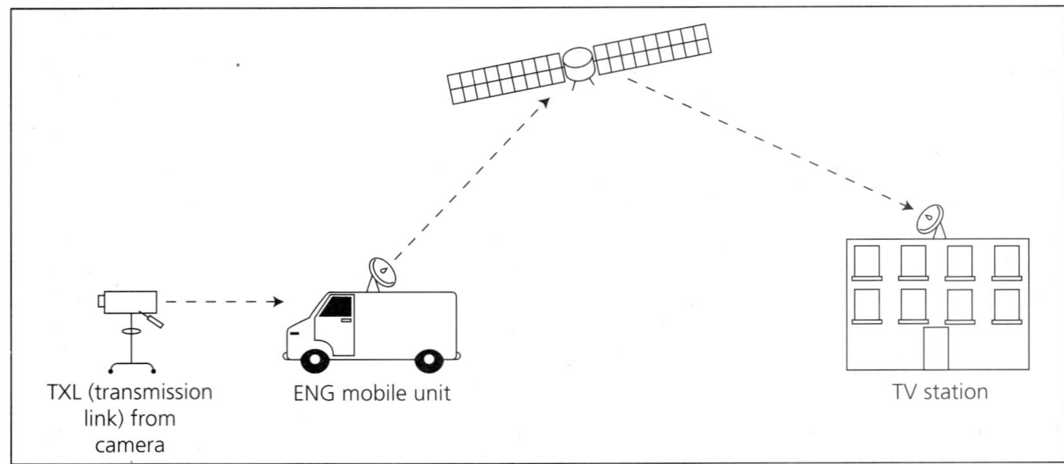

Figure 10.4   *ENG connections using a satellite uplink.*

After using signal-strength meters to determine the best position for the satellite uplink, an engineer in the field immediately begins sending test and tone signals (usually video color bars and a 1,000 Hz audio tone) so that the home base can establish and adjust its connection. In addition to providing a live television feed to the home base, satellite transponders are used to set up voice communication between field crews and selected personnel at the home base. This connection includes both telephone links and an **interruptible foldback (IFB)** circuit, a voice channel that enables the director to break into a program's audio feed to talk directly to the field reporter through the reporter's earpiece. IFBs also permit news anchors and other personnel to talk to field reporters about upcoming segments. In this way, satellites help to establish the interactivity needed for a successful production. However, the time delay in these satellite-transmitted conversations is about a quarter-second, a lag that is sometimes irksome to the participants.

The ENG Preproduction Stage

Before leaving on a field assignment, the ENG crew should be thoroughly prepared. To paraphrase an old adage, *it is better to have gear and not need it than to need it and not have it*. In line with this advice, most serious ENG operations keep a checklist of the most often needed items to make preparation routine. In addition, it is useful for each crew member to have a press pass to enable him or her to cross police lines.

An *information log* listing the addresses and locations of valued conveniences is also handy. The log might include the locations of bathrooms and working phones, as well as names and numbers of key field contacts (the mayor's secretary, the police commissioner, and so on). It also helps to know key locations for getting the clearest line-of-sight microwave relay for the mobile van. In many ways, the information log serves the same purpose as the production book kept by the competent producer.

Cameras and Tape   Before and during the trip to the shooting location, you should discuss with the crew the objectives of the assignment. Upon arrival, unpack and set up the equipment you intend to use. As far as the camera is concerned, if time and the nature of the job permit, use a tripod to steady your shots. Remember to white balance the camera even if you have just done so at a previous location. The most popular cameras and tape formats for ENG operations currently tend to be Betacam (½ inch SP or M-II) and the older ¾-inch formats, although in some cases the S-VHS and even Hi-8 formats are used. Depending on what you have, be sure to bring enough batteries and tapes in the proper format to carry the day.

Before shooting begins, always record about ten seconds of tape before cueing talent to ensure that you have tape up to speed and to provide enough control track for later editing. Shoot an additional ten seconds of tape after each segment to give the editor enough control track to edit. When a tape cassette is finished, immediately label it. Identify each tape (and tape box) by date, time, and event covered to reduce confusion. Bring a marking pen and stick-on labels for this simple accounting task. If you don't want to record anything else on a cassette, remove the "record" tab from its back panel to eliminate the risk of taping over crucial footage.

Lighting   If the shoot is to take place outdoors in daytime using available light, take note of the location of the sun and, if possible, place the camera so that the sun is behind you. If that is not possible, use lens shades to offset less than optimal angles of the sun. If the sun is extremely bright, use flags to reduce or eliminate contrast ratio problems. Reflector boards can help fill in dark shadows. If it is windy, stabilize flags and reflector boards with clamps or with the assistance of utility crew to eliminate accidents and flickering effects. Better yet, try to use an area sheltered from wind and direct sunlight, such as the lee side of a building.

Rain may require the use of raincoats for cameras and other major equipment that needs protection. **Raincoats** are waterproof covers custom-fitted to cameras, recorders, and other key equipment. In the absence of custom-fitted raincoats, plastic garbage bags are a good substitute. Exercise extreme caution when using lights in the rain.

For night and/or indoor shoots where available light is not enough, you will need artificial lights. The main goal is simplicity: provide adequate light for air-quality video. This can often be achieved using three portable instruments either on stands or clamped to available surfaces. The lights should be tunable so they can be used in spot or flood positions.

Most portable light kits feature lensless tungsten halogen lamps with barn doors, scrims, and gels for controlling spill and intensity. Wooden clothespins (the two-piece, wire-loaded variety) are ideal for pinning filters and scrims to lamps. Pack a pair of heat-resistant gloves for safe handling. Bring lights that run on batteries in case AC power is not available. In some cases, you may have to settle for a **speed light,** a single light mounted on top of the camera. Although this alternative is not the best, it is sometimes all you have.

If possible, set lights according to the principles of good lighting design discussed in Chapter 4. Remember to light to create texture, depth, and acceptable contrast ratios. Lighting from above rather than below avoids unnatural, unflattering shadows. Use key, back, and fill light to achieve proper lighting effects.

If you can power only one or two instruments, you may be able to stretch your resources by using **bounce light,** that is, light reflected ("bounced") off a wall, ceiling, or other reflective surface. Bounce light increases the amount of fill and even backlight on a subject and reduces contrast range. To use bounce light, position the subject near a white-colored wall or a similar surface. If a wall is available but it is not white, you can tape white paper to it.

For outdoor night shoots, it is often desirable to find settings with illuminated backdrops. In reporting about the federal government, the White House and Capitol buildings are frequently used as backdrops, not only because they are important but because they are mostly white and are already lit. In the absence of prelit backgrounds, place the subject close enough to some background that your lights spill light onto it. If none of these choices is available, expect the subject to look cameo lit (not necessarily bad).

If you shoot indoors in the daytime using tungsten halogen lights, they will be incompatible with natural light in terms of color temperatures, and you will need to decide how to deal with the natural light streaming in from windows. Either draw the drapes, pull the shades, cover the windows with opaque paper, or coat the windows with filters that color correct the natural light.

Finally, for daytime outdoor shoots, it is sometimes desirable to augment natural light with artificial lights, for example, to boost flat light conditions caused by cloudiness or to offset the light variations that occur when a shoot takes place over several days. Under such conditions, use HMI lights to match the color temperature range of natural light.

Audio   Take audio levels as soon as possible, and shoot some test tape to make sure all systems are go. Use simple audio design—perhaps a single hand-held mic for the reporter's speeches. For two-person interviews, the latest production approach uses a wireless lav for the reporter and a wireless hand-held for the interviewee, but you can still do an adequate job sharing a single conventional hand-held mic. For longer interviews, another approach is to have reporter and interviewee each wear a lav.

You can also mount a shotgun mic on the head end of the camera and record all audio from there, but sound recorded in that way often lacks presence, and ambient noise can be distracting. Use this method only as a last resort, when crew or equipment is not available for other configurations.

For outdoor shoots, always use windscreens. Either before or after taping, record room noise or nat sound for twenty or thirty seconds so that you have some available for editing purposes, if needed. Nat sound can also provide nice background for audio sweetening, which will often include postproduction voice-overs from studio announcers or news anchors.

Finally, prepare to monitor audio through a headset and to ride the gain throughout the taping to capture the best possible sound.

The ENG Production Stage

Of course, no matter how sophisticated telecommunications technologies are on land or in space, nothing can replace the intelligent use of facilities. At the site of a newsworthy event, the reporter's and crew's journalistic experience, production and writing ability, editing knowledge, values, ethics, and people skills, and even the organization's policies, all influence the finished product. What principles and techniques can guide you in gathering air-quality footage? This section briefly outlines some key practices for single-camera field production to help you capture your story with the most professional results.

Sequential Thinking   Shot sequencing and visual continuity are natural by-products of standard multicamera studio production. The same action is viewed simultaneously by several cameras from different angles with different shot compositions, enabling you to cut from shot to shot without losing the normal flow of action. In contrast, in single-camera field production, matching action from different camera angles, positions, and compositions is not a natural by-product. Instead, segments must be shot at different times from different vantage points and edited into final form in postproduction to simulate matched action. Simulating matched action is an essential ingredient of **continuity,** or the smooth flow of uninterrupted action from shot to shot. In single-camera field production, continuity between shots must be fabricated; it is an illusion.

Creating matched action is possible only with careful planning. You must think sequentially, planning the entire sequence of shots ahead of time.

Jump Cuts   Without sequential thinking, one problem you may create is a **jump cut,** an unnatural transition showing an abrupt change in the subject's location or appearance. Imagine a sequence beginning with a wide shot of a woman wearing a hat and preparing to sit down in a chair. The shot includes the entire body of the standing talent with a full view of the chair beside her. The next shot shows a close-up of the subject already sitting, no longer wearing the hat (Figure 10.5). This sequence constitutes a jump cut because the viewer never sees the subject sit down or remove the hat.

Jump cuts are jarring to viewers because they telegraph the message that the viewer has missed part of the action. Jump cuts ruin the sense of continuity and smoothness. On special occasions that may be just what the director wants, but usually jump cuts should be avoided.

The way to avoid them is to shoot **overlapping action**. This means reshooting the same action again from a new camera position. For example, after shooting the move to the chair while in the wide shot, bring the camera in for

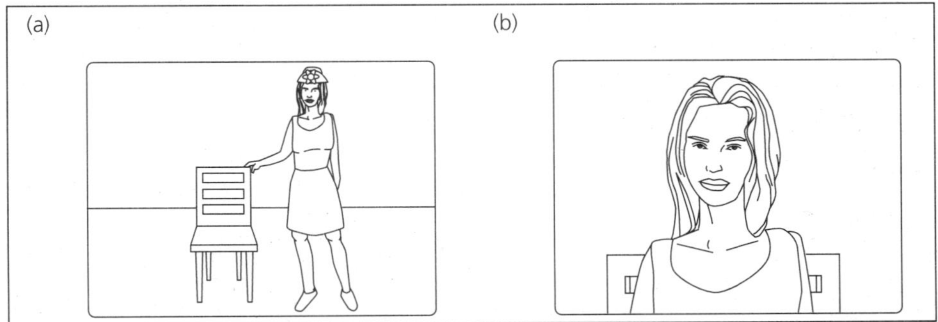

Figure 10.5   *A jump cut. The transition from (a) to (b) involves a sudden "jump" because we never see the person sitting down or taking off her hat.*

Code of Broadcast News Ethics

To help deal with issues of staging and other ethical concerns, the Radio-Television News Directors Association has developed a Code of Broadcast News Ethics, reproduced below.

The responsibility of radio and television journalists is to gather and report information of importance and interest to the public accurately, honestly, and impartially.

The members of the Radio-Television News Directors Association accept these standards and will:

1) Strive to present the source or nature of broadcast news material in a way that is balanced, accurate and fair.
   a. They will evaluate information solely on its merits as news, rejecting sensationalism or misleading emphasis in any form.
   b. They will guard against using audio or video material in a way that deceives the audience.
   c. They will not mislead the public by presenting as spontaneous news any material which is staged or rehearsed.
   d. They will identify people by race, creed, nationality, or prior status only when it is relevant.
   e. They will clearly label opinion and commentary.
   f. They will promptly acknowledge and correct errors.
2) Strive to conduct themselves in a manner that protects them from conflicts of interest, real or perceived. They will decline gifts or favors which would influence or appear to influence their judgments.
3) Respect the dignity, privacy, and well-being of people with whom they deal.
4) Recognize the need to protect confidential sources. They will promise confidentiality only with the intention of keeping that promise.
5) Respect everyone's right to a fair trial.
6) Broadcast the private transmissions of other broadcasters only with permission.
7) Actively encourage observance of this Code by all journalists, whether members of the Radio-Television News Directors Association or not.

Reprinted by permission of The Radio-Television News Directors Association.

the close-up, then ask the talent to repeat the action of sitting down and removing the hat. Then match action in the editing suite by picking the frames in the medium shot and close-up that are most alike, and edit them together.

Of course, in news coverage, it is not always possible to control actions to obtain overlapping footage. Some news groups prohibit **staging,** defined strictly as any act performed specifically for the camera. A more liberal definition permits having subjects repeat actions for the camera as long as those actions would normally have occurred in the camera's absence. Depending on your news outlet's policy, if a repeating action is the subject of a story you are shooting and you need several versions of it from different angles, you may have to wait until it comes around again to get overlapping footage. Or, if you are covering a unique event not likely to be repeated, such as a building demolition, and you wish to record it from more than one angle, you will have to shoot it simultaneously with several cameras.

Cutting on Action   To maintain smooth flow of motion, it is also best to **cut on action**. Think again about our example of the woman sitting down on a chair. Just as she is about to settle into the chair, you could cut to the close-up from the new camera position, capturing the subject an instant before she makes contact. Cutting on action results in smoother transitions because the viewer is more involved with following the action than with the edit itself or the camera's new position.

Cut-ins or Inserts   After using an establishing shot to set the scene, it's a good idea to feature close-ups that carry forward the main action of a story. The **cut-in** or **insert** is a close-up that captures a key moment of visual interest to drive home a story's main point.

Imagine you are covering an airport reunion of a soldier with his family after a long tour of duty (Figure 10.6). You start with a wide shot of the terminal, followed by a medium shot of a specific waiting area. The camera then captures a group shot of the soldier's anxious family as the arrival is announced, closing on his daughter playing with a rag doll. When he arrives at the gate, you carry a two-shot of the first hug between the soldier and his daughter. At this point, you *insert* (or *cut in*) a brief close-up of the rag doll carelessly slung behind the soldier's back to symbolize the child's joy in reuniting with her father. Then you cut back to the two-shot once again. If the story is carried forward to the next tour of duty, a cut-in of the departing soldier might include a close-up of his hands snapping a suitcase shut.

Cutaways   Cutaways are shots that lead the viewer's attention away from the main scene, often to related action outside of it. Cutaways provide bridges or transitions to subsequent scenes. For example, after the shot of the soldier and his daughter hugging, a cutaway might be a shot of a smiling flight attendant watching the action from nearby. Later in the story, when the subject is getting ready to leave on another tour of duty, after the cut-in of his hands snapping the suitcase shut, a cutaway shot might show his mother phoning the neighbor who will drive him to the airport.

Cutaways such as these, which relate closely to the story content, are sometimes called *motivated* cutaways. In contrast, *unmotivated* cutaways feature more neutral content. In general, motivated cutaways are more interesting than unmotivated ones. But even unmotivated cutaways can provide transition to the next scene. A simple example of an unmotivated cutaway used for transition is a wide exterior shot of a building where the next scene is to take place.

In addition to supplying transition, both cut-ins and cutaways do several things: they add pertinent visual information, they drive the story forward, and they compress time. Further, from an editing perspective, cutaways provide a way to hide jump cuts. For example, cutaways can be used to connect interview segments that were recorded at different times.

The Reverse-Angle Shot   One of the most common cutaways is the **reverse-angle shot**. In a single-camera shoot of an interview, for example, the camera may be set up to capture a shot of the subject over the shoulder of the interviewer. From this position, the camera can zoom past the interviewer to feature a one-shot close-up of the subject. When the interview is done, it is often helpful for editing purposes to shoot additional footage of the interviewer

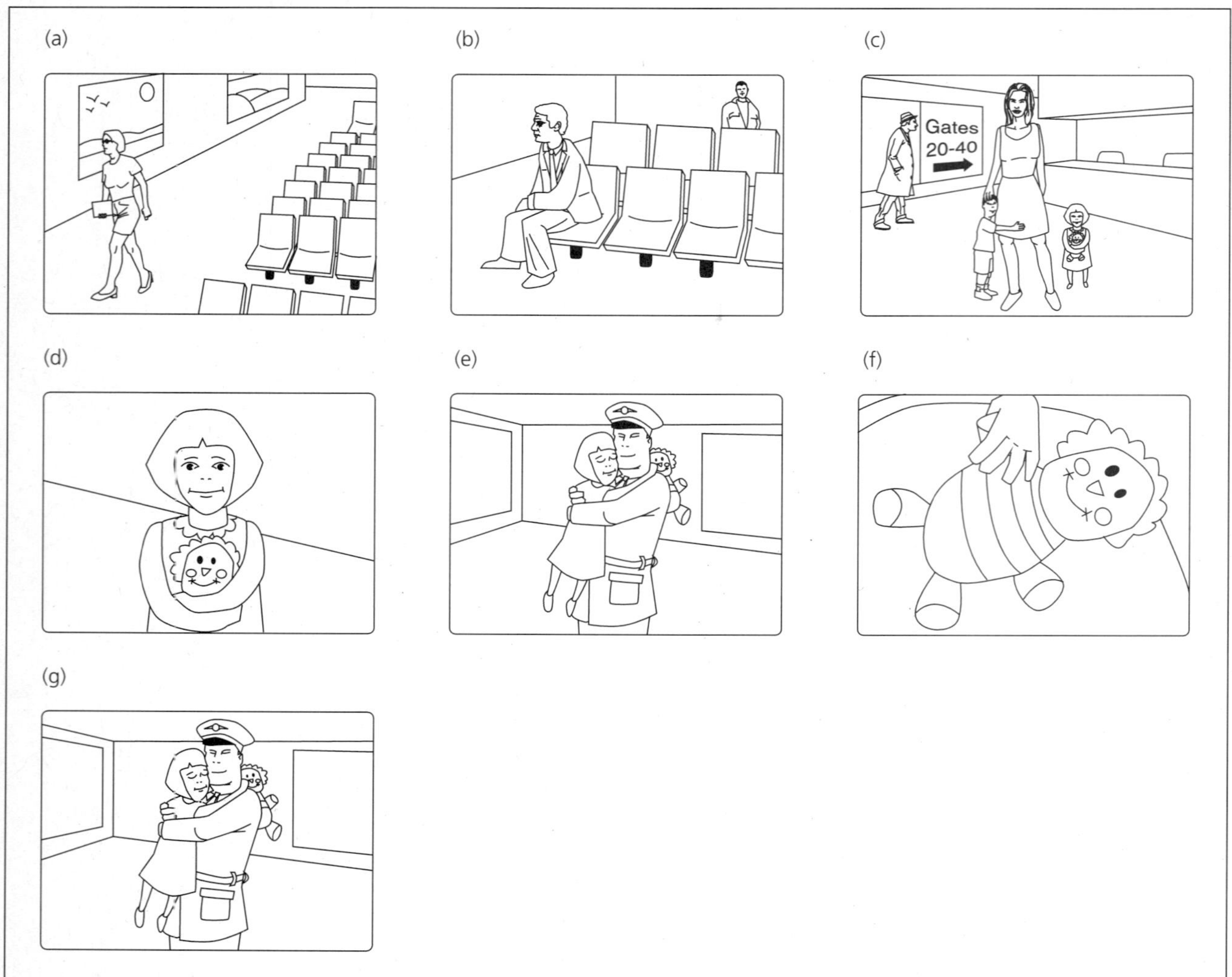

Figure 10.6 *A shot sequence illustrating a cut-in or insert. (a) Start with a wide shot of an airline terminal. (b) Trim to a medium shot of a specific waiting area. (c) A family waits in an airport terminal for a soldier returning from duty. (d) Close-up of the daughter playing with a rag doll. (e) As the soldier arrives, a two-shot of the first hug between the soldier and daughter. (f) Insert of a brief close-up on the doll. (g) Return to the two-shot of the hug.*

from the opposite perspective, or reverse angle—that is, from behind the subject (Figure 10.7). From this position, you can zoom past the subject's shoulder to get a one-shot close-up of the interviewer, either pretending to listen to the subject's answers or repeating the questions as they were asked during the actual interview. In editing, these reverse-angle shots can be used to provide smooth transitions between different portions of the interview. They give the editor the flexibility to arrange responses in a different sequence, if desired, and to eliminate portions of the interview that are not wanted. Of course, from an ethical standpoint, you should ensure that the views of the subject are not misrepresented. Journalistic accuracy should be a constant concern.

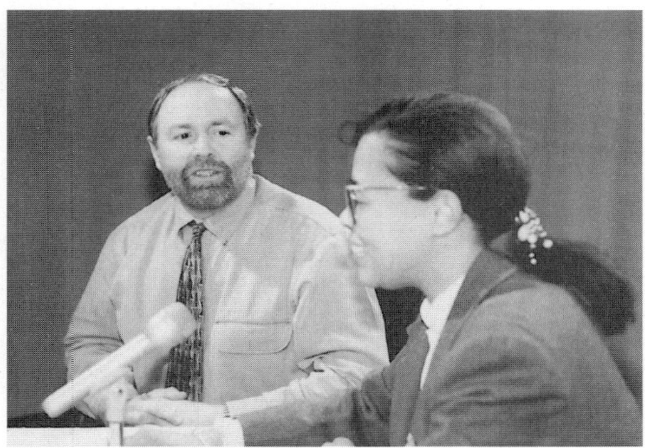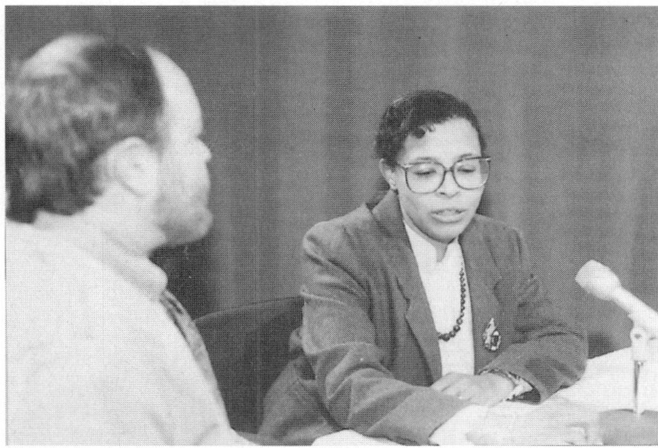

Figure 10.7   *Sequence illustrating a reverse-angle shot in an interview. (left) Shot of the subject over the interviewer's shoulder. (right) Reverse-angle shot of the interviewer from behind the subject.*

Directional Continuity   Nothing confuses an audience more than watching footage of subject movement that illogically changes direction from shot to shot. The most common examples of this shoddy production practice occur in covering horse races or parades. The action moves across the screen in one direction, followed immediately by footage of the same subject inexplicably moving in the opposite direction (Figure 10.8). To eliminate such *false reversals*, as they are called, follow the axis-of-action rule.

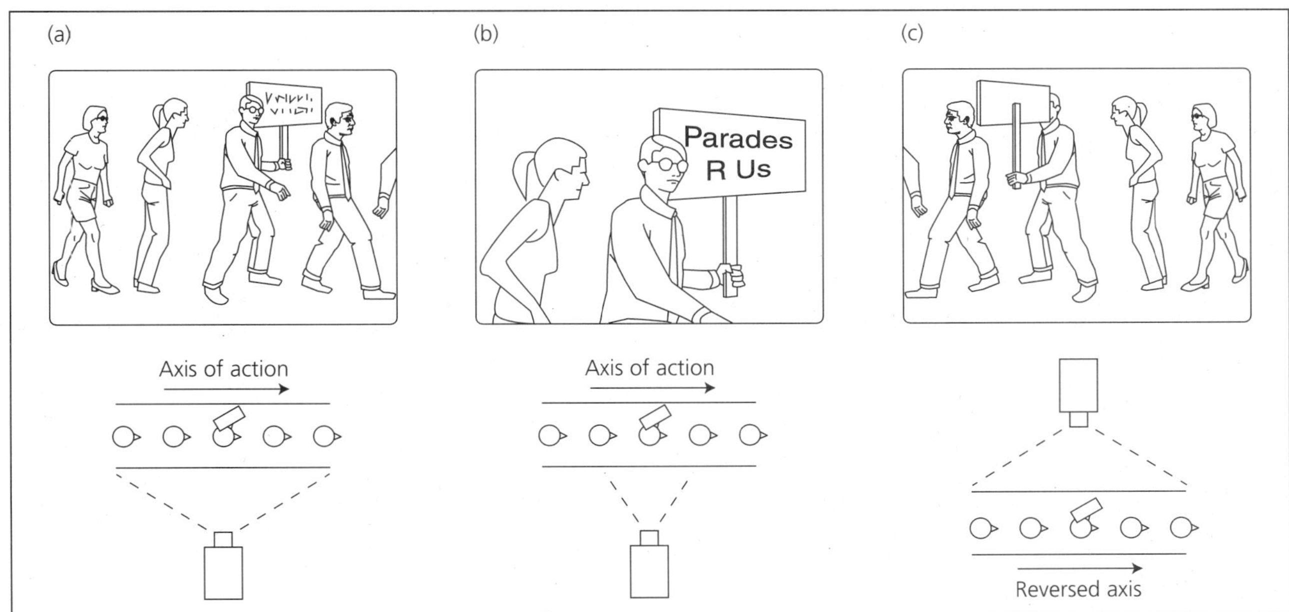

Figure 10.8   *Sequence illustrating the axis-of-action rule. To cut from the wide shot in (a) to the close-up in (b) is fine, because the axis of action has not been violated. To cut from (b) to (c), however, would be a false reversal; the subjects would seem to have changed directions, when they have not.*

According to the **axis-of-action rule,** you should establish an imaginary line along which the main action flows and *keep the camera on the same side of the line for all shots*. If you are shooting a parade with marchers moving from left to right in your viewfinder, the axis of action is parallel to the plane of your camera lens. You can move the camera to a new position as long as you do not cross that line. Similarly, in an interview, the axis of action may be thought of as the imaginary line connecting the subject's and interviewer's mouths. To avoid awkward pictures when shooting over-the-shoulder shots and reverse-angle shots, keep the camera on the same side of that line.

Of course, sometimes it is not possible to restrict all camera shots to the same side of the axis of action. In our parade example, what happens if the police make you move to the other side of the street? In such cases, there are several ways to soften transitions between shots that change direction. One is to inject an intermediate shot (a cut-in or cutaway) that distracts the viewer and softens the change in direction. Handy cut-ins are the **head-on shot** and the **tail-away shot,** which in the parade example are shots of the parade group either approaching or moving away from the camera, respectively (Figure 10.9).

Another remedy is to take the viewer along for the ride. For example, in the case of a horse race, when the horses round the turn (thus changing direction), you may be able to follow them with a high-angle shot so the audience sees the change happen and accepts it readily.

Matching Camera Angles   The ENG camera operator must be sharply aware of the need to match camera angles when shooting segments that will be cut together later. Matching camera angles means following the line of action between two related shots so that when they are edited together, they appear consistent. For example, if you shoot a basketball player being interviewed by a shorter reporter, it is important to match the shot of the player with the reverse-angle shot of the reporter. If the camera angle on the subject is low, making the shot appear as though the interviewer is looking up at the subject, the reverse angle of the interviewer should be set so that the subject appears to be looking down at the interviewer along the same axis (Figure 10.10). If such consistency is lost, segments cut together can look jarring to viewers.

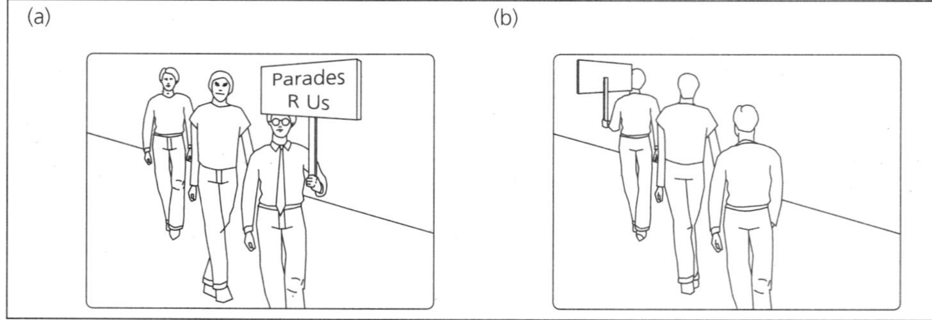

Figure 10.9   *Shots that can help smooth out a transition involving a change of direction. (a) A head-on shot. (b) A tail-away shot.*

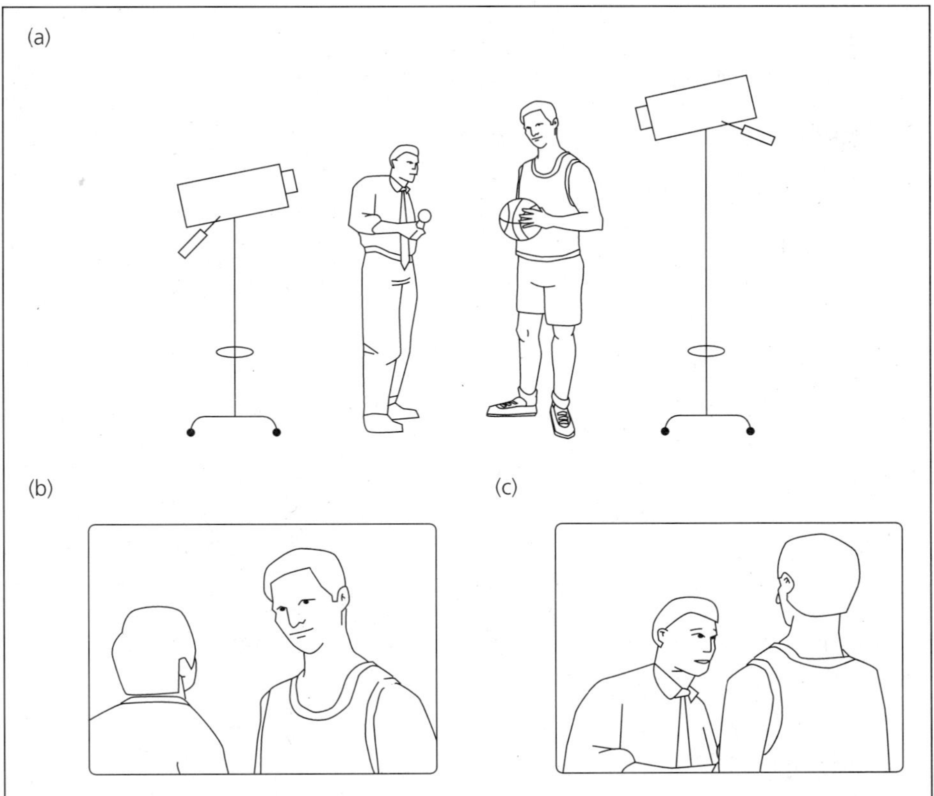

Figure 10.10   *Matching camera angles. (a) Positioning of two cameras to capture a stand-up interview between a reporter and a taller basketball player. Note that one camera is pedded high, the other much lower. (b) The OS shot of the player being interviewed. (c) The reverse-angle shot of the reporter, which is angled down to the same degree that the player's shot is angled up.*

Panning the Camera   Rough, unmotivated camera panning irritates viewers. Excessive panning takes attention away from the focus of your story and leads to dizziness when done in the extreme. In general, the rule is: *pan the camera to follow movement.* Very rarely is panning on a static scene justified. It is especially annoying to pan back and forth as though painting a fence. *Without question, it's better not to pan at all than to pan badly.* For more specific guidelines, check the accompanying list of Professional Pointers.

Safety First   For all camera work in the field, if you bring the camera close to the action, train yourself to shoot with both eyes open. For example, if you are stationed on the sideline of a football game, use one eye to monitor the viewfinder and the other to watch the area around you. This way you can anticipate where to shoot next. More important, you will be better able to tell when a 300-pound defensive tackle is about to crash into your camera.

ENG in Extreme Weather   Cold, wet, and windy weather conditions are a challenge to the ENG field crew. Covering major storms, hurricanes, tornadoes, and floods therefore requires special planning. If everyone has evacuated a dangerous weather zone and you are assigned to cover the impending

PROFESSIONAL **POINTERS**

Panning the Camera in Field Production

- *Avoid shaky, uneven movement.* Work with a tripod whenever possible. If the camera is hand held, use your body as if it were a tripod, keeping your elbows tightly at your sides and your feet spread slightly wider than normal for added stability. To smooth out your panning motion, practice!

- *Begin your pans by placing your feet in the position your body will be at the end of the pan, and twist your body to bring the camera to the beginning point.* Then, when the shot begins, slowly untwist your body along with the action you wish to cover (Figure 10.11).

- *To increase stability, lean on a steady object such as a tree or car roof.* You can even sit down.

- *Slow down.* No matter how slowly you think you are panning, it will always appear faster than you thought. Therefore, whenever you can, make the pan even slower than you think you need.

- *Work at making the panning motion truly horizontal.* If pans of the camera must leave the horizontal (such as in following a plane taking off), keep the subject matter oriented in roughly the same portion of the screen space throughout the pan.

- *Begin and end panning shots with static footage of the subject.* This static footage will give the audience a chance to grasp the subject matter in the frame and recover from the motion before a new shot or scene begins. It is jarring to the audience to cut from a still shot to a panning shot of different subject matter, and vice versa.

- *When possible, allow the subject to enter the frame before you pan to follow the action.* This is known as *anticipation*. Likewise, when you finish covering an action, it is perfectly acceptable to steady the camera and let the subject exit the frame. Editors love this because it gives them a natural edit point for creating neat transitions between sequences. This approach is possible even when subject motion cannot be controlled, such as in an airplane takeoff.

- *Build your pans.* Let the high point of subject action occur as the subject fills the screen. For example, you might position the camera at the finish line of a race. Also, begin covering action at a sharp angle to the subject rather than at a right angle. This gives viewers a chance to recognize the subject and orient their attention.

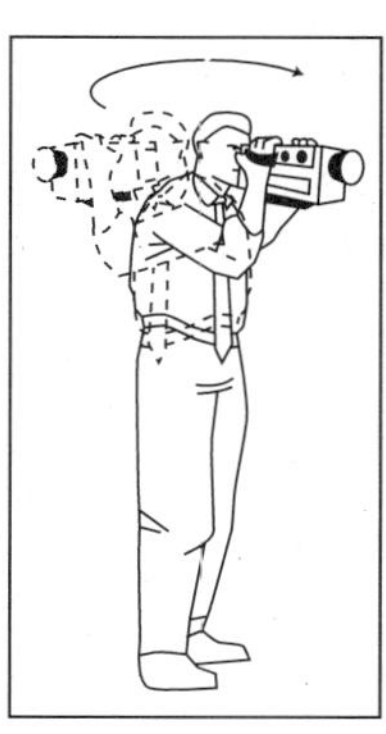

Figure 10.11   *To pan a field camera with minimal jerkiness, start with your body straight, feet spread, elbows close to your sides, and the camera in the position it should be at the end of the pan. Then twist your body to move the camera to the starting position. To accomplish the pan, slowly untwist yourself.*

storm and its aftermath, you will need to adapt to extreme conditions while keeping yourself and your equipment safe. A good first principle is: *Use common sense, and don't be a hero.* Walking through a three-foot-deep puddle in semidarkness to get a key shot may be a bad idea if your next step is onto a hidden power line.

If the power is out, you will have to function without conventional electricity. You may also have to do without access to food or water for several days. Under such conditions, the ENG shoot soon begins to resemble a rustic camping trip or even a military operation. Plan accordingly. In addition to food and several changes of warm clothing, pack enough blankets, pillows, towels, and toiletries for several days. Depending on the area you are in, a snakebite kit may be a useful addition to standard first aid supplies. Flashlights and extra batteries are always useful. A multiband radio (battery powered) that can receive weather channels is also an obvious asset.

As for equipment, the main objective is to *keep it dry*. Besides the raincoats already mentioned, additional waterproof barriers are invaluable. Plastic, waterproof storage containers for tapes (and food) will keep things dry until you get back to the station.

When shooting in rain and wind, put the wind at your back to keep water off the lens. Keep soft, dry towels handy to wipe the lens if necessary. If gale forces get too rough, shoot from inside the van for stability. Again, don't be a hero if the wind starts tossing street signs around as though they were Tinkertoys; at that point, you may already have gotten enough storm footage to permit you to retreat to a safe zone. A shelter is a good place to shoot additional storm footage and an even better place to begin shooting aftermath segments by interviewing people who are waiting out the storm. The human-interest segment of the story can begin or continue there.

In extreme cold, keep camera batteries warm to extend their usefulness. Low temperatures can cut battery life to less than half. Store them under your clothes, next to your skin, to keep them warm until right before you use them. Use the electrical generator in your van to recharge them.

Going from a cold to a warm environment can promote condensation, which can paralyze recorder heads and other mechanical parts of equipment. Some recorders have sensors that shut them off when the moisture level gets too high. If this happens, you simply need to wait until the machine dries out.

Your job does not end when the storm ends. You still must shoot dramatic exterior aftermath footage to illustrate the storm's effects. Then you must get the story back to the station. If power is out, or if you are out of range of a microwave link, you will need an alternative method of reaching the station with your story. You may need to use video relay equipment at a nearby telephone company outlet.

Live ENG Communication Systems   As we saw in the earlier examples of the Gulf War and the golf tournament, various communication systems are used to coordinate live field productions. These systems are best understood in terms of how their reach, range, and interactivity aspects connect different groups of relevant parties. The most obvious receivers of live video from the field are the end users, namely audiences, who see broadcast or cable feeds as part of regularly scheduled programming and special reports. However, earlier in the process, various production personnel exchange messages to help them produce and deliver finished programs. These include producers, directors, reporters, and crew members in the field, as well as directors, news anchors, and others at home bases, who are often out of direct earshot and line-of-sight of one another.

Table 10.2 describes some of the communication devices used to connect relevant parties to one another, and Figure 10.12 diagrams the connections they establish. Without these systems, smooth delivery of field coverage would be impossible. When working, these systems are invisible to the end user. However, when breakdowns occur during live transmissions, such failures quickly become apparent to viewers in the way the end-product looks; viewers experience dead air, mismatches between audio and video, uncoordinated or missed coverage of key moments, embarrassing cutaways with apologies, and so forth. It is therefore little wonder that good field producers are committed to building redundancy into their communication systems to avoid catastrophe. Though these systems are especially important in live shoots, they are also used in many taped productions.

The double-headed arrows in Figure 10.12 indicate real-time, interactive (two-way) communication between relevant parties with a given communication system (voice, video, or data). The single-headed arrows indicate one-way message flow between parties. For example, the cell connecting off-air home base directors with on-camera field reporters contains an IFB entry because directors at the home base use IFBs to talk to on-air talent in the field. This link has a single-headed arrow because field reporters can hear messages from the directors but cannot use the IFBs to talk back to home base personnel. However, field reporters can send voice messages to home base personnel via the audio portion of the video signal. Hence, it is through two separate channels (therefore some redundancy) that interactive voice communication is established between these two relevant parties. If one of the two channels is lost, some communication is still maintained, a comfort to the field producer when Murphy's Law kicks in.

TABLE 10.2

Communication Devices for Field Production

Device	Function
Walkie-talkies	Establish voice contact among crew members who are out of earshot and/or line-of-sight of one another.
Scanners	Permit news crews in the field to monitor police and fire department activities.
Cellular phones, two-way radios	Connect members of field crews and home bases with one another.
Headset intercoms	Link field director and crew members.
IFBs (interruptible foldback circuits)	Allow directors and other personnel to talk to on-air talent through an earpiece worn by the talent during a live telecast. The IFB system is commonly called a *program interrupt (PI)*.
Video line monitors	Enable field personnel (both on and off camera) to see and hear a live feed of the program being transmitted from the home base.
Battery-powered televisions	Serve the same function as line monitors.
Camera-cable and wireless private lines (PLs)	Allow remote truck personnel and production crews to communicate with one another, as well as with other remote control rooms and home bases during all production phases.
Pagers (beepers)	Provide data communications via radio transmission to alert personnel to contact someone.
Megaphones	Enable field directors to communicate with nearby crew in the field during preproduction phases.

Tape Logs   Much ENG-style raw footage is composed of interview segments, cover shots, cutaways and reverse-angle material, event coverage (also called *actuality* footage), and *stand-uppers* (direct-to-camera shots of the field reporter talking). Therefore, editing is often required to complete the program. Since time is of the essence in news, it is best to be poised for the postproduction editing from the start. For taped segments, this means creating an accurate log of each tape as it is being shot, if possible, or shortly thereafter. The log should identify each piece of footage in order, including its length in minutes and seconds, with SMPTE time code information if available. These accounting procedures are an invaluable aid to the postproduction process.

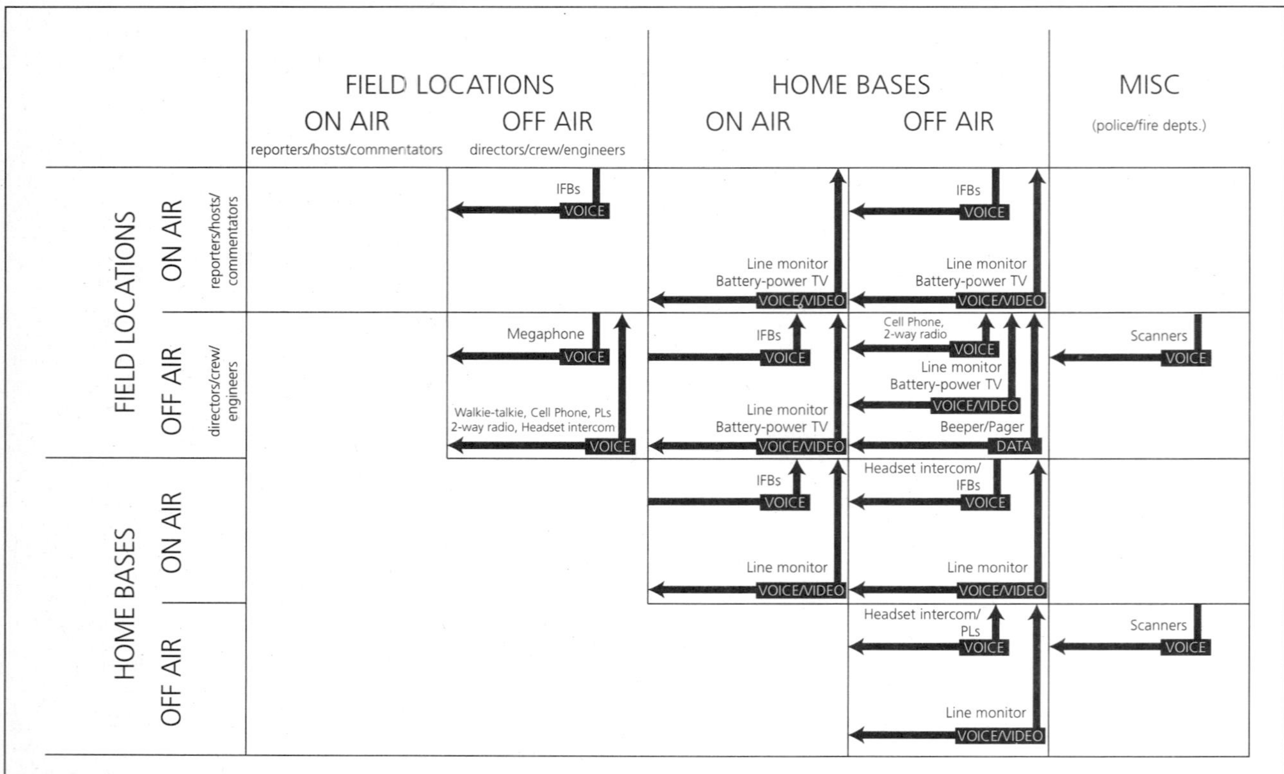

Figure 10.12 *ENG communication systems. The diagram shows how selected communication devices establish links, either one-way (single-headed arrows) or two-way (double-headed arrows), between field locations and home bases.*

The ENG Postproduction Stage

Once production is completed, you will need to strike all of the equipment quickly, efficiently, and safely before moving on to the next location. You may wish to hold a postproduction meeting (a *debriefing*) to discover ways to do better the next time. Usually the producer, director, technical manager, and assistant director meet to discuss any problems that were encountered. Sometimes the technical manager's job includes feeding a daily "trouble report" to the home base. Besides helping to solve problems, the debriefing and reports build morale among the crew.

Other postproduction chores include thanking all of the personnel in the field. In addition to the crew, others who deserve thanks (either in person, by phone, or by letter) include all support personnel and contacts who helped you set up and execute the shoot. It is not only the polite thing to do but also wise, because the people you have just worked with may work with you again in the future.

Finally, for taped footage, the editing can begin, a process we will discuss in detail in the next chapter. If you have created an accurate log of each tape, the editing process will be greatly simplified.

ELECTRONIC FIELD PRODUCTION

Even stories that begin as breaking news can evolve into a different kind of field production. In 1994, on the night when football star O. J. Simpson was chased by the police and then arrested for murder, viewers saw many marginally lit, shaky pictures shot from helicopters. Later, however, when Simpson's trial began, much of the televised news coverage was not nearly so spontaneous or rough in production values. On the contrary, much of it was relatively routine and predictable. News crews camped for months, filing daily stories from the Los Angeles courthouse, the county jail, and the crime scene. Reporters presented carefully crafted feature stories for later broadcast. For example, some feature stories integrated retrospectives of Simpson's football career along with interviews with friends and family members of the victims.

For such stories, a more filmic approach is often taken, using a single camera with multiple setups. In particular, greater care is paid to such production values as talent rehearsals, camera and subject placement, blocking, lighting, sound, continuity, scriptwriting, and editing. Roughly speaking, the shift away from spontaneity and rough production values to more planning and higher production values distinguishes ENG from what has come to be known as **electronic field production (EFP)**.

Because of the advance-planning aspect of EFP, shoots can be better designed and more leisurely in their execution than ENG shoots. Also, EFP uses more sophisticated equipment. All in all, EFP aesthetics often match or exceed those in the studio. In this chapter, EFP refers to recorded single-camera production done in nonstudio locales. The multicamera remote (MCR), both live and on tape, is discussed later.

The types of programs produced using EFP techniques range from simple interviews done in people's offices to complex presentations shot in numerous locations. Commercials, corporate meetings, magazine programs, instructional, educational, and industrial videos, promotional and public relations campaigns, and even feature documentaries may all employ EFP techniques. EFP productions may be done in one location or many. They may feature almost no postproduction work, or they may use a wide variety of editing and sweetening techniques.

The EFP Mobile Unit and Equipment

The EFP mobile unit and equipment resemble those of the ENG operation, with some notable exceptions. For example, since there is no need for a live feed from the field, the EFP mobile unit dispenses with microwave and satellite transmission hardware. Instead, the surface of the van's roof can be used as a camera platform to get high-angle shots. Without the transmission hardware, more room is available for additional camera, recording, lighting, and audio equipment. Of course, the well-equipped van has customized storage rigs to store and secure all equipment during transport. AC power connections and an electric generator that runs on diesel fuel are added advantages.

Cameras and Tape   Cameras for EFP can be heavier, higher-end models for added picture quality, with tripods and shoulder-mount accessories to match. Additional equipment can include an array of lenses and filters, a jib for high, sweeping camera movements, a portable dolly, and tracking equipment for sophisticated camera movements on the ground.

Higher-quality EFP productions currently tend to use Betacam cameras with Betacam SP or M-II recorders. However, some work is done using S-VHS and even Hi-8 technology, though the latter two are often limited to in-house corporate and educational (closed-circuit) settings. The older ¾-inch videocassette recorders are still in use in some settings. All these formats can run on batteries. However, it is always better to use AC power when available.

Lighting and Filters   Depending on the nature of the shoot, EFP lighting equipment can be as sophisticated as that found in a fully equipped studio. In fact, if the shoot takes place outdoors, the equipment can go beyond what the studio environment requires.

Filters are particularly important for EFP. Video cameras are made to operate without the need for color correcting under lighting with color temperatures in the 3,200 K range (as produced by tungsten-halogen lights). Under all other conditions, some filtering is necessary to maintain proper color. Table 10.3 lists the most common filters used to match color temperatures with

---

TABLE 10.3

Common EFP Filters

Type of Filter	Function
Camera-mounted filters	
Neutral-density	Reduce the quantity of light entering the camera. Reduced exposure permits the use of wider lens apertures, thus cutting depth of field.
Fluorescent	Correct for the greenish cast of fluorescent lighting, bringing it to 3,200 K. FLB filters correct for fluorescent lights with color temperatures of 4,500 K. FLD filters correct for "daylight" fluorescent lighting in the 6,500 K range.
Amber	Correct sunlight to 3,200 K. Also correct artificial light in the range of 5,600 K (HMIs) to 3,200 K.
Light-mounted filters	
Dichroic	Correct quartz-halogen lights from the 3,200–5,600 K range to make them compatible with sunlight.
Window-mounted filters	
Amber filter sheets	Correct incoming sunlight to the 3,200 K range.

lighting conditions you are likely to encounter on field shoots, including camera-mounted filters (mounted either behind or over the camera lens), light-mounted filters (mounted or clamped in front of the lighting instrument itself), and window-mounted filters (mounted on windows to correct to 3,200 K the color temperature of sunlight entering the production area).

Besides controlling the color temperature, filters allow you to control the amount of light that enters the camera. By controlling the light level, you can vary the aperture size you wish to use, thus making it possible to control depth of field. When using filters to limit the amount of light entering the lens, you can use a **filter factor** to determine the amount of light lost due to filtering. For example, a filter factor of 2 cuts the amount of light entering the lens in half, which is the same amount that would be cut by reducing the lens aperture by one full f-stop. Using filters with known filter factors allows you to regulate exposure levels and thereby control depth of field, even under extremely bright conditions. For example, by using a neutral-density filter with a filter factor of 4, you can shoot at a lens aperture two f-stops wider than would be possible with no filter, resulting in shallower depth of field. For filters used in combination, remember to multiply filter factors to determine how many f-stops you have jumped.

Audio The audio needs of an EFP production can go beyond the simple arrangements of the typical ENG shoot. In addition to the requirements outlined for ENG, you may need to include booms, wireless mics with RF transmitters, fishpoles, and shotgun mics. It is also common to use audio mixing boards with headsets for riding the gain during EFP productions. Further, if you want audio foldback or playback, you will need to bring portable speakers as well as audiotape recorders. Of course, you must also supply enough cable to hook everything up, as well as gaffer's tape to secure cables.

The EFP Preproduction Stage

Perhaps one of the biggest differences between ENG and EFP shoots is the preproduction phase. The EFP crew has the luxury of scouting locations and planning all phases of the production before any shooting begins.

An EFP site survey should answer a wide range of questions, including the following: What is the site like? Can we get a location sketch? Is it indoors, outdoors, or both? If outdoors, where is the sun? How will that affect the shots we need to get? How will time of day affect the nature of light during each shooting day? Is it always this quiet? Is the ebb and flow of people through this area uniform? It is always an advantage to conduct the site survey close to the day the shoot will begin to simulate the conditions as accurately as possible. If the site survey takes place three months in advance, much may have changed by the time the shoot occurs.

As for equipment concerns, the following questions are important: Where are our camera locations? What audio and lighting equipment do we need? Is there sufficient electrical power? Do all the outlets actually work? Which cable runs are shortest? Does that door open all the way?

In terms of scheduling, ask the following: What are the shooting dates? Can we get everything we need in the allotted time? What do we do if bad weather (or a similar problem) throws off our schedule? What kind of cooperation will we get from the jurisdiction or the property owner for additional use of space and facilities?

Practical issues also need to be worked out: Is parking available? Lodging? Food? Bathroom facilities? Telephones? Who pays for all that? Do we have the necessary security and insurance coverage to ensure that our equipment will be properly cared for and protected from damage and theft?

Finally, in terms of legal issues, ask what contracts, permits, clearances, and release forms are needed, and then plan accordingly. These and countless other questions should be answered during the preproduction phase of the EFP shoot.

As for voice communications among the crew, the EFP operation can make good use of cellular and car phones to coordinate activities among crew members who will be arriving in separate vehicles. Once on location, crew members can communicate with one another with walkie-talkies. Battery-powered megaphones are also useful during rehearsals. Pagers, headset intercoms, and private lines may also become important during different phases of the production process.

The EFP Production Stage

An EFP shoot can mix all of the challenges of studio production with all of the unknowns of the field. Because of the varied types of programs handled in EFP, it is impossible to specify all the production situations you will encounter. In this section, we will look at some of the more common production considerations for several leading program genres.

Meetings and News Conferences   Government hearings, news conferences, and similar indoor gatherings present unique problems. The subjects you are there to cover may be quite expert in their professional fields, but they may lack on-air experience and may not appreciate your production needs. On the other hand, if they are poorly covered, their insights may be lost. It is your job to keep this from happening.

More specifically, the subject may wander from a fixed microphone position, causing you to lose key audio. Or the subject may refer to graphic materials that are simply not air quality, such as thinly lettered charts on white reflective cards. Speakers may sweat profusely under hot lights after refusing to wear makeup or forget they have promised to stay in the lighted area for proper video coverage. Or they may turn away from the camera altogether.

The area where the coverage is to take place may be inadequately lit and lack adequate electrical outlets. The room may be too large to be lit properly with the instruments you have. There may be room set aside for chairs to accommodate audience members but little or no space for cameras and lights. The room's decor—busy backgrounds, for example—may present extra problems for video coverage. The room may be echoey and boomy because of hard wall surfaces. The audience may be noisy. Other sounds, such as from plumbing and nearby traffic, may present additional audio problems.

Further, audience participation in, say, a question-and-answer period during or after speakers' presentations may require lighting in two directions, creating potential glare problems. To solve such problems, it may be necessary to find alternative camera positions. Moreover, you will need to decide how speeches by audience members will be captured. You can hang an area mic over the group if it is small, but if that is not feasible, you may need to use a fishpole mic carried by a production assistant. Another alternative is to use a mic stand for audience members to step up to when they ask questions. Or you can have someone at the speaker's podium field questions and then repeat them to obtain adequate audio coverage.

For all of these potential problems, preproduction planning is critical, but you will also need to arrive early at the site to begin working on problems that crop up at the last minute. If graphics will be used, try to prepare air-quality versions of them in advance. To integrate materials into the program coverage, get a sequence of the events or discussion topics in advance and ask the on-air presenters to inform you of any significant changes.

Documentary and Magazine Features   Feature stories for television are infinitely varied in content. Material can range from a hard-news, in-depth documentary, such as an exposé on police corruption, to a soft piece concerning the latest grooming trends at a dog show. Length can also vary, from a one-minute segment designed to fit into a larger show to an hour-long format. Regardless of the topic or length, here are some things to keep in mind to make the production run smoothly:

- Conduct a site survey and pre-interview with all key personnel— clients, on-air talent, production staff, and field contacts—before the first day of shooting.
- Finalize the script to reflect the approach you will take. Get it approved in advance by the client, and make sure it is understood by both talent and crew.
- Note the running time for each segment, and be careful to log each one for later editing. Label boxes and tapes so you won't confuse them with others you are using. As with ENG operations, remove the "record" tab from the backs of finished tapes so they don't accidentally get recorded over at later shoots.
- Make a quick on-site review of each segment to be sure you have air-quality program material for each. This means reracking some tape for every segment and rolling it in the camera or tape machine to ensure you have in fact gotten the footage you think you have gotten. Also, monitor the audio to confirm that sound was successfully collected.
- If you need to reshoot segments, be careful not to "burn out" your talent with too many retakes. There comes a time when you must decide that a certain level of performance is all you are going to get from someone. If possible, show the client the footage at a private meeting and offer your professional opinion. That is, in part, what you are being paid for.
- Don't forget to shoot transitional material for later editing, including master and establishing shots, close-ups, reverse angles, and cutaways. Also, record nat sound for later editing.
- Check footage you have collected against a checklist to make certain you have indeed gotten the things you need.
- If you need to return for more shooting at a later date, take snapshots of the sets and costumes, if any, so that they can be replicated the next

time. Similarly, if shooting outdoors, note the time of day and the weather conditions so that later shoots can match lighting conditions.

The EFP Postproduction Stage

At the conclusion of the production, your editing task may include additional audio and video sweetening, such as adding titles, credits, graphics, music, narration, and sound effects. However, the editing task is not the only postproduction concern. There may be promotional chores to get the finished piece exhibited.

As with ENG operations, thank everyone involved. Finally, hold a debriefing meeting with the crew and other staff members to iron out problems that came up so you can improve your future performance.

MULTICAMERA REMOTE PRODUCTION

Multicamera remote (MCR) production is, as the name implies, a production process done in nonstudio locations that uses more than one camera at the same time, permitting the director to cut between cameras exactly as is done in a conventional studio operation. Everything that can be done in a conventional studio can be done in an MCR operation. A mobile truck serves as the control room. Using ENG-type transmission facilities and intercom systems, the MCR operation delivers coverage to a home base for transmission to virtually any audience, potentially worldwide. Multicamera remote productions include live coverage of the Olympics and other sporting events, such as professional baseball and football and NCAA college basketball. MCR techniques are also used for political conventions, concerts, theatrical events, and major awards ceremonies.

Like EFP productions, MCR operations sometimes cover events staged primarily for television, making it possible to produce supporting material ahead of time. In these cases, event coverage can be integrated with prepared segments, feature stories, biographies, and interviews (*backgrounders*). Coverage is often carefully timed to accommodate cutaways for commercials and promotional announcements.

The MCR Mobile Unit and Equipment

In addition to using all the equipment already described for ENG and EFP operations, the MCR expands its arsenal to include banks of monitors and a production switcher. In short, at the high end, all of the control room hardware found in the most advanced studio facilities is available. Some high-end units (trailer trucks over 40 feet long) contain even *more* equipment than most studios do (Figure 10.13). For example, units covering major sports have a number of slow-motion replay recorders and laser disk machines in addition to sophisticated transmission facilities, camera control hardware, character

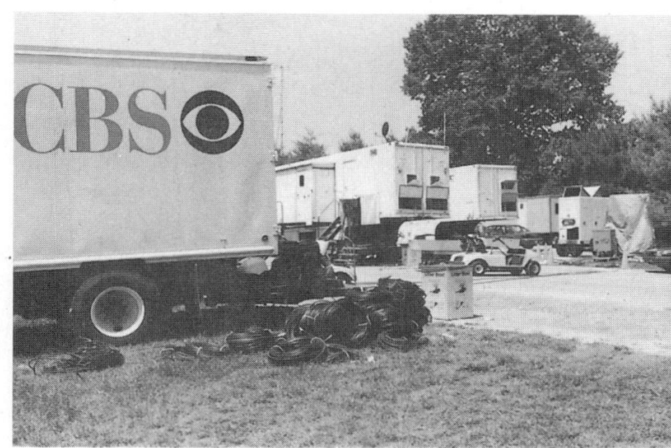
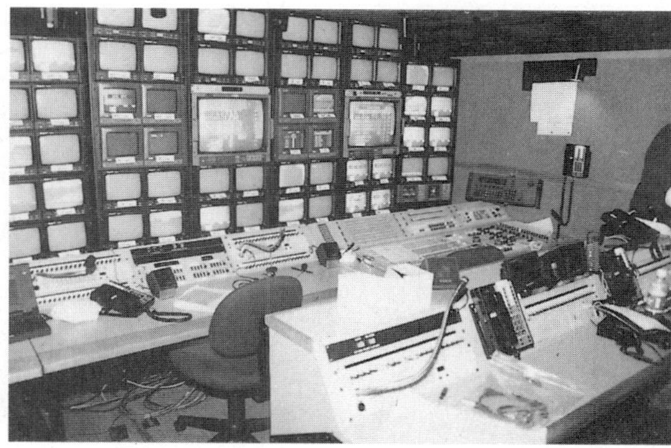

Figure 10.13 *MCR facilities used at the Kemper Open golf tournament. (left) The back of a 46-foot MCR trailer truck, shown in a village of trailers and vans containing technical, maintenance, production, and transmission facilities. (right) Some (not nearly all) of the equipment inside the truck.*

generators, still store machines, and a separate audio console. Power requirements for such facilities can easily exceed the load limitations of generators. For this reason, power needs may be supplemented with outside connections from local utility companies.

The MCR Preproduction Stage

The same concerns outlined in the preproduction phase of EFP operations are relevant to the MCR operation. In addition, the MCR preproduction phase should also be concerned with the following:

1. Since you will likely not rely solely on your own power and communications lines for service needs, be sure to establish reliable contacts for electricians, telephone company personnel, and any other maintenance services. During the site survey, power needs should be clearly established, and a decision should be made to go with either all generator-supplied power or all land-supplied power to eliminate phasing problems caused by two different sources.

2. Parking for the mobile vehicle may require special permissions from the local authorities. To ensure access to the remote site, arrange to get permits or reservations for entrance and parking.

3. Review the paperwork generated from the site survey ahead of time to establish the best location(s) for the vehicle(s) and equipment. Work out a plan that requires the shortest cable runs. For many events, such as established sporting events and concert venues, fixed (buried) cables may already be available. Inquire ahead of time to take advantage of them.

4. In addition to noting the location of each camera, be especially aware of the number of cameras you need to deploy, the mobility their locations afford, and the lens requirements of each position as a function of the shots you want each camera to get.

I N D U S T R Y
voices

John McCrae
Head of Field Maintenance and Technical Personnel,
CBS Sports Remotes

Q: Tell my readers about what you do for CBS.

A: I am in charge of all the technical aspects and technical personnel that keep our mobile trucks on the road. We have five trucks. Four are double sets of tractor-trailers; the other is a small (35-foot), self-contained unit. I've got sixteen technicians, who are probably the most highly trained technically outside of our engineering people.

Q: You have some background in electrical engineering?

A: I went to Stevens Institute of Technology, where I took electrical engineering. My high school had both a radio and television station, where I did a local news broadcast five days a week and spent some time in radio as an on-air DJ. In 1977 I got my first paying job as a DJ at an ABC affiliate radio station. But I quickly realized that I probably would not make it as a top-market DJ, and I kept to my engineering studies. When I started job searching for real, I was very interested in the technical aspects of both radio and television, and one of the outfits looking for new blood was CBS.

Q: When did you learn the craft of large, multicamera remotes?

A: I started in 1979 as an assistant. I worked the political conventions in 1980 in Detroit and New York. I worked with a guy named Jim Patterson, a very good teacher. Soon I had learned enough to be a studio manager. I was assigned weekend sports and the evening news just as Dan Rather took over. I did those shows until 1983.

Q: How do you handle internal voice communications among personnel?

A: Today we use digital programmable communications systems, and we program who can talk to whom through a computer. If we have sixty people working a remote, they don't all have to talk to one another. The producer may have sixty keys to talk to everyone, but other people may have as few as six. The important thing is that the system can be changed on the fly. If someone decides they need to talk to someone they haven't been given, then *boom*—it's a couple of keystrokes on the panel and they have communication.

If we lose communication with one another, we're down. It's the most important problem to avoid. If I lose a camera, we can get by, and you won't even notice it because we have other cameras to cover a key shot. But if we lose internal communication, the director can no longer tell cameras what shots to get, he can no longer tell them that they're on the air, he can no longer tell them what he'd like them to shoot.

Q: How is the communication system set up?

A: We use a map with the location of all personnel marked off and lines designating who has to talk to whom. Then we set up the system accordingly. We have a TX truck, which is essentially a master control room on wheels that ties all our facilities together.

As far as production goes, the building blocks are the same no matter what shape the ball is. If it's an arena, somewhere there's going to be an *announce* position. Those people need both headsets and hand mics. They generally use the hand mics for on-air shots and headset mics when they are not on camera. They need to use headsets because the noise levels in arenas are very high, and they have to hear what's happening in the truck. They also need to hear their partner. Without headsets, they may not be able to hear each other even if they're only two feet apart. Typically they get program audio in both ears, with an interrupt in one ear. The interrupt carries directions from the truck, and it's extremely important, because when we're going to commercial, when we're showing graphics, the announcers need to be warned.

So, you've got an announce position. Then you've got camera positions. Now if we're doing eight football games, we use standard positions for the cameras across the board so they all look like they're coming from the same network. We try to make the camera positions equivalent. And we make the graphics from game to game consistent so we can cut in exciting moments from other games during regional coverage and still look consistent.

Q: What about regular telephone connections?

A: To this day, we still use standard telephone trunk lines. The phone company gives us special consideration. They know there are other groups like radio and print folks with similar needs, so they install lines where we need them overnight. We also have backup phone lines for the announcers in case the camera coverage goes down. In such cases, you'll see a graphic that says "Please stand by," but you'll still hear the announcers. We also use plain old telephone lines for our field connection back to CBS in New York.

There's a lot of communication from the truck to folks in the arena or tournament site. Every camera wired to the truck by triax cable has communication circuits with phone capability built in. We also use wireless private lines, kind of like hand-held radio with a couple of extra goodies added to it.

Q: How many CBS personnel are working here at the Kemper Open Golf Tournament [the site of the interview]?

A: There are forty-eight technicians: camera operators, utility, tape operators, technical director, audio personnel, audio assistants, graphics. Then there are fifteen management and logistics people, myself, and runners. The producer, director, assistants, and spotters equal about forty. All told, that's over a hundred CBS employees to make the "Kemper Open" show.

Q: How does the whole operation unfold?

A: The steps we go through are roughly as follows. Step one is the *survey*. We create a schedule of events. We line out the production aspects, including camera shots. We decide on auxiliary areas for interviews. We also decide on the graphics at this point.

Next, we decide on the *technical* aspects. This includes camera locations, cable runs, truck locations, power considerations. We have separate generators for power if we need it. None of these units has

power on board. We need to set up telephones and communication links. We also need storage for equipment. Finally, we need to set up transmission: how we plan to get out with our signal. Will we use fiber or satellite to get out?

Next are the *logistical* considerations. Where do we park? Do we have hotel accommodations? Has someone taken care of airport schedules, food, toilets?

Next, *personnel*. Staffing must be scheduled so we don't wind up paying people for days they are not needed.

Finally, we deal with the issue of *contacts*. This includes tournament personnel, greenskeepers, parking attendants, etc. We also have to deal with *seat-kill*, which means paying for seating that is killed off by our camera placements. No paying customer wants to sit behind a camera platform.

After all this is worked out, you go back to the office and plan the budget so that you get it all done for the lowest cost. What days do I need what people for what purpose and for how long? Where am I getting all the equipment and personnel I need, and how much will it cost?

Q: My last question: What do you look for in an intern?

A: I want someone who listens, who is not afraid to get his or her hands dirty. This is a tough industry. Sometimes the language you hear is not always so pleasing. People get frustrated. I've been called worse things than you've ever heard in your life by a person and fifteen minutes later sat down at a bar and had a beer with him. If you hold a grudge, you're going to be done in this business in a short time.

To be a good intern, you must listen and learn. Get it the first time. Follow directions and take criticism constructively. Remember to ask for constructive criticism. Also, try to pick your area of expertise and focus on that.

5. Audio needs are just as critical as video needs. Arrange to deploy mics near all the sounds you intend to collect. For example, for live coverage of an NCAA championship basketball game, separate microphones may be used to pick up sneaker squeaks, ball swishes through the nets, grunts of players, reactions of spectators, and speeches of announcers and commentators. These needs may require the use of headset mics, shotgun mics, parabola mics, and wireless lavalieres with a variety of pickup patterns. You may also want to accept a direct audio feed from the arena's PA system.

6. In addition to intercom voice links among relevant parties via camera headsets, additional private lines may be added at selected locations for floor managers, assistants, and other production personnel. As with ENG operations, IFBs should be used to send voice cues from control room personnel to on-air talent.

7. Feeds for both program and preview video can be provided wherever needed by running cables from the truck's control room to selected field locations.

8. Establish security measures to protect equipment from theft and destruction.

The MCR Production Stage

Once the preproduction phase is completed, it is time to set up and rehearse. To ensure a successful production, arrive early to solve any last-minute problems. Review the schedule of events with the crew, and make sure the sequence of tasks is correct. For example, if special platforms are needed to support cameras and other equipment, make sure they are ready when the cameras arrive. Many shops tack a handwritten schedule to the control room door of the remote truck.

Once everything is set up, the rehearsal phase parallels that in a regular studio production. Verify that all systems are up and running. For example, are transmission facilities functioning properly? Do you have reliable intercom connections and adequate video feeds for all crew and on-air personnel? Are all camera control units working properly? Are cameras color-balanced and shaded? Are microphones deployed where you need them? Have levels been taken? A facilities check is done by a broadcast associate and assistants who put on all the headsets and check all the monitors and audio lines.

Once all systems are go, conduct run-throughs. For events that cannot be thoroughly staged in advance, such as sporting events, assign responsibilities so that camera and other production personnel know roughly what they will cover for a number of different scenarios. In baseball, for example, the camera located in center field can practice zooming and panning to capture the pitcher and batter, pop-ups behind home plate, and so forth.[2] During rehearsals, the crew should become accustomed to strange or awkward surroundings and any limitations the site presents.

At showtime, if you are broadcasting live, be sure to maintain contact with the studio, especially for time cues for commercial cutaways and other taped

[2]Standard camera positions for most major sports events are described in Catsis (1996).

inserts. Start times are strictly followed and must be coordinated carefully with the rest of the broadcast day. Promotional announcements and possible schedule changes should be monitored from the remote site. A line monitor in the remote truck carrying the station's broadcast feed is the best way to coordinate field coverage with the home base.

The MCR Postproduction Stage

The same postproduction chores noted in the ENG and EFP sections apply to MCR as well. In MCR productions, with their large amounts of expensive equipment, it is especially important to keep a checklist to make certain nothing has been left behind.

If the program is taped rather than live, the editing tasks are likely to be larger and more complex than in ENG or EFP. In the next chapter, we will focus on the full range of editing processes that can be used for both studio and field productions.

KEY TERMS

flyaway video satellite uplink *(232)*
reach *(233)*
range *(233)*
interactivity *(234)*
electronic news gathering (ENG) *(234)*
repeater station *(236)*
satellite news gathering (SNG) *(237)*
transponder *(237)*
footprint *(237)*
interruptible foldback (IFB) *(238)*
raincoat *(239)*
speed light *(239)*
bounce light *(240)*
continuity *(241)*
jump cut *(241)*

overlapping action *(241)*
staging *(242)*
cut on action *(243)*
cut-in (insert) *(243)*
cutaway *(243)*
reverse-angle shot *(243)*
axis-of-action rule *(246)*
head-on shot *(246)*
tail-away shot *(246)*
electronic field production (EFP) *(253)*
filter factor *(255)*
multicamera remote (MCR) production *(258)*

QUESTIONS FOR REVIEW

1. How do the concepts of reach, range, and interactivity help explain the way communication systems are set up to make field productions run smoothly? Give examples of each concept.

2. What similarities and differences exist among ENG, EFP, and MCR field productions? How does SNG differ from ENG in terms of equipment and transmission methods used to cover news?

3. Why is satellite transmission called distance insensitive?

4. What preproduction planning might you do to execute a field shoot of a local neighborhood PTA school meeting?

5. How can you maintain continuity when editing footage from an EFP field shoot? What techniques can you use to avoid jump cuts and distracting sound variations from one segment to the next?

6. What function do shooting inserts and cutaways serve in producing footage for field productions?

7. Why is the axis-of-action rule important for preserving directional continuity? Give some examples.

8. What kinds of filters can help maintain proper color temperature when shooting under varied lighting conditions?

11 Editing: Aesthetics and Techniques

You are watching "The Nesters," a sitcom about title characters Bill and Carol Nester, affectionate young newlyweds. Both work in different companies, not far from their home. Normally, they wake up and have breakfast together, then drive to work in separate cars. At lunchtime, they always meet at their favorite restaurant. They return to work at 1:00 and drive home at 5:00. Such is their typical day.

But not today. In this episode, entitled "The Surprise," their afternoon routine is disrupted by Carol's plan to throw a surprise birthday party for Bill at his office. To stall him while his pals get the office ready, Carol calls the restaurant from the neighborhood gas station, feigning car trouble, and asks Bill to come get her. While polishing off two baskets of his favorite bread sticks, Bill has been worrying about her lateness and also about the big report his boss has demanded by the end of the day. He rushes to get Carol, but then she convinces him to wait while their mechanic, who is in on the deception, fakes an attempt to diagnose the car problem. After much delay, both call their offices to explain why they're late. Of course, Carol is really calling Bill's office to see if it's time to bring him back yet. Bill finally insists they go to his office together to "button up a few details." At 3:00 they finally arrive at Bill's office, where the party comes off as planned. He is relieved to learn his buddies have already printed his report and delivered copies to his boss. Along with other presents, Bill receives an extra-large basket of his favorite bread sticks.

If you were shooting a video of this episode, how much screen time would it take? Certainly not the six or seven hours of clock time that elapse in the story. In terms of *screen time,* conveying all the details of this episode would require just minutes.

Imagine what viewing would be like if all television programs presented their subject matter in real time. Most of them would be too expensive to make and too tedious to watch. Instead, largely through *editing,* television programs routinely present stories that cover long time spans in a much shorter period.

On the simplest level, editing is the process by which video segments are assembled into a desired sequence after they have been shot. Aside from condensing the time span, this process permits scenes to be shot out of order, according to the most convenient schedule, and rearranged afterward. Most important, editing is used to enhance clarity and add impact to programs. This chapter focuses on both the aesthetic principles and the practical techniques of editing. The topics covered include:

EDITING AESTHETICS

continuity editing • classical editing • dynamic editing: thematic montage • pictorial complexity and mise en scène • editing "The Nesters" • editing sound • technical advances influence editing decisions • throwing out all the rules

EDITING TECHNIQUES

control track editing with the basic two-deck system • logging footage and making an edit decision list • frame counting • assemble and insert editing • SMPTE/EBU time code • the expanded meanings of on-line and off-line editing • analog and digital recording systems: tape and disk formats • nonlinear editing with digital video

EDITING AESTHETICS

It should be clear from our opening discussion that editing spares audiences the insult and inconvenience of having to watch the tedious aspects of real life. But life cannot be chopped and compressed arbitrarily. What are some of the principles videographers follow to make compressed time seem as convincing as real time?

Continuity Editing

To condense time without confusing the audience, editing must maintain *continuity*, which we defined in Chapter 10 as the smooth flow of uninterrupted action from shot to shot. Editing that condenses exposition while preserving continuity is called **continuity editing** or *cutting to continuity*. Successful continuity editing depends largely on preserving cause-and-effect relationships and unity in space and time.

Shot Order and Cause-and-Effect Relationships   To achieve good continuity, the editor must understand what to include and in what order so that the audience can comfortably follow the action. Imagine the meaning you might derive from the following sequence: a low-angle shot of a safe plummeting to the ground from an office window, followed by a shot of a child's frightened

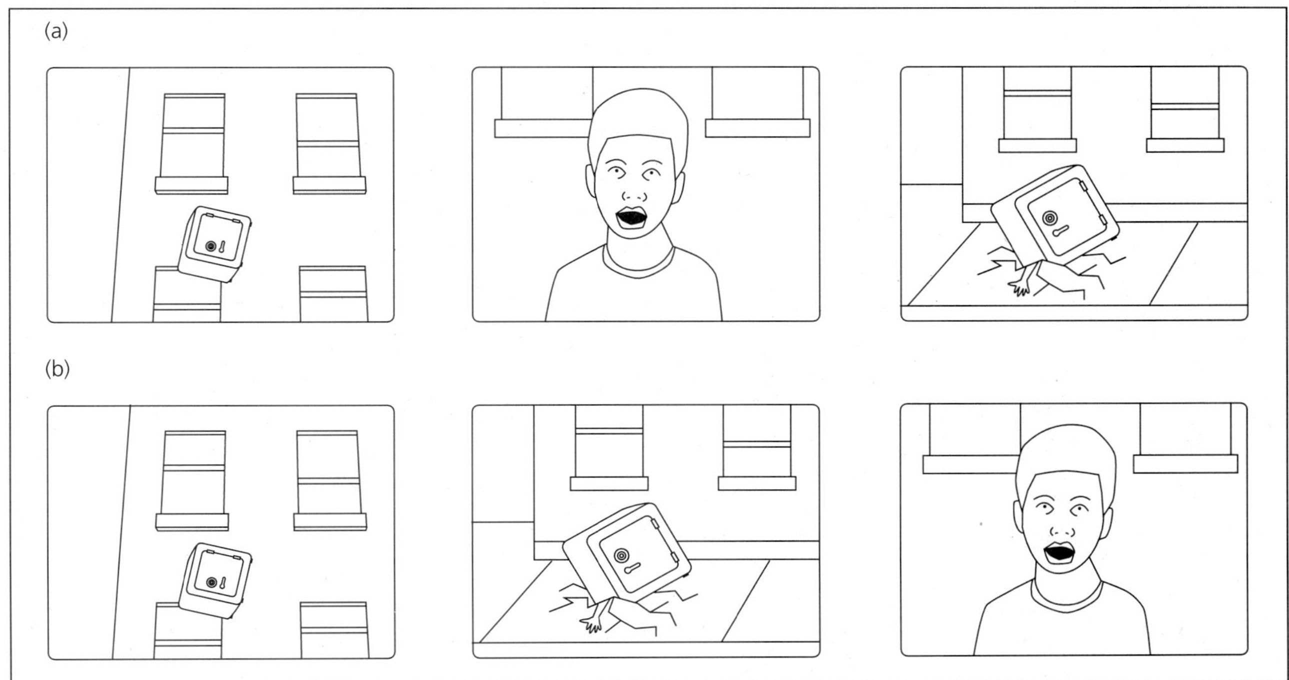

Figure 11.1   *Using shot order to influence the viewer's perception of cause and effect. (a) A sequence of a falling safe implying that the child has been crushed by the safe. (b) Exactly the same shots, but rearranged so that the child appears to be a witness rather than a victim.*

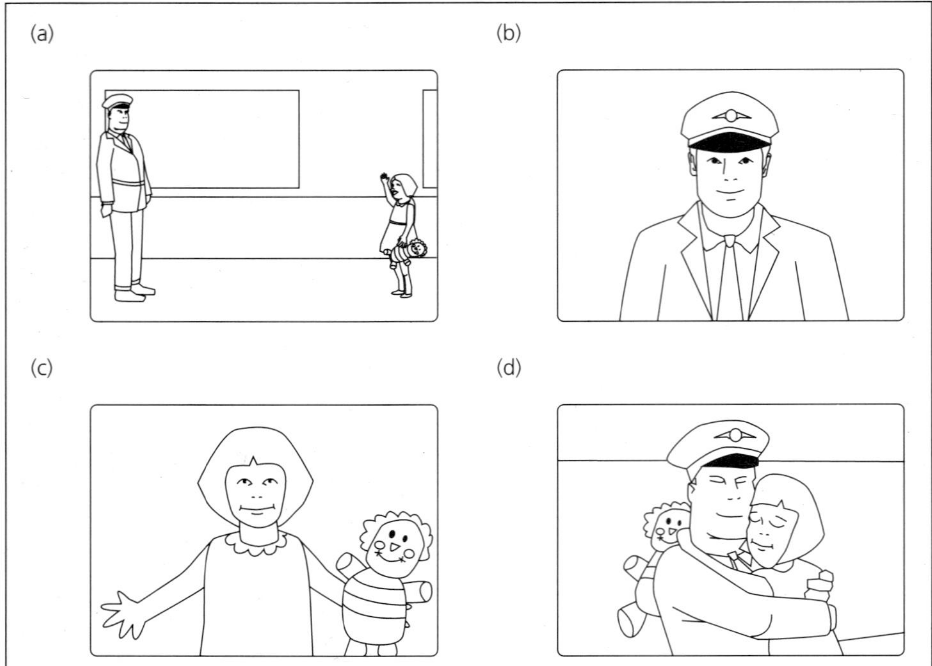

Figure 11.2   *Creating unity in space and time. (a) Establishing shot of father and daughter waving to each other shows their relationship in space and time.. (b), (c) Close-ups convey their emotions as they run toward each other. (d) A final two-shot as they meet rounds off the sequence in both space and time.*

face seen against the background of the same building, followed by a shot of the safe, now on the ground, with damaged concrete around it and a partial view of a child's limb protruding from underneath (Figure 11.1). The implication of this sequence is that the child whose face you have just seen has been crushed by the safe. But if we rearrange the order of these shots so that the cause-and-effect relationships seem to be altered, you get a very different meaning. For example, if the shot of the child's face came *after* the others, you would assume the child had just witnessed the tragedy rather than become the victim of it.

Unity in Space and Time   Recall the example we used in the last chapter of the soldier returning from his tour of duty to be greeted by his daughter in the airport. To maintain *unity in space,* we might cover their reunion hug by starting with a wide establishing two-shot of the father and daughter seen from the side, waving to each other from opposite ends of a large airport waiting area, then cut to head-on close-ups of each subject running toward the other, followed by a final shot of the two seen from the side once again, this time in tighter close-up, just as their bodies meet (Figure 11.2). In this way, the establishing shot captures the spatial relationship of the subjects for the audience. Having the subjects wave at each other in the same shot also telegraphs the message that the subjects are waving at the same time. Thus, *unity in time* is also preserved.

In contrast, if we used close-ups without an initial establishing shot, the subjects' relationship in space and time would remain unknown, and the exposition would be more confusing. Why are we seeing two different people in

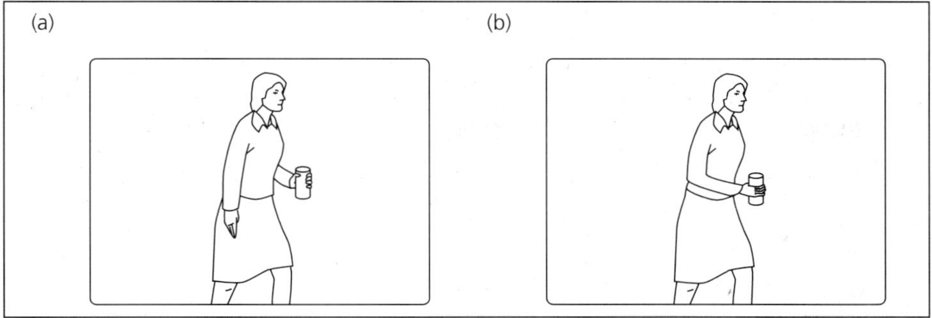

Figure 11.3   *Example of a continuity problem. If (b) directly follows (a), viewers will wonder how the glass suddenly jumped from one hand to the other.*

different places, the audience might wonder, and what do they have to do with each other?

Additional Continuity Concepts   In Chapter 10 we discussed cutting on action, shooting overlapping action when cutting from wide shots to close-ups, following the axis-of-action rule, and matching camera angles when shooting reverse-angle shots. All of these techniques, it turns out, are standard principles of good continuity editing.

A general goal of continuity editing is to make edits as invisible as possible so that the audience is not distracted from the story. The accompanying Professional Pointers list provides some further tips. These pointers are not intended to be applied rigidly, though. Deciding how they should be used in different situations should depend on your program objectives.

Of course, besides preserving continuity, editing determines timing and pace in a story. In general, the timing of cuts can add significant dramatic intensity. This leads us more deeply into the question of editing aesthetics.

PROFESSIONAL   POINTERS

Continuity Editing

- *Match actions and subject matter between two consecutive shots in a single scene.* In a given scene, the background, position, and wardrobe of the talent should be the same from shot to shot. For example, if a long shot of a room has a lighted chandelier but the next shot shows the chandelier turned off, there may be a continuity problem. Similarly, if a talent is holding a drink in the left hand in the first shot but has it in the right hand in the next, or has no drink at all, that can be a jarring distraction (Figure 11.3). Have a continuity person make notes on such details to avoid such gaffes.

- *Images cut together should be of sufficient difference in size, composition, and subject matter to appear motivated.* Motivated cuts make a point, justifiably shifting the

(continued on the next page)

Continuity Editing (continued)

attention of the viewer from one place to another (Figure 11.4). Avoid cutting between images that have small differences in size or composition. Small changes will only irritate the viewer. If there is not sufficient reason to warrant a cut, don't make it.

- *When cutting from wide to tight shots, and vice versa, within the same scene, keep the relative positions of objects in successive shots consistent.* For example, if a wide shot of a subject contains a telephone pole on the subject's right, keep the telephone pole on the same side when you cut to a tighter shot (Figure 11.5).

- *For moving talent, maintain motion vector continuity.* That is, keep the screen direction of the movement constant, even when the talent moves in and out of frame. For example, if a talent moves out of frame to the right, it is generally most sensible to see the talent enter the frame in the succeeding shot from the left, moving in the same direction (Figure 11.6).

- *In addition to motion vectors, consider graphic vectors.* For example, when cutting from a wide to a tight horizon shot, the less jarring cut is the one that presents the horizon line at the same height in the close-up as that in the long shot (Figure 11.7). This is because the horizon line presents a strong graphic vector that is quite noticeable to the viewer from shot to shot. *Graphic vectors* include any strong lines that clearly define the screen space.

- *Pay attention to index vectors.* The term *index vector* refers to the screen direction implied by the physical orientation of people and things in the frame. For example, if two people are playing catch, the normal orientations of the bodies give clues about (indexes to) the action taking place (Figure 11.8). Matching eye-lines when cross-cutting reaction shots is an example of an index vector concern.

- *Preserve continuity as a function of developing narrative.* When locales change, or when a new character is added to those already in a shot, it is generally advisable to establish such changes immediately through the use of wide (establishing and re-establishing) shots. This results in less jarring and more understandable story development.

- *Edit scenes to emphasize significant events and eliminate trivial details, even if minor physical inaccuracies result.* If a scene is cut right, the audience will ignore small physical inaccuracies in favor of strong dramatic development.

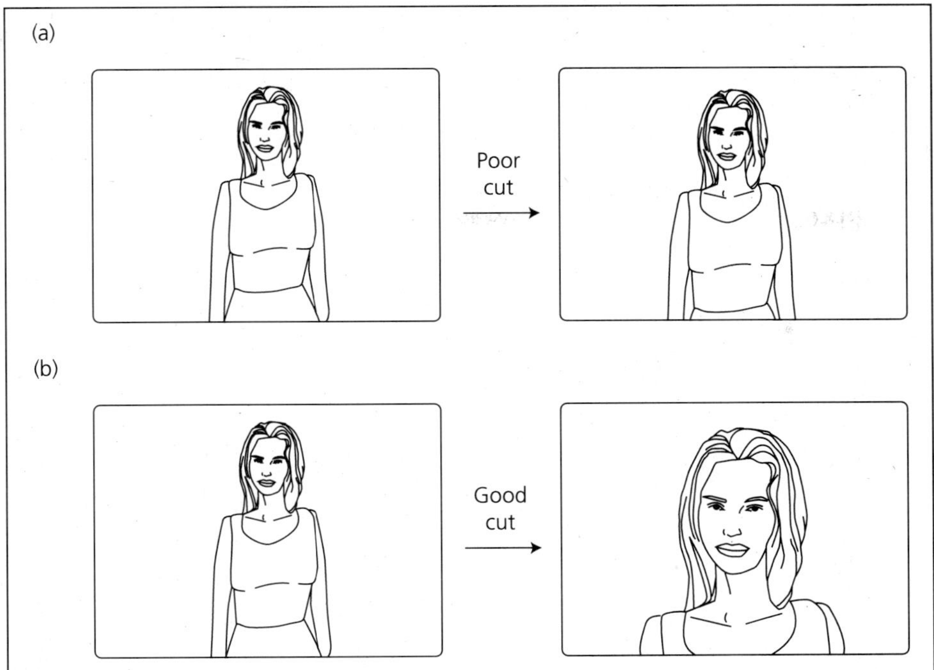

Figure 11.4 *Using motivated cuts. (a) Cutting from the left image to the right would be an unmotivated cut, annoying to the viewer, because the two images are only slightly different. (b) A motivated cut. Here the two shots differ enough to justify the cut.*

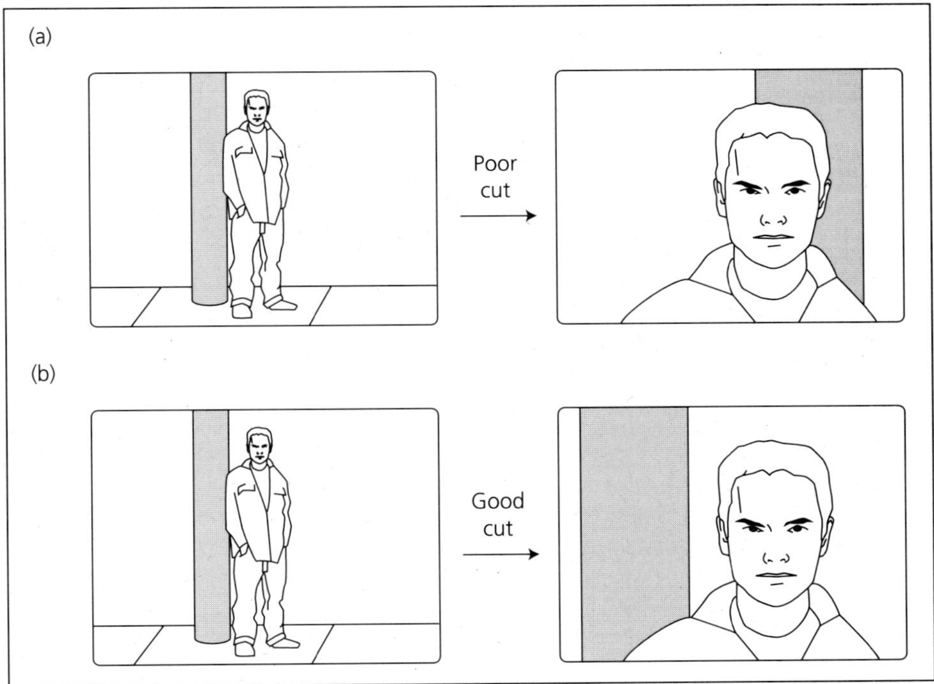

Figure 11.5 *Keeping positions of objects consistent. (a) In this cut, the telephone pole appears to move from one side of the subject to the other, jarring the viewer. (b) A better cut: the pole stays on the same side.*

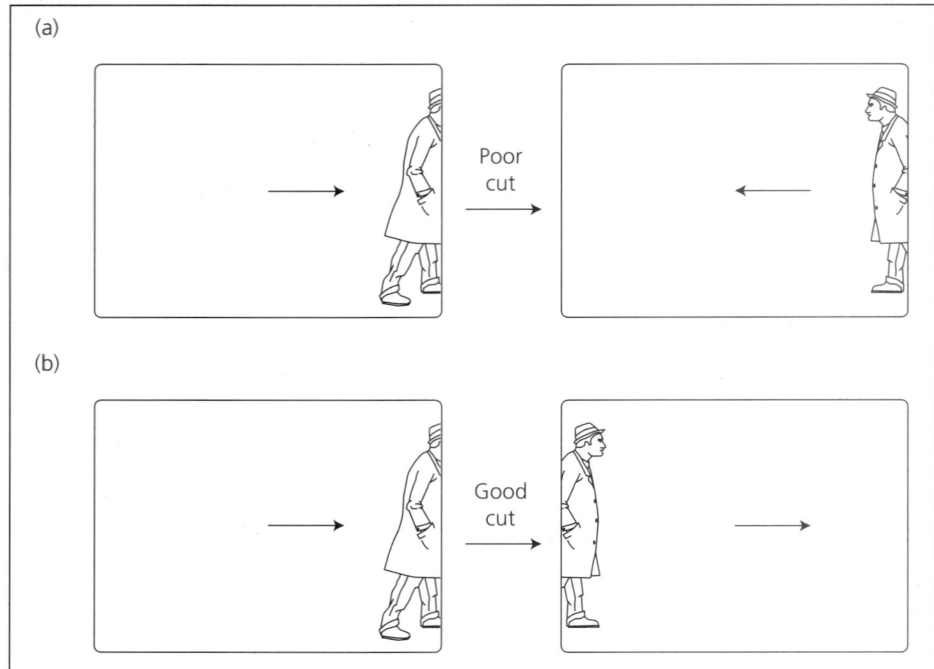

Figure 11.6 *Maintaining continuity of motion vectors when talent moves in and out of the frame. (a) A cut in which the vector continuity is broken because the talent seems to reverse direction. (b) A better cut, in which the motion vector stays constant.*

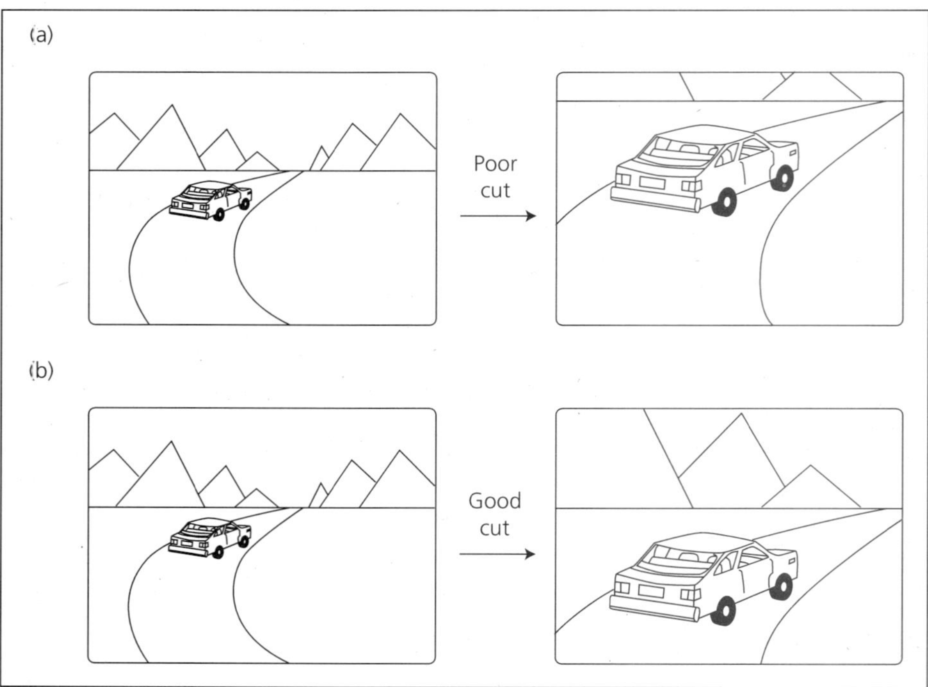

Figure 11.7 *Maintaining continuity of graphic vectors. (a) Because the horizon changes level from one shot to the next, the cut jars the viewer. (b) A cut that keeps the horizon at the same level so that this graphic vector remains constant.*

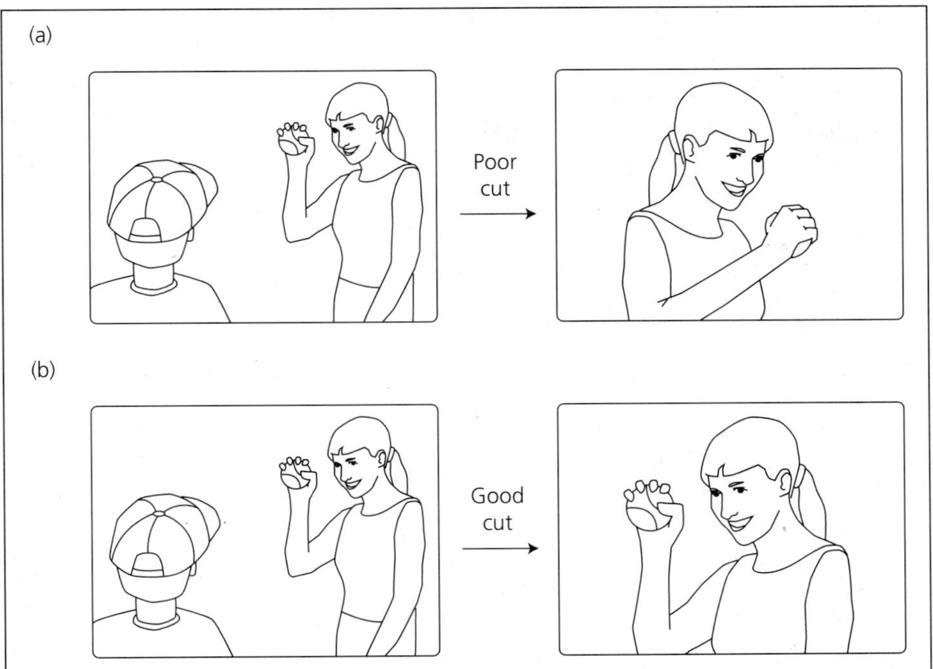

Figure 11.8 *Picture elements that convey a direction of activity establish index vectors. (a) In the wider shot, the positions of the people's bodies and the angle of the throw create a clear index vector. But the cut to a close-up changes the screen orientation and thus breaks the index vector. (b) In a good cut, the index vector is preserved.*

Classical Editing

Editing can do much more than provide clear, economic exposition. By changing camera location and framing, directors can accentuate parts of a scene and vary the intensity of the narrative accordingly. For example, it is possible to cover a boxing match with a single wide shot of the entire ring from a fixed camera position. That would provide lucid exposition. But imagine the same coverage accented with tight close-ups of each fighter's face as punches are landed. Then intercut close-ups of ringside fans reacting to each barrage (Figure 11.9). Clearly, such editing would intensify the dramatic, physical, and psychological aspects of the story. Editing that intensifies the story in this way is called **classical editing**.

Reaction Shots and Parallel Editing Many classical editing techniques were popularized by American film director D. W. Griffith in the first few decades of the twentieth century, and they are still in wide use today. For example, Griffith intercut tight close-ups of talent reactions—*reaction shots*—to heighten the emotional impact of stressful or disturbing situations.

Griffith also popularized the practice of intercutting between events supposedly occurring at the same time but in different places. The most common example of such *parallel editing,* as this technique is known, is a sequence beginning with a shot of a damsel in distress, Little Nell, being tied to the rail-

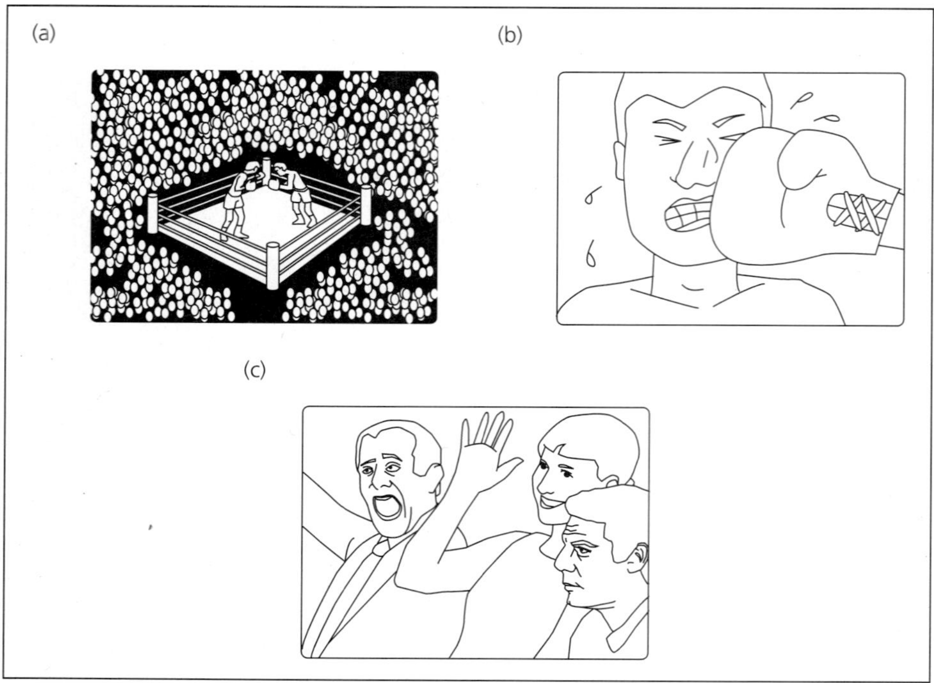

Figure 11.9   *Classical editing often uses close-ups to intensify important elements of the story. In this sequence, to convey the drama of a boxing match, a wide shot of the entire scene (a) is intercut with close-ups showing the boxers' faces and (b) the screaming, gesticulating fans (c).*

road tracks by the evil villain, Snidely Whiplash (Figure 11.10). The next shot shows a train approaching from right to left in the frame, getting progressively larger, followed by a shot of the hero, Dudley Doright, racing on horseback from left to right, also getting progressively larger. Shots of Little Nell struggling to get free are then intercut with further shots of the train and the hero, both approaching ever closer. In successive cuts, each image is presented for shorter and shorter lengths of time. Ultimately, of course, the hero arrives to rescue Nell in the nick of time. In this way, three separate pieces of footage that may have been shot weeks apart are combined to increase the story's dramatic intensity and greatly heighten audience involvement and interest.

The Traditions of Formalism and Realism   Classical editing permits the director to create and control dramatic emphasis. Through the use of close-ups and multiple camera setups, the viewer is forced to see only the aspects of a scene the director wants seen, and no others. This approach takes much control of dramatic intensity out of the hands of the on-camera performer. The style of editing that emphasizes selected elements of scenes in this way to heighten audience attention is known as **formalism**.

Some view formalist editing as too controlling, too manipulative, and even undemocratic. Critics say that formalist editing deprives audiences of the chance to make their own decisions about what to look at in the image. In contrast, offering the audience scenes edited for continuity only—or, better yet, one long master shot—leaves more choice in the audience's hands.

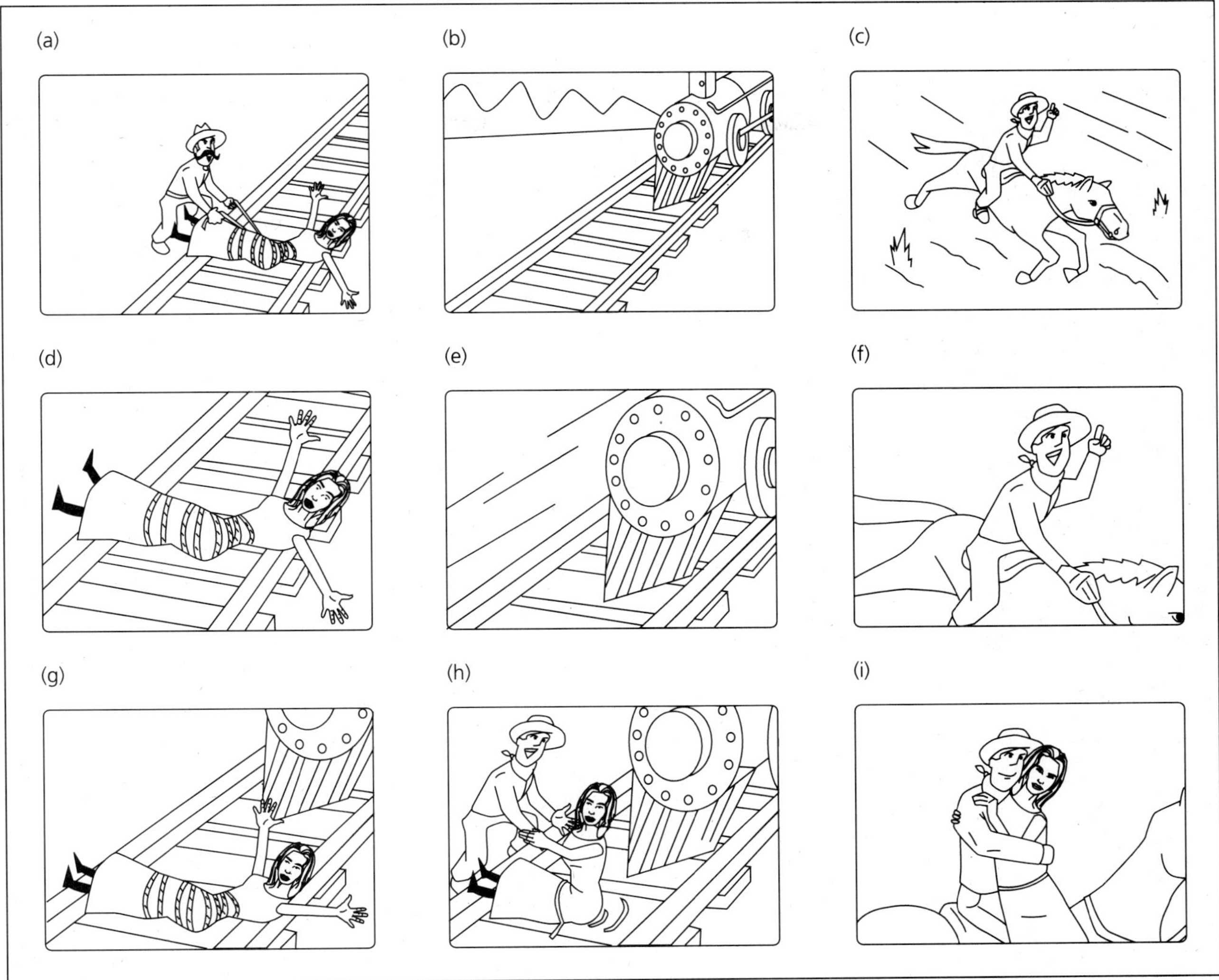

Figure 11.10 *Parallel editing, a type of classical editing, cuts back and forth between events that are occurring in different places at the same time. (a) The stereotypical villain, Snidely Whiplash, tying Little Nell to railroad tracks. (b) A train approaching. (c) The hero, Dudley Doright, riding to the rescue from the screen direction opposite to the train. (d) Close-up of Little Nell struggling to free herself. (e) The train getting ever closer. (f) Dudley getting ever closer—can he outrace the train? (g) The locomotive almost upon poor Nell. (h) The hero arrives in the nick of time! (i) Hero and heroine reunited at last.*

One leading critic of formalism was French film critic André Bazin. In sharp opposition to the formalists, Bazin took an approach often called **realism**. Bazin's theories stemmed from his view that one of the strongest elements of the photographic art (including film and video) is its ability to capture realistic images of the world. Bazin thought editing violates a connection to physical reality and could destroy a scene's impact. In place of editing, he advocated the use of deep focus, long takes, wide shots, and camera movement to follow key action through panning, tilting, and tracking.

If we agree with Bazin's views, it is easy to think of a situation where *not* editing adds dramatic impact to a scene. For example, imagine a story about a pilot who crash-lands in a desert. He hikes from the crash site in search of water, but his thirst becomes overwhelming, and he collapses on a sand dune. Now the camera carrying the scene of his collapse slowly tilts up, and a tracking shot reveals that just on the other side of the sand dune is an oasis with plenty of fresh water. Obviously, the irony and dramatic impact of such a scene would be lost if edits were made between the collapse and the oasis.

The extent to which the formal control of classical editing is praised or denigrated is probably a function of how it is used. It is up to you, as the director, to determine which edits are warranted and when it is better to leave the master shot alone so that audience members can choose which parts of the picture they want to see.

Dynamic Editing: Thematic Montage

In addition to continuity and classical editing, a radically different technique called *dynamic editing* is widely used to shape program content. Largely developed by Russian filmmaker Sergei Eisenstein shortly after Griffith experimented with it in his 1916 silent-film epic *Intolerance,* **dynamic editing,** also known as **thematic montage** (from the French *monter,* meaning "to assemble"), stresses the association of ideas but *without regard to unity in space and time.*

In *Intolerance,* Griffith addresses the theme of "man's inhumanity to man" by juxtaposing scenes of the crucifixion of Jesus with conflicts taking place in ancient Babylon, sixteenth-century France, and twentieth-century America. The segments that he intercuts are spatially, physically, and/or psychologically discontinuous, but are related in concept and theme.

Eisenstein further developed the technique of thematic montage in his 1925 movie masterpiece *Potemkin,* a film originally intended to paint a sweeping picture of the Russian revolution of 1905. During production, Eisenstein decided to limit the film's scope to a mutiny on board the battleship *Potemkin* and a massacre of peasants by the Russian army in the port of Odessa. During the mutiny scene, close-ups of the crucifix of the navy priest (portrayed as a slovenly officer and an enemy of the abused sailors) are intercut with close-ups of a sword to underscore the ironic clash between the ship's religious authority and the rebellion. Similarities in the shape and the metallic appearance of cross and sword in these closely juxtaposed shots help Eisenstein make his ethical and political statements.

Since Eisenstein, thematic montage or dynamic editing has become a regular part of the filmmaker's and videographer's repertoire. The psychological associations induced in audiences can vary depending on subject matter and the way images are juxtaposed. Sometimes the effect can be powerful and engaging. Some montages cause tension, whereas others build complementary ideas.

One factor that influences how a montage is apprehended is the eye of the beholder. For example, propaganda pieces often use montage to associate political figures with one another, thereby making critical statements about them. Imagine juxtaposing shots of a presidential candidate giving a speech to a throng of supporters with similar shots of Hitler addressing Nazi crowds.

To the candidate's supporters, such an association could be quite incendiary, whereas opponents might view the material as accurate and agreeable. In this way, besides making commentary on a subject relatively easy for the director-editor, montage can profoundly affect members of the audience, but in different ways.

Pictorial Complexity and Mise en Scène

In addition to audience-related factors, content factors influence the effectiveness of editing. One of these is the **pictorial complexity** of the shots involved. The complexity of the image is a function of how much new subject matter is presented to the audience in a given amount of time. Editing pace should therefore be determined in part by the amount of new material being presented to the audience. Many experts believe audience involvement seems to peak when viewers have been given enough time to absorb most *but not all* of the information in a shot before another shot of different subject matter is introduced. With the right pace of delivering new information, the audience remains on a satisfying curve of rest and excitement.

Another content factor to consider when determining editing pace is the arrangement of objects in the frame, known by the French term **mise en scène** (pronounced roughly "meez on sen"). The editor should consider what and where the most important objects are in each shot and how easy they are to see so that editing pace will aid rather than harm story clarity and development. For example, to show that a main character is depressed and may attempt suicide, you might start with a wide shot of the subject in bed and then cut to a close-up of the bed and night table, revealing a gun on the table. If the gun is in clear view in the close-up, you will not need to stay on the shot as long as you would if the gun were partially hidden.

Editing "The Nesters"

To see how editing enhances storytelling, let us apply the editing techniques discussed so far to our imaginary episode of "The Nesters." We will edit some scenes for continuity simply to preserve lucid exposition and give others more complicated treatment using both classical and dynamic editing techniques. As you read these suggested treatments, think about other approaches you might take.

Continuity Editing   Let's edit the opening scene to continuity. Our objective will be to present lucid exposition, as shown in Figure 11.11. We fade up on an opening shot of our young couple awakening in their bedroom as their clock-radio erupts with the 7:00 a.m. news, time, and traffic report (six seconds). The next series of medium shots shows Bill in the shower, Carol brushing her teeth, Bill drying himself with a towel, and Carol rushing past him into the shower stall (fourteen seconds). The next shot is of Bill in his bedroom mirror, finishing tying his tie, and Carol entering the bedroom wrapped in a towel (six seconds). A cut to the clock shows the time to be 7:25 (one second); then we see Bill in the kitchen hastily pouring coffee and a fully dressed Carol slapping two plates of scrambled eggs and toast on the table (four seconds). They

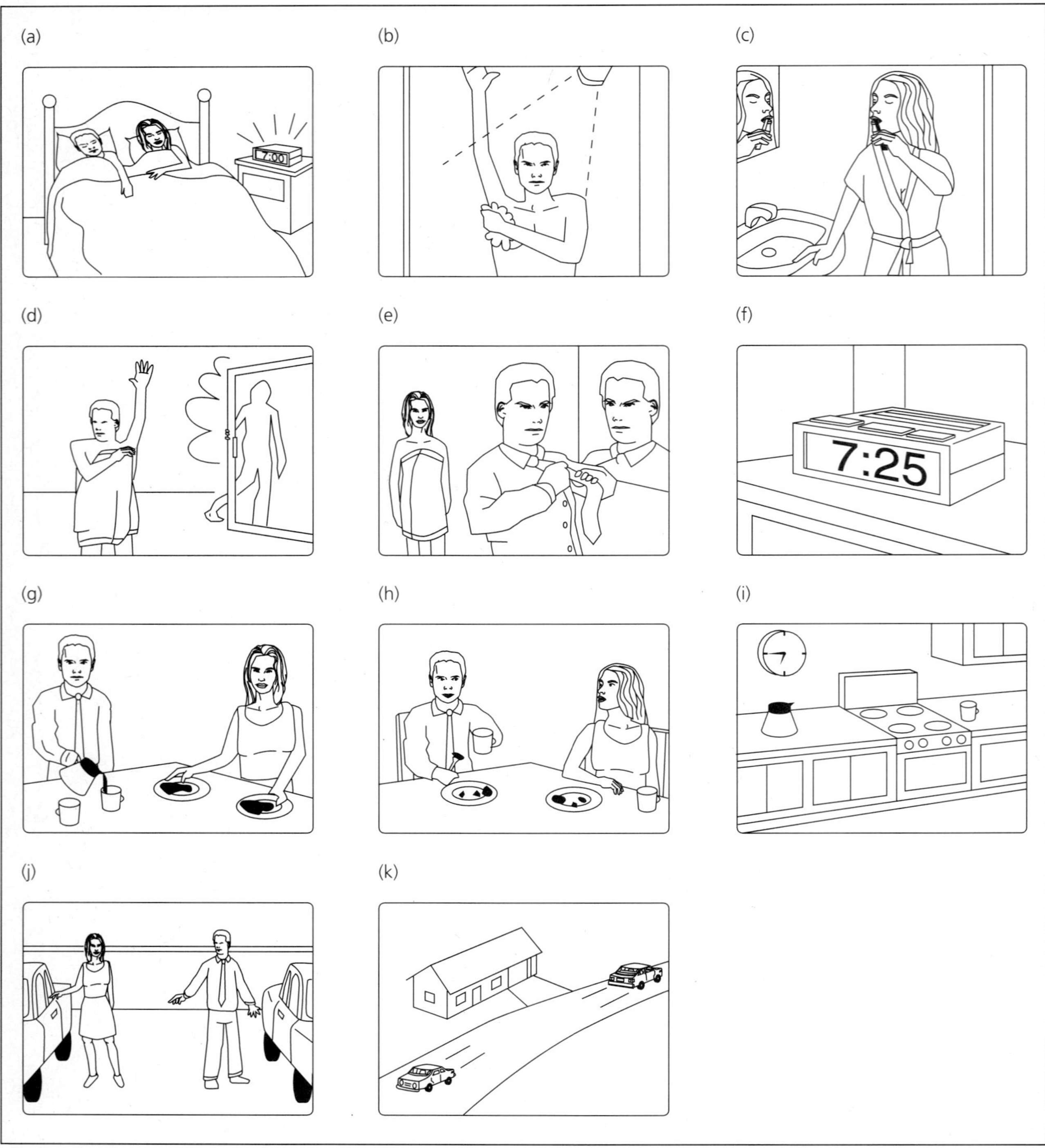

Figure 11.11   *A sequence showing continuity editing in the opening scenes of "The Nesters."*

sit at the breakfast table, gobbling their food and chatting in quick bursts (fifteen seconds). A glance at the kitchen clock reveals that it is now 7:45 (one second). Next, we see Bill and Carol in an exterior medium two-shot at eye level, snatching a goodbye kiss and jumping into their respective cars (four seconds), followed by a wider aerial shot of their cars zooming out of the driveway and heading in opposite directions (four seconds). When written out, the exposition sounds tedious, but the total screen time of this sequence is just fifty-five seconds. In less than a minute, we have established many of the basics of the Nesters' life together and allowed ample opportunity for comedy (Bill singing way off key in the shower, Carol's glasses falling off during the hasty kiss—whatever corny or original bits the scriptwriter might imagine).

Classical Editing   Classical editing may be used to emphasize more dramatic and psychological aspects of our story. For example, at the restaurant, to convey Bill's concern about Carol's unexplained lateness, we can do more than present a medium shot of Bill waiting alone at the table eating bread sticks. We can emphasize Bill's growing impatience by inserting close-ups of his wristwatch from his point of view (POV) (Figure 11.12). Such cut-ins drive home Bill's increasing concern about Carol and about his own tight schedule. Similarly, to convey Bill's growing discomfort, we could include increasingly frequent cutaway shots of the restaurant door from Bill's POV each time it opens.

Dynamic Editing   Dynamic or montage editing could help us focus attention on Bill's feelings of loneliness and his concern about Carol's lateness. While he waits in the restaurant, along with the POV shots of Bill nervously glancing at his watch, we might add shots of him looking out the window, noticing geese nuzzling each other in a park pond as well as several couples strolling together hand in hand (Figure 11.13). These shots would drive home our theme of two young newlyweds deeply in love.

Editing Sound

Thus far, we have discussed editing without attending to the role of sound. Sound can motivate on-screen action, and cutting decisions, as much as any

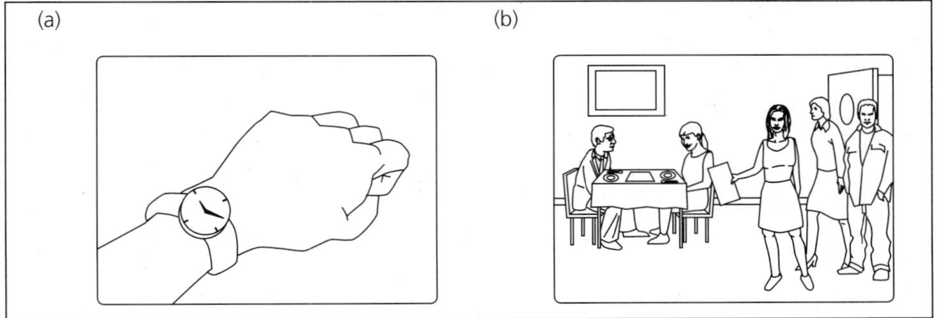

Figure 11.12   *Classical editing of Bill's restaurant scene from "The Nesters" might use shots from his point of view to help convey his anxiety as he waits for Carol. (a) A close-up of his watch, as he sees it, can stress his concern about her lateness. (b) A shot of the restaurant's entrance, where the hostess is leading other people in, can emphasize that he keeps looking for Carol.*

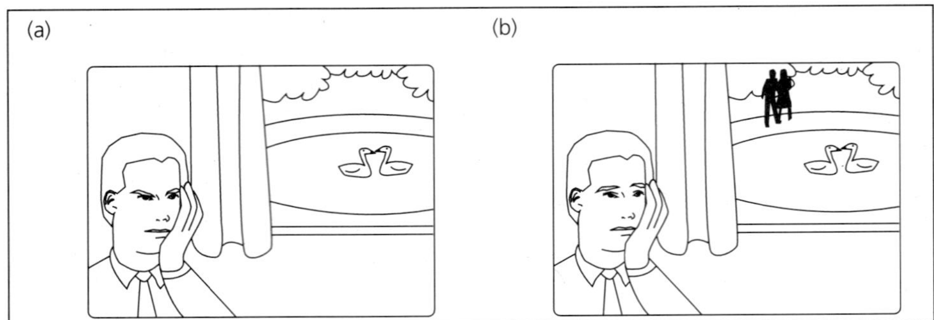

Figure 11.13 *Dynamic editing of Bill's restaurant scene in "The Nesters" might accentuate his apprehension with apparently unrelated shots that fit the theme of lovers being together. For instance, he could look out the window to see (a) two geese nuzzling each other and (b) a couple strolling by, hand in hand.*

other production or story element. In "The Nesters," the sound of the clock-radio erupting with the news, weather, and traffic report motivates the characters to awaken and get out of bed. Clearly soundtracks can help enormously to set the scene in many different ways. But if sound is so important to the editing process, why have we not mentioned it until now?

The soundtrack is treated separately because the mode in which we apprehend sound differs from the way we see, and these differences have consequences for editing. In terms of our visual sense, we can justify the cut as the basic type of edit because cutting imitates important aspects of normal visual function. For example, if I am in bed reading a book and someone knocks on the door, my eyes (and perhaps my head) shift abruptly from the book to the door. I do not sweep my eyes gradually from book to door, focusing with equal sharpness on the entire transition. Rather, my eyes "cut" to the door in much the same way an edited video cuts between images.

In contrast, when sound enters my perceptual field, I need not shift my head to hear it or identify the direction it is coming from. Further, I can select individual sounds for special attention from those presented to me. As I am reading quietly in bed, I may become especially aware that my clock is ticking. And if a train approaches from a distance, gaining in volume gradually, my attention to the clock does not instantly "cut" to the train. Rather, it gradually shifts from one sound to the other, similar to the usual fading in and out of sound in a video. I can also shift my attention back and forth between sounds of roughly equal volume (in this case, the ticking and the distant train sounds), depending on which one I choose to concentrate on. Hence, "cutting" sound in and out of a program is unnaturally abrupt; it does not accurately simulate the way we normally attend to sound. Therefore, when editing, we often need to treat sound differently from the way we treat visuals.

Synchronous and Asynchronous Sound   Although sound is often a natural accompaniment to visual action, it is purely a matter of choice whether or not to include it with its corresponding visual action on screen. When actions and corresponding sounds coincide, the sound is said to be **synchronous**. Seeing a gun being fired, with muzzle flash and smoke, and hearing the gunshot *at the same time* is an example of synchronous sound.

PROFESSIONAL POINTERS

Editing Sound

- In the absence of compelling reasons not to, use gradual fades rather than cuts to integrate sound into a program. Harsh, noticeable audio cuts draw attention away from the illusion of a continuous, smooth stream of action and invite undue attention to technique.

- To enhance realism, it is not necessary to represent all sound as it occurs in real life. Rather, record only the significant sound. For example, the audience will not care if a fire in an on-screen fireplace does not crackle as long as the fire is not a significant part of the story. If the fire later becomes a main part of the action—say, because an important document has been thrown into it—you can use a crackling sound at that time. Use the same principle for handling sound for phone calls. Decide whether to have the audience hear the voice on the other end of the line on the basis of dramatic motivation, not realism. The audience will not question the choice you have made if the action on screen motivates the decision or attracts enough attention.

- Nat sound used to set a scene need not persist if it is too distracting. For example, if dialogue is to take place at an oil-drilling site, natural sound may be used to establish the location. But as the conversation gets going, it is acceptable to fade down the background noise. Of course, don't cut sharply from the noise of the drilling equipment to the dialogue track. Such abrupt change is unnatural, and the viewer will be sensitive to its lack of realism.

- Sometimes asynchronous sounds are more important than synchronous ones because they make the audience aware of something that was previously unknown.

- Use sound quality, in addition to volume, to add dramatic depth to a program. Even the simplest sound has many complex associations. For example, a series of toots of a car horn, in addition to volume, has duration, presence (the quality of being on-mic or off-mic), tempo, rhythm, pace, pitch, timbre, overtones, attack and decay, dynamics, and even cultural and historical associations. If the sound you are dealing with is music, you may have the added dimensions of melody, harmony, and counterpoint.

- When adding music to dialogue, avoid using lyrics under voice-over, unless this technique is motivated by the program. Lyrics under dialogue can be distracting. If you are convinced that a certain lyric must be used with dialogue, make the volume levels different enough to offset audience distraction.

In contrast, hearing someone's voice while seeing the same person on camera with his or her mouth closed, indicating that the person is thinking, not speaking, is an example of **asynchronous** sound. In asynchronous sound segments, there is a discrepancy between what is seen on screen and what is heard. Often asynchronous sound is used for dramatic effect, to fill in exposition, or to make ironic or psychological statements.

Beyond the artistic impact of using synchronous or asynchronous sound, knowing in advance whether you will use one or the other (or both) can have a significant effect on your editing decisions. For example, if a segment features a quick succession of cuts to build tension or excitement, you may decide that the abrupt synchronous sound changes that result from such a series of hard cuts would distract the audience. Instead of synchronous sound, therefore, you may decide to insert music, dialogue, additional sound effects, nat sound, or just plain silence. You may even decide to use a sound montage created from the original synchronous soundtrack but sweetened with fades to soften or eliminate the abrupt changes. Whatever you decide, being aware of the difference between how we process visual and audio information will make the job of fabricating effective soundtracks easier. See the accompanying list of Professional Pointers for more suggestions for editing sound; again, these principles are offered as general guidelines, not as rules to follow slavishly in all cases.

Technical Advances Influence Editing Decisions

Technical advances continually extend the mobility and capability of cameras. Today's cameras are lighter and smaller and can operate under a wider range of lighting conditions than ever before. The result is that cameras can now routinely capture actions that were once unavailable to television (skydiving footage is just one example). These advances in technology can influence your editing decisions.

For instance, to show the flight of an arrow, typical editing would cut from a shot of the archer drawing back the bow and letting the arrow fly to a shot of

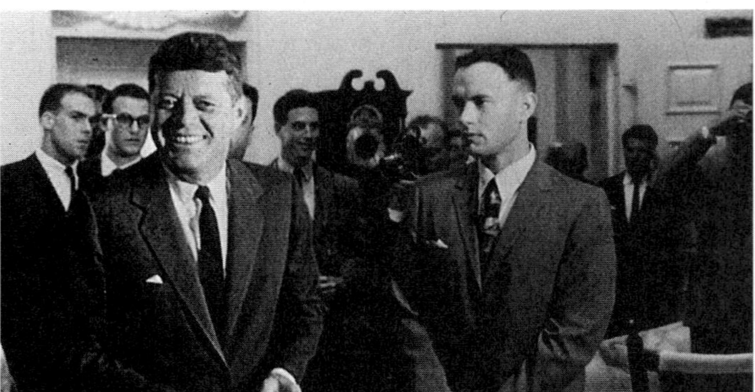

Figure 11.14   *With digital editing, images from entirely different sources can be edited into the same shot, creating a composite. This composite shows Tom Hanks, in the title role from the movie "Forrest Gump," meeting with President John F. Kennedy.*

the arrow hitting the target. The interim flight is not normally shown because it is too fast. However, if it were possible to mount an extremely durable, lightweight, and remarkably small battery-powered camera, equipped with radio signal transmission capability, on the arrow itself, imagine how we might shoot the sequence. Now we could cut from the shot of the archer to that of the arrow-mounted camera just as it leaves the bow. We could televise the entire flight to the target. Someday such shots will probably be possible. When camera technology reaches this stage, the editing process will change accordingly.

In addition to miniaturization, the digital effects used in postproduction are influencing editing aesthetics. With digital technology, composite images from different sources can be edited *into the same shot,* not just in series with one another. Digital technology extends the editing craft to the planning of shots within the frame. Many music videos are good examples of digital editing (Figure 11.14).

Throwing Out All the Rules

Ultimately, editing technique must be judged in terms of the subject matter being presented, the context of the technology available at the time, and the goals of the programmer. Often the principles set forth in this chapter are purposely violated, not only in music videos but also in avant-garde and experimental videos and even in numerous commercials. In the case of music videos, rather than exposition, narrative, or storytelling, the aesthetic goal is often to provide perceptual stimulation and sensual excitation. In such cases, smooth continuity may be neither desirable nor necessary. Yet it is still important to understand conventional editing principles. If you decide to ignore them, you need to know what you are doing and why.

EDITING TECHNIQUES

Before the advent of videotape in the mid-1950s, the only editing performed in television production was done live, in real time, during the broadcast. The director watched a bank of monitors and called for shots from several cameras in the studio, and it was the technical director's job to cut, fade, or dissolve to the next shot while the show was airing. This type of editing, which is still practiced for live broadcasts, is called **on-line editing**.

Videotape, of course, has greatly expanded the editing dimension of television, making video more like film in that footage can be shot, edited, and sweetened in postproduction before being shown. Editing done this way is called **off-line editing**. As we will see later, the terms *on-line* and *off-line editing* have both undergone changes in definition as postproduction editing practices have evolved.

Off-line editing enables you to shuffle recorded program segments like a deck of cards. Further, since all common videotape formats feature a video track and two or more audio tracks, all separate from one another, it is possible to manipulate any one of them independently from the others. This permits the videographer to select, mix, insert, or assemble either audio, video, or both for any program. Add to this the ability to digitally alter single frames of video with computer graphics software and the postproduction suite offers nearly endless editing possibilities.

In addition to separate video and audio tracks, videotape has a **control track** or **pulse track** consisting of a series of electronic sync pulses recorded near the edge of the tape. This track enables you to locate specific frames of video for viewing and editing.

Control Track Editing with the Basic Two-Deck System

The basic videotape editing system (Figure 11.15) consists of two tape decks. One is used to run *source* tapes, that is, unedited tapes. The other is used to run the *edit* tape, or the tape onto which you *dub*, or copy, the edited program. Monitors and speakers are provided for viewing each tape's video track and hearing its audio track(s). An **edit controller** enables you to coordinate the decks, search through tape, and execute edits.

A **shuttle mode** on the edit controller allows you to quickly locate and view video segments on either source or edit tapes as they are run back and forth through their videotape decks. Since all segments are stored in series on tape, they cannot be randomly accessed; instead, individual segments are retrieved by "shuttling" back and forth through the tape until the desired segment is found. For this reason, the use of any tape-based editing system (whether analog or digital format) is referred to as **linear editing**.

Most systems also allow tape to be slowly *jogged* to locate edit points more precisely within segments. Even single frames of video can be further trimmed or added using *trim* buttons. All of these electronic functions key off the tape's control track.

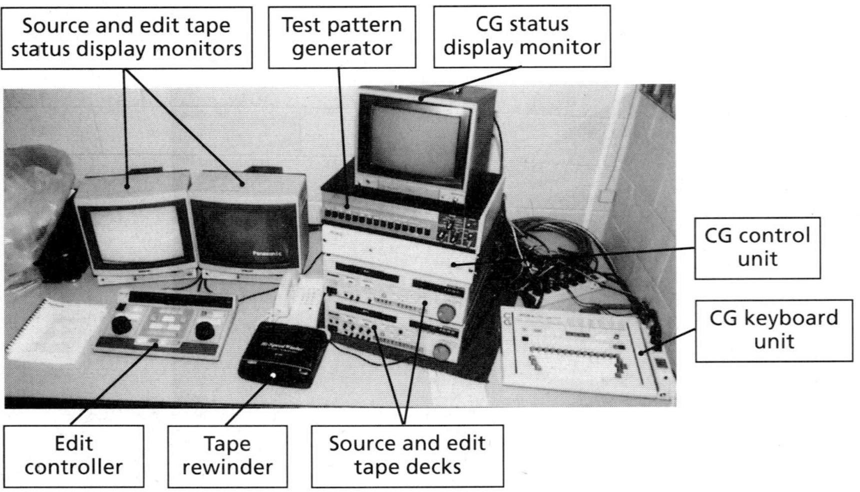

Figure 11.15   *The components of a basic videotape editing system.*

RAW FOOTAGE TAPE LOG			
Counter/Frame #	Time	Description/Comments	A/V Quality
00:00:00:00- 00:00:30:00	30 sec	Nesters waking up (take 1)	audio noise from blankets; video good
00:00:30:00- 00:00:38:06	8 sec	Nesters waking up (take 2)	good A/V*
00:00:38:06- 00:00:42:19	4 sec	Close-up of clock (7:25 A.M.)	good (no audio needed)*
00:00:42:19- 00:00:52:11	10 sec	Carol Nester brushing teeth (take 1)	no toothpaste on brush (no good)
00:00:52:11- 00:01:04:12	12 sec	Carol brushing teeth (take 2)	good A/V*

Figure 11.16 *Sample of a log of the raw footage from "The Nesters." Here the person who created the log indicated the best takes with asterisks in the "A/V Quality" column.*

Logging Footage and Making an Edit Decision List

During the search process and while replaying tapes, you can keep a written log of the raw footage, including the location and duration of each segment (Figure 11.16). You can also formulate an edit decision list before performing actual edits. The **edit decision list** or **EDL** (Figure 11.17) is a written, rough-cut version of the proposed edited program, listing in order the beginning and end points for every segment. Many production houses use systems that accept EDLs on disk. With an EDL, you can do most of your tinkering with edit choices without having to sit at an editing console.

Writing an EDL is usually not optional except in the most basic editing situations. For simple assignments that skip the EDL phase, edits may be executed as soon as edit points have been selected. Time constraints often dictate whether or not you generate an EDL before performing actual edits.

Frame Counting

As tape advances or rewinds through each deck, a **frame counter** advances or reverses for each tape wherever control track is recorded. The frame count is displayed as an eight-digit code arranged in four pairs of numbers referring to hours, minutes, seconds, and frames (Figure 11.18). For example, if the source readout on the edit controller, starting from zero, counts frames from the beginning of the control track, the digital display after one minute of con-

EDIT DECISION LIST

PROGRAM: **THE NESTERS**   DATE: **November 15, 1998**

Segment	Description	Tape #	Hr	Min	Sec	Fr	Hr	Min	Sec	Fr	
1	Nesters waking up to clock radio (take 2)	1	00	00	30	00	00	00	38	06	
2	Bill Nester in the shower	1	00	17	21	18	00	17	25	18	
3	Carol brushing teeth (take 2)	1	00	00	52	11	00	00	55	10	

Figure 11.17   *An edit decision list for the opening shots of "The Nesters." From the raw footage log shown in Figure 11.16, particular shots have been selected and placed in order as an initial guide to the editing.*

tinuous control track will read 00:01:00:00, where the first pair of numbers (on the far left) designates the hour, the second pair the minutes, the third pair the seconds, and the last (on the far right) the frame number. Assuming a stable and continuous control track, the readout of a continuous frame count after one hour, twelve minutes, and forty-five-and-a-half seconds would be 01:12:45:15.

Of course, the digital counter can be zeroed out at any time by the operator with the press of a button, and a new frame count can begin at any point that has control track. The major advantage of the digital display is that it makes it easy to log the location and duration of any segment of video. From written logs, you can construct a rough EDL even without access to a videotape recorder.

Assemble and Insert Editing

Once you have chosen edit points for a series of segments, you may perform edits a couple of different ways. Depending on your needs, you may choose to transfer *all* tracks from the source tape to the edit tape, including the video track, both audio tracks, and the control track. This type of editing, called

01:12:45:15

Figure 11.18   *A frame counter.*

assemble editing, is most often done when raw footage must be put into final form quickly, as is often the case in ENG operations.

To perform assemble edits, you must first delegate the edit controller to do such work by pressing the appropriate button on the console. Once in assemble mode, the edit deck erases all material (including the control track) on the edit tape ahead of the signal being laid down by the source deck.

Instead of assemble editing, you can transfer *selected* tracks of audio or video from the source tape to the edit tape. Transferring selected tracks is called **insert editing**. Insert editing is done when more extensive postproduction work is needed. For example, you would use insert editing to add asynchronous audio to a video track. You would also use it to add video inserts and/or cutaways (called *B-roll* or "beauty" shots) to an interviewee's long speech (called primary or *A-roll* footage).

After selecting insert mode on the edit controller, you need to specify the particular tracks of video and audio that you wish to transfer to the edit tape. For this task there are usually three buttons, designated "video," "audio 1," and "audio 2," which set up the edits so that only the requested tracks transfer when you give an execute command.

An important difference between assemble and insert modes concerns the way the invisible but critical control track is handled. In assemble mode, all audio and video tracks, *plus the control track,* are transferred from the source tape to the edit tape. In contrast, in insert mode, selected tracks of audio or video, *but no control track,* are transferred. This means that in insert mode, you must first lay down a control track on the edit tape for either the length of the tape or the length of the program (whichever is shorter). Unfortunately, laying down control track can be done only in real time, and this can become a burden, especially when working on a tight schedule. For a one-hour show, it would take an hour to lay down a complete control track on the edit tape. Because of this time requirement, some shops keep a supply of **preblacked tapes,** tapes with continuous control track already recorded on them.

Assemble mode has its own problems. Although the control track from the source tape should transfer to the edit tape whenever editing occurs, sometimes the source deck does not match the new control track exactly to the control track that was laid down in prior edits. In addition, some dropout of control track can occasionally occur between edits. When mismatches and gaps occur, some **tearing** of the video signal can result, causing picture roll, momentary glitches, loss of color, or other problems. That is why assemble mode is best used when only a few simple edits must be performed for relatively long segments.

In summary, assemble editing is fast, but sometimes results in an imperfect control track. Insert editing takes more time, but the unbroken, prerecorded control track avoids discontinuities or gaps in the video signal.

Executing Assemble Edits To execute assemble edits, begin by delegating the edit controller to assemble mode. Then specify an *in* point on the source, and edit tapes by pressing the appropriate buttons. You can also specify an *out* point on either the source or edit tape, but it is not required. If an *out* point is not specified, the assemble edit will simply continue until you either stop the machine or the control track runs out.

Important: When performing assemble edits, the edit tape must have a little bit of control track recorded on it, just for start-up purposes. Usually about a minute of control track is enough to get the process going. Control track is then transferred from the source tape to the edit tape as each edit is performed. To put initial control track on the tape, simply record one minute of black onto it from a camera, or dub it from a preblacked tape. The system will then have enough information to execute assemble edits.

Executing Insert Edits   To perform insert edits, begin by delegating the insert mode on the edit controller. Be sure you have properly striped the edit tape with a continuous control track. You can confirm that you have a continuous control track by watching the action of the frame counter as either the source or edit tape advances or rewinds through the edit decks. If you see moments when the frame counters stop counting, you know the tape contains segments that lack control track. If there are gaps, you will need to insert continuous control track on the tape before continuing.

When the tape is ready, specify which parts of the video signal you wish to dub. Remember, the insert mode permits you to choose just the video track, one or both audio tracks, or any combination of these. Locate the beginning of the desired segment by shuttling through the tape, watching the monitor and listening to find it. Then jog and trim, if necessary, to locate precisely the frame you want to use as an edit point. When you find the desired frame, mark an *in* point for the source tape by pressing the appropriate button on the console. Then repeat the process to locate the end of the segment, and mark an *out* point. Next, locate and select *either* an in point *or* an out point (not both) for the edit tape. Do not specify the remaining out or in point, because the editing system calculates it from the information you have provided.

If you like, you can specify in and out points for the edit tape rather than for the source tape. Then specify just one such point on the source tape. The machine does not care which way you do it.

Previewing and Performing Edits   In either assemble or insert mode, you can preview the edit you have requested by pressing the *preview button* on the edit controller. Previewing gives you a chance to see what the edit will look like before you accept it. If you decide you like it, you can execute the actual edit by pressing the *edit/perform button*. In edit/perform mode, the machine dubs a copy of the selected material onto the edit master, and when the preselected out point is reached, the tapes come to a halt and the edit is complete.

Prerolling Tapes   When previewing or actually performing edits, the system automatically *prerolls* both tapes and then rolls them forward to their *in-points*. Usually a three-to-five-second preroll time is programmed into the edit controller by the operator beforehand. Prerolling permits the tapes to reach proper speed before passing the designated *in-points* so that edits are produced cleanly.

SMPTE/EBU Time Code

Although control track editing is quite precise in that edit points can be fixed at the beginnings and ends of frames rather than somewhere in between, it is not perfectly frame accurate. This is because videotape can stretch or slip as reels of tape are spun back and forth during searches. In addition, during high-speed rewinds or fast-forward searches, so many frames are speeding past the tape heads that the machine may sometimes simply miscount frames. As a result, strict correspondence between frame numbers and specific video frames may be lost. In short, control track editing gives you no guarantee that a particular video frame will have a permanent address throughout the editing process.

Even bigger discrepancies arise from the fact that the counter can be zeroed out at any time or can begin on a tape that has not been fully rewound. When this occurs, the frame numbers appearing in early written logs or EDLs may be so different from subsequent counts that searches based on them are no longer accurate. To surmount these difficulties, it is necessary to give every frame a unique and permanent address.

As we saw in Chapter 7, a system for performing frame-accurate coding of videotape was developed by the Society of Motion Picture and Television Engineers. It is called *SMPTE time code*, or, since it was also adopted by the European Broadcasting Union, *SMPTE/EBU time code* (usually pronounced "simply time-code" for short).

Two types of SMPTE time code can be recorded on videotape. One, called *longitudinal time code (LTC)*, is recorded lengthwise on either the audio or cue track. The other, *vertical interval time code (VITC*, pronounced "vitsee"), is recorded vertically in each frame of the video track.

For some grades of tape, LTC may be found in a special address track. On VHS tape, for example, LTC is located on an audio track. When LTC time code is recorded on an audio track, it must be recorded at a high enough level to be read clearly but not so high that it causes audio problems such as cross-talk or squeals. You must also guard against the possibility of losing it by overdubbing audio.

An advantage of VITC time code over LTC is that it cannot be lost when audio tracks are added in postproduction. A disadvantage is that VITC records only when video is recorded and cannot be read when the tape is fast forwarding.

Time code may be recorded at the time tape is shot, or it may be added later. In-studio productions often record it during production. In field work, time code may be recorded as tape is shot if the camera is equipped with a built-in time-code generator or if an external one is available.

Setting Time Code Before a Shoot   In setting time code for a shoot, it is generally advisable to start from zero and move forward continuously for each and every tape. This way, every frame of video will have a unique address, and all frame numbers will be arranged sequentially without gaps. This is especially good for computer editing, because any large gaps between segments or repeats of frame numbers may confuse the computer.

It is also possible—and, under some circumstances, desirable—to tie time code numbers to the time of day. This can be helpful if you wish to coordinate parts of your shoot with a studio clock or some other time source. It can also make it easier for production personnel to log segments and prepare EDLs.

Drop-Frame and Non-Drop-Frame Time Code   The original standard frame rate of 30 frames per second, adopted by the National Television System Committee (NTSC), was later changed to 29.97 frames per second to accommodate the chrominance signal needed for color television. This revised NTSC standard was adopted throughout most regions of North America, Japan, and South America. Other countries, where 50 Hz AC power is the standard, use a different system called *Phase Alternation by Line (PAL)*. The PAL system, which uses 625 lines, 50 fields, and 25 frames per second, is most prevalent in Great Britain, Europe, Africa, the Middle East, and the Far East. A third system, called *SECAM* (for *Système Electronique Couleur avec Mémoire*), which also uses 625 lines at 25 frames per second, is in place throughout much of the rest of the world.

As far as U.S. producers are concerned, the odd frame rate of 29.97 becomes important in long programs because it can cause time discrepancies in edited material. For this reason, time code can be altered slightly to provide more accurate timing. **Drop-frame time code** compensates for the difference between 29.97 and 30 frames per second by ignoring frames 00 and 01 each minute except at every tenth minute. Actual video frames are not lost, but the frame numbers are dropped, thus eliminating discrepancies between actual time and time code values. For each hour of video in drop-frame time code, 108 frames of video are uncounted, amounting to a net correction in time code display of 3.6 seconds per hour.

For short programs, where the small discrepancy is hardly noticeable, **non-drop-frame time code** can be used. For each hour of non-drop-frame time code, one hour and 3.6 seconds of actual time elapses.

Time Code Readers   After time code has been recorded on tape, it can be read several ways. A *time code reader* must be used to see the frame numbers during playback and editing searches. With a time code reader, the time code may appear either on the top or bottom of the screen along with its corresponding video footage, or it may be fed to a separate digital readout.

Time code numbers can also be "burned in" to a video dub of footage and then viewed on a regular videotape recorder (VTR). This type of arrangement is called a **window dub** because the time code numbers appear in a small window on the screen (Figure 11.19). Window dubs are useful for preparing logs and EDLs that can then be used to plan the editing. Working from a window dub keeps expenses down because you can search and view the footage on a regular VTR almost anywhere, without paying editing equipment rental fees. Of course, window dubs cannot be used for actual editing because the footage contains time code display that cannot be erased.

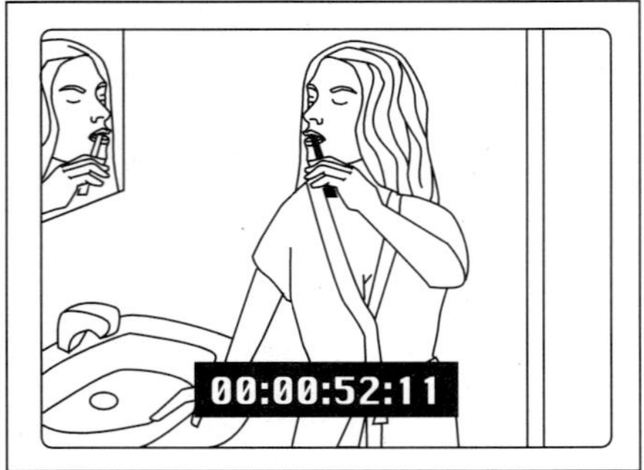

Figure 11.19   *A window dub showing the time code numbers "burned in" on a copy of the video footage.*

The Expanded Meanings of *On-Line* and *Off-Line* Editing

The practice of generating detailed and accurate EDLs from viewing window dubs before entering the actual editing phase of a production has expanded the meanings of the terms *off-line* and *on-line editing*. *Off-line editing* now generally refers to the entire preparation phase after raw footage has been shot, when master tapes are viewed to prepare logs and/or EDLs and are searched for edit points prior to entering the editing suite to execute the final copy. Similarly, the term *on-line editing* has expanded to include more than just live cutting among several cameras during studio broadcasts. Now it includes editing on high-end equipment to bring the program into final form. Generally, the more editing work is done off-line, in the new sense of the term, the less expensive and pressured the on-line phase is.

Multisource Editing   Expanding your editing facility by connecting more than one source VTR to the editing system through a video switcher enables you to use more sophisticated transitions than just cuts (Figure 11.20). With two or more source VTRs, each running a separate source tape, you can dissolve and fade between sources.

This arrangement also permits you to perform **A-B rolling,** creating an edit master from two video sources. For example, if two tapes (the A and B sources) share a common time code, you can roll both tapes at the same moment from the same point and then treat them as if they were two video sources in a live studio production, cutting, dissolving, and wiping between them as you wish. During the entire time, the edit master takes the feed you send through the video switcher and rerecords it. The product that results is the mix of video signals you sent to the record unit. Of course, it is possible to use more than two sources. If three sources are used, the term is *A-B-C rolling*.

Another term to be aware of is **A-B roll editing**. If, instead of doing the shot selection during an A-B roll, you delegate it to an editing control unit or a computer, you are performing A-B roll editing rather than simply A-B rolling. Why is this distinction necessary? In the case of A-B rolling, both sources roll from a common time code point, whereas in A-B roll editing, the sources do not have to be in sync with each other. One advantage of this latter system is that changing tapes on the source machine involves less work.

Analog and Digital Recording Systems: Tape and Disk Formats

Since the advent of videotape, the industry has continually moved toward developing smaller formats to replace the 2-inch version first introduced in the mid-1950s. Miniaturization reduces costs and makes shooting and recording easier, especially in the field. Today high-quality video recording is possible with tape formats small enough to fit on cassettes nearly the size of a standard audiocassette.

One major technical innovation was **helical scanning**, which made it possible to use narrower tape widths to record analog video. In this arrangement (Figure 11.21), video information is recorded *on a slant* on the videotape rather than in perpendicular fashion, thus providing greater space for each frame of video on a given tape width. Since helical scanning was introduced, several narrower tape formats have been developed.

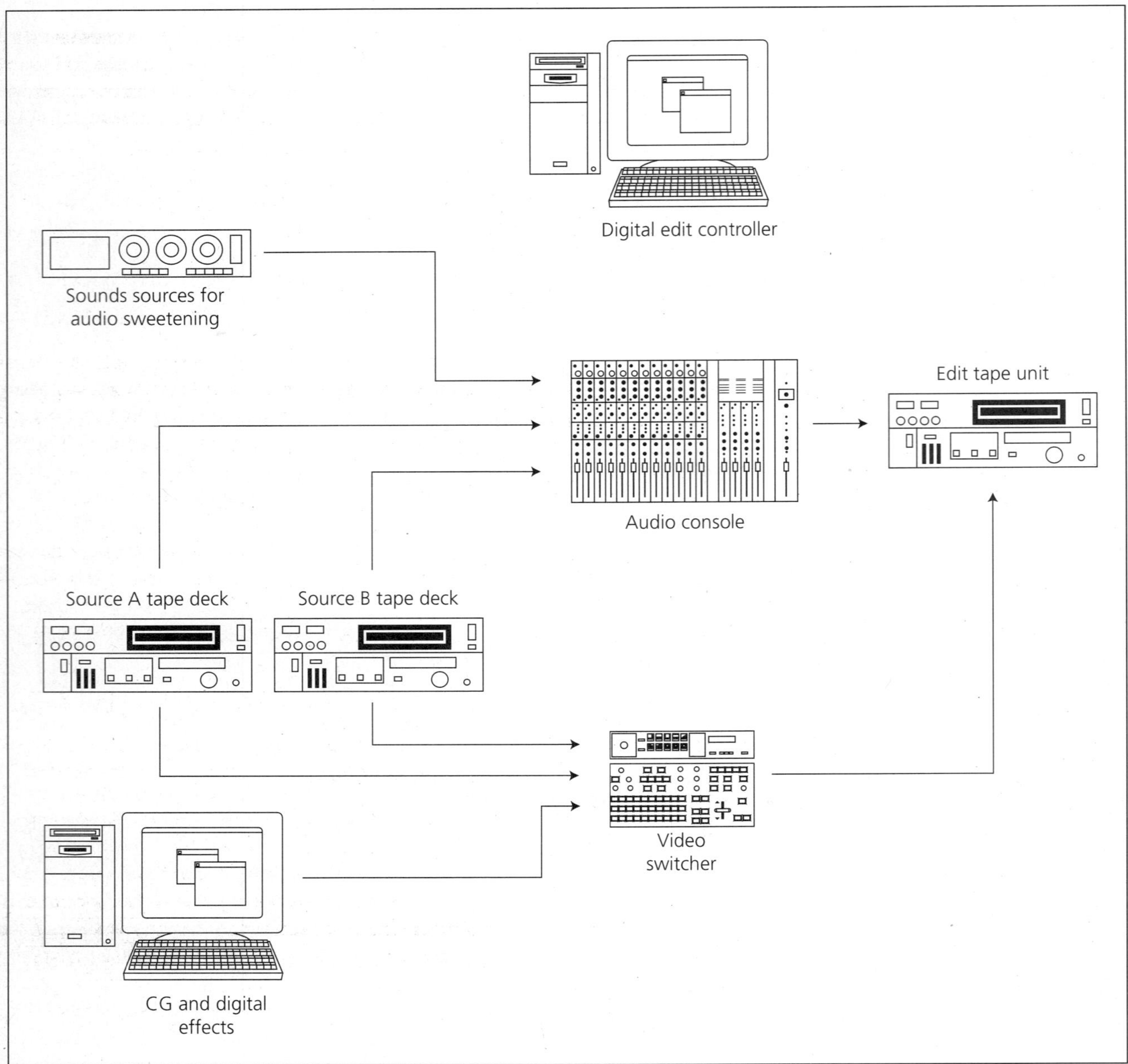

Figure 11.20   *Schematic of typical components used in multisource editing. The edit controller, shown at the top of the diagram, mediates signal flow among all the other components.*

Another trend is the move from analog to digital recording, which has brought about a number of newer digital tape formats. In addition, disk-based recording has arrived, making tapeless recording of digital video directly onto computer disk possible, thus eliminating the time-consuming step of transferring video from tape to disk before editing with a computer. One example of this technology is the *CamCutter,* a digital newsgathering (DNG) camera made by Ikegami that uses a digital hard drive produced by Avid

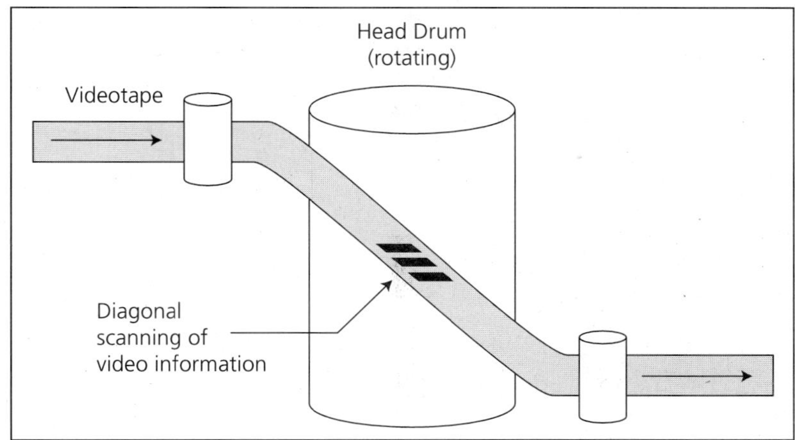

Figure 11.21 *The helical scanning system. Because the tape runs at an angle across the head drum, the video information scans in a diagonal pattern on the tape. This way, video information can be recorded on a greater area of the tape, which enables the tape to store more information.*

Technology for recording video and sound. Because it stores video on a digital hard drive, the system allows the user to edit shots right in the camera. Another advantage of tapeless digital recording is that it saves time during the editing process, since the editor no longer needs to search source tape for desired segments. The random-access feature permits you to select segments in any order and to call them up immediately. This kind of system should eventually eliminate tape-based recording, though that reality is still some years away.

Whether recording is done on tape or disk, or is in analog or digital form, the video signal may be recorded in either a component or a composite format. In **composite recording** systems, both color (chrominance) and brightness (luminance) parts of the television signal are recorded together using only one wire. As the original recording method standardized by the NTSC, composite recording is referred to as the *NTSC signal*.

In contrast, two newer **component recording** systems, called either *Y/C* or *RGB system* recording, have been developed. In Y/C system recording, the luminance (the *Y* part of the signal) and the chrominance (the *C* part of the signal) are recorded separately using two separate wires. In RGB system recording, the red, green, and blue portions of the video signal are recorded separately using three wires. The advantage of using separate circuits to collect the different parts of the video signal is that it yields a higher-quality signal that loses less image quality when undergoing repeated dubbing during editing.

However, there is a downside to separating video signal components from one another. When the signal components are kept apart, they require more expensive circuitry in all the signal processing equipment, including monitors, switchers, and editing decks. In addition, to be broadcast, the final product must eventually be made NTSC compatible through a conversion process.

Since many of the recording systems just described are incompatible with one another, it is important to know the different capabilities and limitations

TABLE 11.1

Analog and Digital Tape and Recording Formats

Tape Type (name and width)	Recording Format	Production Applications	Comments
Analog			
VHS (1/2 inch)	NTSC composite	Home video, movie rentals	Not for broadcast use
S-VHS (1/2 inch)	Y/C component, recorded as a composite signal	ENG, corporate and closed-circuit applications	"Downward compatible" with VHS (lower-quality VHS can be played on S-VHS equipment, but not vice versa)
BetacamSP (1/2 inch)	Y/C component	High-end studio, ENG, and corporate video	SP stands for "superior performance." RGB signals are separate. Also has four audio tracks. Tapes cannot be played on other systems.
Hi-8 (8 mm)	Y/C component, but can be rigged for NTSC output	Home video recording and professional use	High-quality recording, but should be transferred or "bumped up" to S-VHS or BetacamSP for editing work. Has three audio tracks.
U-matic (3/4 inch)	NTSC composite	ENG and corporate video	Outmoded by 1/2-inch formats but still in use in some shops
M II (1/2 inch)	Y/C component	Same as BetacamSP	Overshadowed by competition
Digital			
Digital Betacam (1/2 inch)	RGB component	Studio and field production, professional quality	May be used for intensive editing and postproduction work. Also able to handle *compression* of the video signal, by which video information is digitally reduced for storage.
DVCPRO (1/4 inch)	RGB component	EFP, professional editing	Handles compressed video. Has stereo audio. Provides excellent quality, even after complex editing jobs.
DVCAM (1/4 inch)	RGB component	Same as BetacamSP; extensive editing capability	Performs editing and dubbing with compressed video. As with most digital video, virtually no loss of signal quality.
Digital-S (1/2 inch)	Y/C component	Professional-level video, comparable to BetacamSP	Downward compatible with S-VHS. Like DVCAM, permits editing with virtually no loss of signal quality.

(continued on the next page)

TABLE 11.1

Analog and Digital Tape and Recording Formats (continued)

Tape Type (name and width)	Recording Format	Production Applications	Comments
D-1 (3/4 inch)	RGB component	Excellent professional quality	Excellent for postproduction editing. Often used in creating complex effects and animations.
D-2 (3/4 inch)	NTSC composite	Excellent quality for broadcast and in-house, corporate, and closed-circuit setting	Since compatible with NTSC, can be used with other broadcast sequipment without extensive reconfiguring.
D-3 (1/2 inch)	NTSC composite	Also excellent quality, useful for both studio and field settings	Compatible with NTSC, so good for broadcast use. Small format makes it good for field production. Like other digital formats, permits extensive editing with virtually no loss of picture quality.
D-5 (1/2 inch)	RGB component	Superior quality	Excellent in postproduction. Contains four audio tracks, capable of CD-quality sound. Small format makes this an excellent choice for field production.

---

of the available recording formats before shooting and editing begin. Table 11.1 should help you recognize which tape formats are compatible with which recording systems.

Nonlinear Editing with Digital Video

As Table 11.1 makes clear, a number of digital formats are now readily available. Videographers can record on digital videotape, or digitize analog video, and then store the information on tape or computer disk (or on large-capacity laser disks) and edit it via computer.

Once on disk, digital video can be manipulated like files of computer data. Digital editing offers major advantages over the linear process described so far in this chapter. For one thing, it enables you to access segments randomly, the way a word processor accesses words and paragraphs in a written manuscript, rather than searching sequentially along a tape. In contemporary jargon, this random-access feature allows us to call the process **nonlinear editing**.

In nonlinear, digital editing, segments can be located more quickly than is possible in linear formats. Digital editing is also frame accurate, and it allows you to move segments around repeatedly (and even make copies of copies) without the slightest loss of signal quality. This last characteristic is a great improvement over analog video, which loses some quality with each successive dubbing. Digital editing allows you to work with first-generation signal quality no matter how many corrections you make during the editing process.

Furthermore, digital editing makes it much easier to revise your editing decisions by inserting or deleting segments. On an analog tape, once the signals are recorded in series, you cannot move them to make room for additional segments or expand them to fill space left by segments you may want to delete. Last-minute changes, then, can be done only by re-editing the entire program, a time-consuming prospect. Time pressure might move you to consider using a partially edited program as the new master and rerecording it with new material. But this approach bumps already edited portions of the final program another generation away from the original, further degrading overall signal quality. At times, the only solution is to scrap the edited work and start over.

Digital video solves this problem. Just as a word processor can move words or paragraphs to make room for new material without forcing you to retype the entire manuscript, digital video technology allows you to shift material around repeatedly to accommodate new additions or deletions without re-editing. This shifting around of material to make room for late additions or deletions is called **rippling**. Not only is it relatively easy, but it can be done without any loss of signal quality.

Together the random-access and rippling features of digital video editing provide major advantages over linear systems. They give you more technical freedom to create the finished program in any order you wish without having to worry about where things will fit.

KEY TERMS

continuity editing *(267)*
classical editing *(273)*
formalism *(274)*
realism *(275)*
dynamic editing (thematic
   montage) *(276)*
pictorial complexity *(277)*
mise en scène *(277)*
synchronous sound *(280)*
asynchronous sound *(282)*
on-line editing *(283)*
off-line editing *(283)*
control (pulse) track *(284)*
edit controller *(284)*
shuttle mode *(284)*
linear editing *(284)*

edit decision list (EDL) *(285)*
frame counter *(285)*
assemble editing *(287)*
insert editing *(287)*
preblacked tape *(287)*
tearing *(287)*
drop-frame time code *(290)*
non-drop-frame time code *(290)*
window dub *(290)*
A-B rolling *(291)*
A-B roll editing *(291)*
helical scanning *(291)*
composite recording *(293)*
component recording *(293)*
nonlinear editing *(295)*
rippling *(296)*

QUESTIONS FOR REVIEW

1. What are some practical and aesthetic reasons for using editing in video production?

2. Define *continuity editing, classical editing,* and *dynamic editing.* Compare and contrast them, and give examples of each.

3. How does shot juxtaposition affect audience interpretation of cause and effect in an edited sequence? Use an example to illustrate.

4. Describe the formalist and realist traditions and how they influence editing decisions in different situations.

5. Generally speaking, how do pictorial complexity and mise en scène affect editing pace?

6. In what ways might technology development affect editing aesthetics? What recent developments have had the most impact on editing practices?

7. Why is the control track important in both analog and digital editing?

8. What advantages does time code editing offer over control track editing?

9. What are the differences between assemble and insert editing?

10. What advantages does nonlinear editing offer over linear editing?

PART THREE

12 Writing and Script Formats

The art of writing is the art of applying the seat of the pants to the seat of a chair.
Mary Heaton Vorse

How vain it is to sit down to write when you have not stood up to live.
Henry David Thoreau

An author can hold correct opinions and yet not be able to express them in polished style. To put one's thoughts on paper without being able to organize them or to express them clearly, or without being able to hold the reader with some kind of charm, means one is making an inexcusable mistake of both his leisure and his pen.
Cicero

To write well, there are required three necessaries—to read the best authors, observe the best speakers, and much exercise of his own style.
Ben Jonson

If you wish to be a writer, write.
Epictetus

The quips and counsel from these veteran authors offer invaluable advice to video scriptwriters. They know good writing is best learned by doing: by living, observing, reading, and writing. Further, it helps to be realistic about one's abilities, as well as tenacious and able to take rejection. But what in particular do you need to know to write for video as opposed to writing a play, a novel, or an essay?

Since video writers may deal with a great number of topics and quite different types of subject matter, both fiction and nonfiction, this chapter cannot cover all the situations you may encounter. However, it will offer some valuable guidelines for video writing and prepare you for using basic script formats. The topics covered include:

PRINCIPLES OF VIDEO WRITING

characteristics of the medium • characteristics of audience and genre • characteristics of the competitive market

BASIC SCRIPT FORMATS

the rule of pragmatism • the director's working script • news scripts • scripts for commercials • full-page scripts • scriptwriting software

A FINAL NOTE

PRINCIPLES OF VIDEO WRITING

Certain features of video writing make it different from other types of writing. The nature of the medium itself, the audience and genre, and the competitive market all influence the scriptwriter. In the most obvious sense, the work of the television writer never comes to the audience on paper. Rather than being read, the writing is heard, after being embedded in a program along with other sounds, images, and actions. This has strong implications for the writer.

Characteristics of the Medium

Television can deliver sounds and images live—as they occur—such as in news and sports. It can also present performances repeatedly, thanks to storage media such as film, tape, and disk. Either way, video programs convey a quality of natural immediacy and action. Video's strength lies in its ability to convey human action through real-time audiovisual displays. Even complicated computer systems have yet to render human action at a comparable level of picture quality and immediacy.

What writing styles fit best with these characteristics of the medium? Most television writers advise using simple, concise language, in a direct, natural, conversational tone, in the present tense. Such language works well on video because it fits video's active, immediate, and dynamic nature; it is also quickly and easily comprehended. Unlike narrative writing, it emphasizes showing over telling, revealing over describing, and demonstrating over explaining, taking advantage of the medium's ability to showcase processes as they unfold. Imagine a cooking program that squandered its precious airtime talking about recipes but never actually preparing one.

Pictures, Sound, and Words: the Goal of Synergy   Though it may be humbling for the television writer to admit, the dramatic, arresting images of video are likely remembered better than the words that accompany them. Therefore, it makes great sense to let words play a supporting role to pictures. Decide which parts of the story will be conveyed by nonverbal elements, and then write to support the program as a whole. Let the words provide what the other program elements fail to supply.

This does not mean words play a minor role. In fact, in the well-written television program, words and pictures frequently co-determine each other. In radio, you might write a line such as "Why don't you wear this red polka dot dress to the party?" Those words would give the listener a mental image of the dress. However, on television, where the dress might be perfectly visible in a close-up or medium shot, a better line might be "Why don't you wear *this*?" These simplified words demand a picture to back them up, and the picture in turn calls for words that are simple and direct.

Pictures often serve to support the claims made by the words. For example, if an announcer's voice-over claims that microwave oven X brings water to a boil faster than its chief competitor, the claim will be more persuasive if it is accompanied by a visual demonstration of the product's superiority. As another illustration, imagine the difference in impact between merely hearing about a plane crash and hearing about it while seeing graphic video of the disaster. Such *synergy,* in which the combination of program elements has a greater impact than any individual element alone, is one main goal of the television writer.

In addition to visuals, the writer can sometimes rely on sounds to reduce or replace the need for verbal description. In news coverage of an intense forest fire, if location sound features loud crackling and wind, verbal descriptions of the noise may be superfluous. In fact, injecting surplus verbiage over the nat sound would risk drowning out the very noise that generated the description.

The Impact of Editing   Besides the effects of shot content and nonverbal portions of the soundtrack, the writer's task is influenced by the extensive editing often done in video. For example, television writers providing commercial and promotional copy often write scripts for thirty-second spots that reduce them to ten-second versions known as **liftouts**. To achieve coherence in either version, writers work to preserve sense and provide smooth transitions from sentence to sentence.

More generally, since videotape editing often condenses stories into their highlights, the pace and order of shots and sequences may change during the production process. Such changes can affect the sense, flow, rhythm, tempo, and tone of the written material. For this reason, writing changes are often needed to preserve meanings lost or changed by program editing. Experienced video writers know to prepare for this possibility and not become overly frustrated by it.

Fixed Program Time and Order   With newspapers, magazines, and books, time spent with a particular story may differ widely among audience members. Different people have faster or slower reading speeds; they read for different purposes, and they have different amounts of time available. Readers can also skip around if they choose or review items of interest a second or third time. In short, print gives readers a great deal of control over the amount of time they spend with the material and the order in which they attend to different items.

In contrast, viewers of live television are tied to a fixed schedule. Even with taped video, when programs are seen the way they are meant to be seen, viewers are exposed to sequences in a predetermined order, each for a pre-specified amount of time.[1] Once a speech is delivered, it is gone. What are the implications of this for the writer?

As mentioned earlier, it is helpful to keep the language simple and conversational. Avoid long sentences. Try to present ideas clearly the first time. Particularly important, *pace* speeches for television at an acceptable rate for the majority of viewers. Material presented too slowly can bore viewers or, worse, insult their intelligence, motivating them to tune out. On the other hand, if material is presented too quickly, viewers can become confused. Therefore, try to present speeches at a pace that will keep them interesting while not confounding the viewers.

---

[1]Of course, with videotape, it is possible to skip chunks and go back and forth as in a book or magazine; it's just slower and clumsier because of the limitations of the tape. Perhaps when we complete the move to faster random-access storage systems, people will come to use videos more as they do the print media. However, the pace of presentation *within* selected units of material will still be more fixed in video than in print.

Telegraphing what is coming next is often useful. **Telegraphing** involves warning viewers about upcoming topics so that they are prepared to receive the ideas more easily. A common example occurs when the program host tells viewers to get a pencil and paper ready for an important phone number or address. Another useful technique is **recapping,** restating important ideas to refresh the audience's memory. This may be especially appropriate when the material is complex, and it also helps late tuners-in to catch on to what is happening.

To check the pace and the need for telegraphing or recapping, gauge how the script works in real time. Read lines aloud to time them and to assess their sound, sense, and clarity. For more difficult segments, you may need to slow down or recap key statements. Don't be satisfied with merely adequate results. Rework lines to make them more understandable and functional. Then be dissatisfied again. Continually demand more of your writing.

Characteristics of Audience and Genre

As Chapter 1 noted, video programs are usually tailored for very particular audiences. As a writer, you should bear in mind the audience's maturity level, values, opinions, social history, interests, needs, and so forth in deciding how to write your copy.

On the most basic level, a video writer considers how much knowledge of the program's subject matter the audience already possesses. What information can be taken for granted so that program time is not wasted? What information must you provide to bring the audience along at a comfortable pace? Where do you risk boring the audience? Such information may not always be available, but it certainly helps to have as much of it as possible so that you can pitch your script properly.

Consider the U.S. Army's video advertising campaign and the slogan they have used since the early 1970s. The slogan does not say, "Aspire to learn the entire universe of possible roles that will permit you to achieve all of your yearnings." Instead, it says, "Be all you can be." Why is the shorter slogan better? Among other reasons, it is better because it is probably understood by more of the army's target audience, American high school seniors. Similarly, a program about AIDS prevention would be written more simply for a target audience of elementary school children than for an international conference of biomedical professionals and health workers.

Video writers must also be attuned to the needs of the particular *genre*, that is, the basic type or category of the program. There are major differences in the writing of fiction and nonfiction, comedy and drama, variety and quiz shows, sitcoms and soap operas, to name but a few genres. The differences involve not only subject matter but also the overall approach you take with the script, including language usage, level of difficulty, diction, dialect, pace of delivery, and demands for factual accuracy. The audiences for different genres vary in age, socioeconomic status, values, and so on, and audience members have specific expectations for each genre. Thus, to prepare for writing in a particular genre, you should steep yourself in examples of that type of program. For instance, what vocabulary could you use in a period western that you couldn't use in a contemporary police drama, and vice versa? How might the pace of a sitcom differ from that of a tragedy?

Recognizing genre differences does not lock you into imitating exactly what has come before. If that were the case, no innovation would be possible. Although much programming is derivative, there is always room for groundbreaking ideas that develop new program forms. Knowing about genre differences allows you to conform to accepted practice when you want and need to and, conversely, depart from it for special impact. A solid understanding of genre differences permits you to use them to your own advantage.

Characteristics of the Competitive Market

In addition to being influenced by the medium, the audience, and the genre, the video writer must respond to the demands of the competitive market. One obvious effect of this market is that a video script often develops through a number of stages as the financial backers decide whether the program deserves their commitment.

The first stage, as discussed in Chapter 1, is often a **proposal** for a new program, whether it be a comedy, a variety special, a game show, a drama for public television, or a more specialized work for an industry client. The purpose of the proposal is to convince a production company or client that the concept has merit, especially audience appeal. The proposal should describe the program's premise, introduce the central characters or talent, and describe the main program segments.

If the proposal is approved, you may be asked to write a **treatment,** a narrative version of the program designed to give the reader a feel for the characters, setting, and action. Then, once you have described the format and subject matter of the program and the main characters or talent, you may be asked to go beyond the treatment stage to show how they all function in a pilot episode. A **pilot** is a test episode used to gauge the value of the program idea.

In both fiction and nonfiction programs, the proposal and treatment stages are designed to convey an idea of the subject matter and tone of the program. They may also convey typical elements of the action, the pacing, the set design, and any other key features that give the reader insights about the show. The main focus is on identifying the program's audience appeal.[2] In the current television environment, which is rife with choices, the competition is so intense that the need for strong proposals and treatments is greater than ever before.

This competition also affects the way the script itself is written. For example, video writers feel strong pressure to engage the viewers' interest quickly. In past decades, network series invariably started with title, theme, and credits, but many of today's prime-time network programs start "cold," with script material presented even before the credits and opening theme. This **cold start** is designed to grab an audience that is also being wooed by cable and satellite programs, video rentals, and numerous other independent programming sources. Therefore, an arresting lead is crucial: a program's opening words should command attention and create immediate audience interest. Then the follow-up should hold that interest so that the viewer does not start clicking the remote control to search for other options.

---

[2]For a full description and detailed examples of different types of program treatments and proposals, see Blum (1995).

Today's successful television writers know that they must deliver story interest within seconds, not minutes, and that any lag can result in lost audience. The intense market competition influences the writer to cut to the chase, so to speak, without wasting any time.

BASIC SCRIPT FORMATS

Television scripts assign speeches to particular talent. However, scripts that do *only* this are far from complete. The extra-verbal information in television scripts includes cues for music and other background sounds, camera assignments, talent blocking, microphone cues, and notations describing special effects and transitions. Information about lighting and facilities needs, set and prop lists, floor plans and lighting plot diagrams, and other production elements may also be included.

Of course, some programs cannot be fully scripted in advance, because not all of the words spoken during the show are known in advance. Programs that feature ad-lib or off-the-cuff remarks may be written in a **semiscripted format**. Still other programs may be handled using an even less detailed **rundown sheet,** which is little more than a list of a program's major segments in order, with approximate running times. Though semiscripts and rundown sheets do not include all of the words spoken, they can still provide as much of the nonverbal information that full scripts do. In other words, the fact that a program uses a rundown sheet is no reason to omit support documentation such as set and prop lists, a list of facilities, and so on.

Examples of semiscripted shows include interviews, musical variety programs, and sports coverage, for which some (but not all) of the materials have been prepared in advance. Such scripts indicate the order of presentation of segments (if it is known beforehand), and they often contain **in-cues** and **out-cues** for each segment, that is, phrases preceding or ending talent speeches that alert production personnel when it is time to move to the next segment.

News shows may be fully scripted. However, if they feature late-breaking stories or live segments, a semiscripted format may be the most sensible one. For semiscripted news, the script may include notations to alert the director and crew about when and where transitions between stories occur. There may also be notes indicating the technical source for each story (for instance, studio VCR or live feed from the field). Further, the notes can say whether the upcoming story is presented with sound on tape (SOT) or with a reporter's voice-over (VO), and when a segment is to be accompanied by character-generated text (CG) or chroma-key inserts (C/K).

The least structured script format, the rundown sheet, is used for such programs as "The Late Show" and other nighttime talk shows, audience participation programs, and variety formats, in which the order, general content, and approximate running time of each segment is set in advance but the actual words spoken are not known until they are uttered. Figure 12.1 shows an example of a rundown sheet for a three-minute, stationary-talent interview program. Successful programs that start out using a semiscripted format can develop such a regular rhythm over the years that eventually the veteran director and crew become comfortable with just a simple rundown sheet. "The

Richard A. Blum
Television Screenwriter and Professor of Screenwriting, University of Central Florida

Q: Can you give my readers a synopsis of your writing and video career?

A: I have taught screenwriting at various universities and published several books on the entertainment industry. My other credits include serving as senior program officer for the National Endowment for the Humanities Media Division, supervising PBS productions of *Vietnam: A Television History* and *The Scarlet Letter.* At Columbia Pictures–TV, I worked as assistant to the VP for programming, overseeing series like "I Dream of Jeannie," "Bewitched," "The Partridge Family," and "Movie of the Week." I've also worked as senior executive producer supervising productions for Rainbow Programming Services' pay-TV channels, Bravo and Playboy.

Q: That's an extensive background. In terms of writing for television, what principles do you emphasize to young writers just starting out?

A: I always stress that you must write for the production marketplace and that scripts must be producible and castable. If you're going to write for television, you must know exactly what to expect of the pragmatic aspects of the medium.

Q: For example?

A: Take episodic series and TV films as examples. For these forms, your story must have a clear beginning, middle, and end. Often the most difficult challenge is clearly stating the premise of the work as a logline, that is, a sentence or two as you might see them in *TV Guide.* Young writers should also watch series for tips about story structure and character conflicts.

Then, when you develop the story into a script, be certain to use appropriate professional formats.

No one will even read your work if it isn't formatted correctly.

Q: Any professional tips for how to shape characters and plots?

A: I always stress that lead characters must be appealing to audiences, and they must have credible motivations and conflicts. Also, it's important to set up a driving force early in the story that jolts the lead character into action so we care about what happens at the outset.

If you're writing for an existing series, keep the lead characters the focus of the story. If you introduce a new character in an ongoing series, get him or her out by the end of the episode. Otherwise you've created a new set of problems for the producer, who must now build a new character arc and worry about casting.

Q: What can young writers do to get their stuff read?

A: Once you've written a script, get feedback from knowledgeable people so you can revise it. Revise and polish it until it's an acceptable sample of your work. That's when you can finally send it out.

Expect to do many revisions. Remember, television is a collaborative medium, so there are going to be many factors that shape the final product.

Q: When it's in final form, what should the script look like physically?

A: To look professional, it can be very simple. No fancy covers. Just a three-hole punch and metal brads through the margin. Then submit it to an agent, a producer of a similar genre show, or a script competition.

Q: Any specific outlets that you know about?

A: The Academy of Motion Picture Arts and Sciences has a competition called the Nicholl Screenwriting Fellowships, and Lorimar Warner Bros. has one called the TV Comedy Writing Workshops. Both are good outlets for writers just starting out.

Q: What do you look for in a writing intern?

A: Interning is highly competitive in writing. Your best calling card is a script that is a superb sample of your work, in both content and form. You can enter TV and screenplay competitions. The Academy of TV Arts and Sciences runs summer internships that are a wonderful way to get your script looked at by professionals. Two of my students wrote a spec script for a half-hour comedy series, entered it in the Warner Bros. TV writing competition, and were selected for the prestigious comedy writing workshop in Los Angeles. They were hired for script assignments on "Fresh Prince of Bel Air." Through internships, another was hired to do rewrites of some major film productions. So the thing producers look for is a good script in top shape. That's what generates success in the writing field.

Tonight Show" (now with Jay Leno), which ran for thirty years with Johnny Carson as host, is an example of such a program.

The Rule of Pragmatism

The only rule that should govern what script to use and what information to include in it, is the rule of pragmatism: *use what works*. This principle means that the finished script should answer most, if not all, the questions the crew has so that the director can concentrate on the most important functions, including clearing up last-minute problems and conducting run-throughs.

To the extent that the script acts as a proxy for the director, providing answers for the crew, the script is doing what it is supposed to do. Therefore, it is a good policy to ask yourself these questions: Have I included everything I can to make every element of the production clear to all involved? Have I eliminated uncertainty and ambiguity so that everyone knows what is supposed to happen and when? Have I made the script so complete that if, by some twist of fate, the entire cast, crew, and director were suddenly removed from the production, a new group could execute the show from the information in the script? The last question may exaggerate what a good script can provide, but it represents the ultimate goal for every scriptwriter.

The Director's Working Script

If we wanted to convert the rundown sheet in Figure 12.1 into a full script for the three-minute interview program, what might the script look like? Figure 12.2 provides an example. The script shown contains a number of elements designed to reduce uncertainty and ambiguity to a reasonable level, making it possible to execute the program with a minimum of misunderstanding. Let's examine some key features of this script.

Starting at the top of the first page, note that the pages are numbered using an inclusive numbering system: "p. 1 of 4," "p. 2 of 4," and so on. This numbering allows all crew members to see whether they have a complete version of the script in proper order *without having to ask the director*. Next, notice that the script has a title. Every script should have a title, even if it is only for a ten-second announcement. If you are producing a series of spots, the titles permit each one to be identified without ambiguity.

Title: *BOOK LOOK*

Date: November 20, 1998

0:00–0:10	OPENING SHOT: HOST AND GUEST CHAT W/TITLE AND MUSIC
0:10–0:30	HOST INTRO OF GUEST
0:30–0:45	GUEST SAYS HELLO
0:45–2:30	HOST AND GUEST CHAT ABOUT AND SHOW BOOK
2:30–2:45	HOST MOVES TO CLOSE, ANNOUNCES NEXT GUEST
2:45–3:00	TALENT SAYS GOODBYE, MUSIC AND CLOSING CREDITS ROLL

Figure 12.1   *A rundown sheet for a three-minute interview program.*

Similarly, notice the date and client name. Including this information ensures that crew members will know which version of the script they have and who the client is. Some production companies even use different-colored paper to signal when a new version of a script has been written.

Next, notice the **split-page format,** which lays out the audio content on one half of the page and the video content (with other cues) on the other half. This layout makes it easy to link the words in the script to other program elements. In multicamera productions, a split-page format aids in coordinating different camera shots with specific speeches and scenes. This format is standard in most types of video production, although some types, such as single-camera EFP-style production, may use a full-page format such as that used in the film industry. The accompanying Professional Pointers feature lists a number of other characteristics of the split-page format. In reviewing Figure 12.2 and the pointers, consider what other elements you might add to the script to make it better than it is now.

Notice that the script in Figure 12.2 continues after the split-page section for the verbal content. The script goes on to include a floor plan, a light plot,

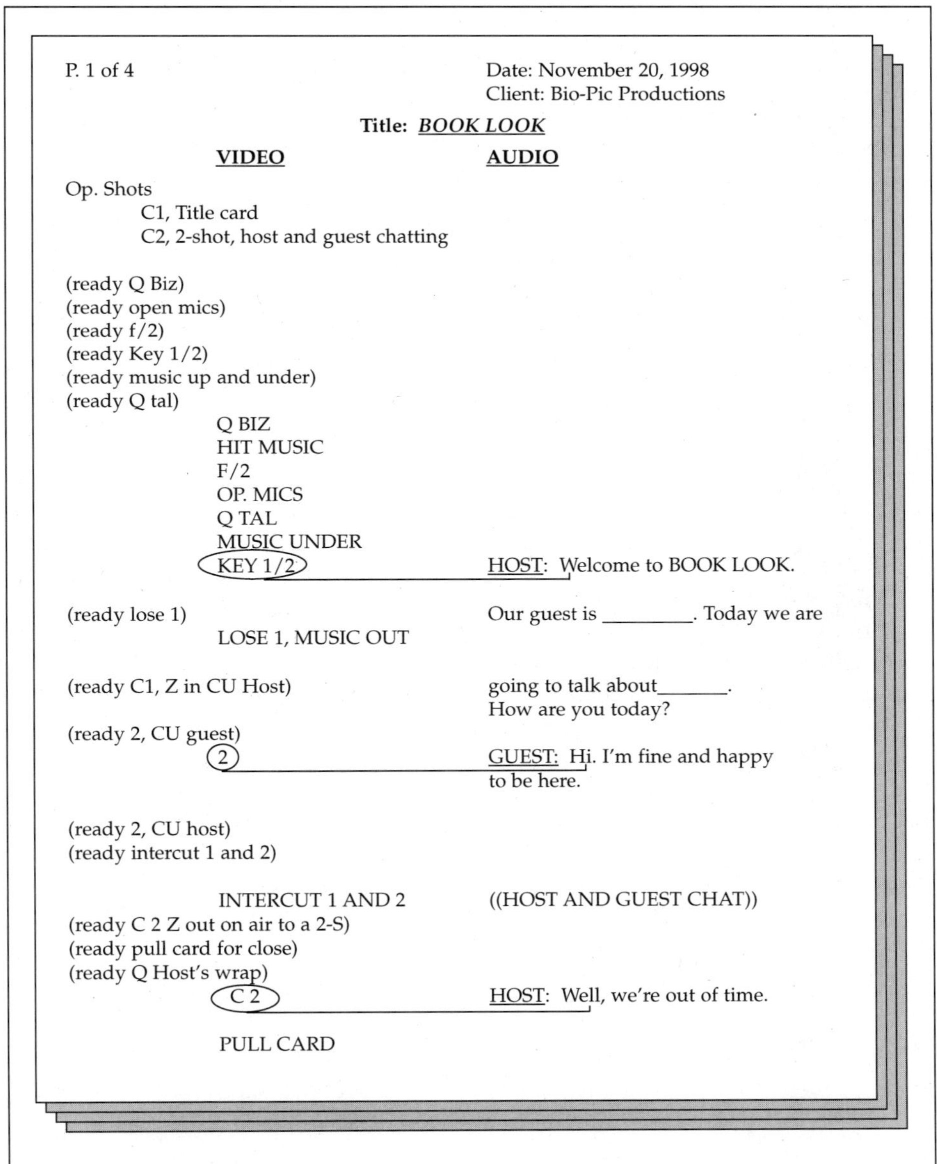

Figure 12.2 *A director's working script for the same three-minute interview program shown in Figure 12.1 (continued on the next four pages).*

KEY TO ABBREVIATIONS

Op. Shots	Opening shots
C1, C2, etc.	Camera 1, camera 2, etc.
Q Biz	Cue business. In this case, host and guest pretend to chat, but in general, a busines cue is any activity talent is given to do before fading up their camera, so that the effect is created of having come in on action (wiping a bar with a rag, dialing a phone, etc.)
Op. mics	Open microphones
Q Tal	Cue the talent to begin the show or some other action
f/2	Fade camera 2
Key 1/2	Key camera 1 over camera 2
Z in	Zoom in
1	Take camera 1 (arrow indicates a specific word or syllable where the take occurs)
F/AUDIO	Fade audio

Figure 12.2   *Continued*

and a facilities (fax) list that indicates set, lights, graphics, mics, props, costumes, and so forth. Set and light lists are provided *in addition to* the floor plan and light plot because the lists make it easy to gather the items needed for a shoot. This portion of the script can list any information that will be important for the crew. For example, notice that the cyclorama on our floor plan is labeled not just "cyc" but "gray cyc." Get as specific as you need to be.

One great advantage of developing a thorough script is that you can place the floor plan and the verbal text side by side during the preproduction phase and confirm whether the shots you have planned are indeed possible. By developing commands with the floor plan as a guide, you guard against the possibility of asking for things that simply cannot be done.

Depending on the company you work for, the crew you work with, and the genre of program you are producing, the level and type of detail provided in the scripts may vary tremendously. You may work with scripts that are quite sketchy compared to the detailed treatment presented here. You may work with some that are even more detailed. In addition, some program genres have unique demands in terms of script preparation. The following sections examine a few different script formats associated with specific kinds of programs.

News Scripts

Television news scripts use the split-page format, clearly indicating the production process for each story. Figure 12.3 shows a typical news script. In this example, a live intro by the studio news anchor is followed by taped ENG video of auction activity with continued voice-over by the anchor. A "font," or CG-generated line of text, identifies the location shown in the video by name. Next, the news anchor is shown on camera again, providing a bridge to the

P. 2 of 4

(ready C1 on card)   Thanks for joining us. Next time, we'll be

(ready K 1/2)

            KEY 1/2   interviewing _________, so don't forget to join

(ready sneak in music)
        SNEAK MUSIC IN   us again for another edition of *BOOK*

(ready lose key)   *LOOK*. So long.
    LOSE KEY
(ready f/blk)   ((HOST AND GUEST CONTINUE
     F/BLK, F/AUDIO

                               CHATTING))

<u>CLEAR</u>
<u>THANKS EVERYONE</u>

Figure 12.2   *Continued*

next part of the story, contained in a news *package,* or prepared tape, featuring an on-the-scene reporter's narration and video of an inventory of jewelry items. Notice that the right-hand column describes the package only in terms of length. The left-hand column identifies the in-cues and the videotape machines to be used. In our illustration, the package uses B-roll and A-roll tapes, each with a separate in-cue. At the end of the package, the reporter appears briefly on camera to sign off, with a font identifying her. Finally, the studio anchor's tag closes the story.

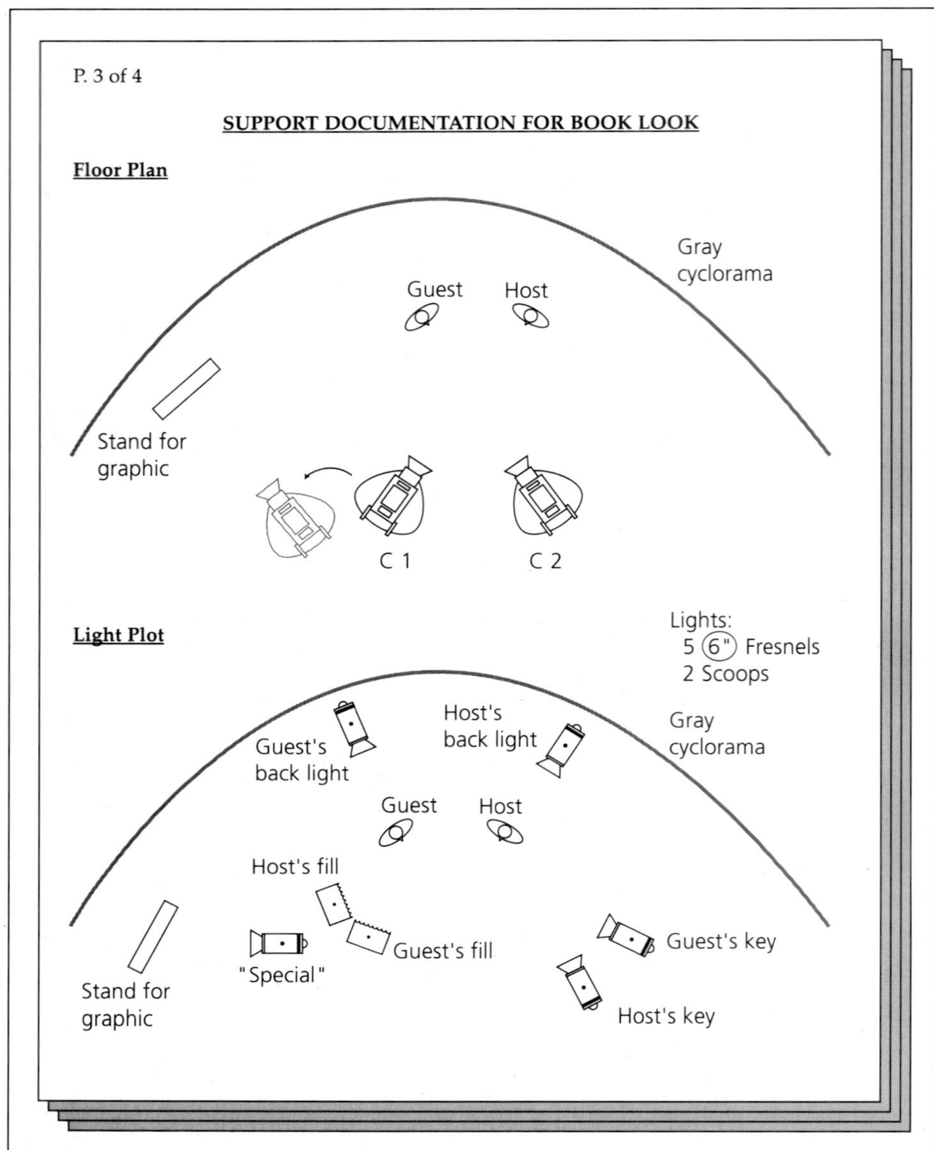

Figure 12.2   *Continued*

Notice these other features of the news script in Figure 12.3:

- Each story is headed by a *slug,* a one- or two-word identifier, in this case "ONASSIS AUCTION." Also included are the date and time, the writer's name, page length for the story, and running time.
- The split-page portion of the script clearly identifies the copy to be read and the reader for each speech. Transitions to new story segments are also clearly indicated.
- Transitions among video sources (in-cues for both A-roll and B-roll tapes) are clearly marked to alert the crew about which sources to cue and when to take them.

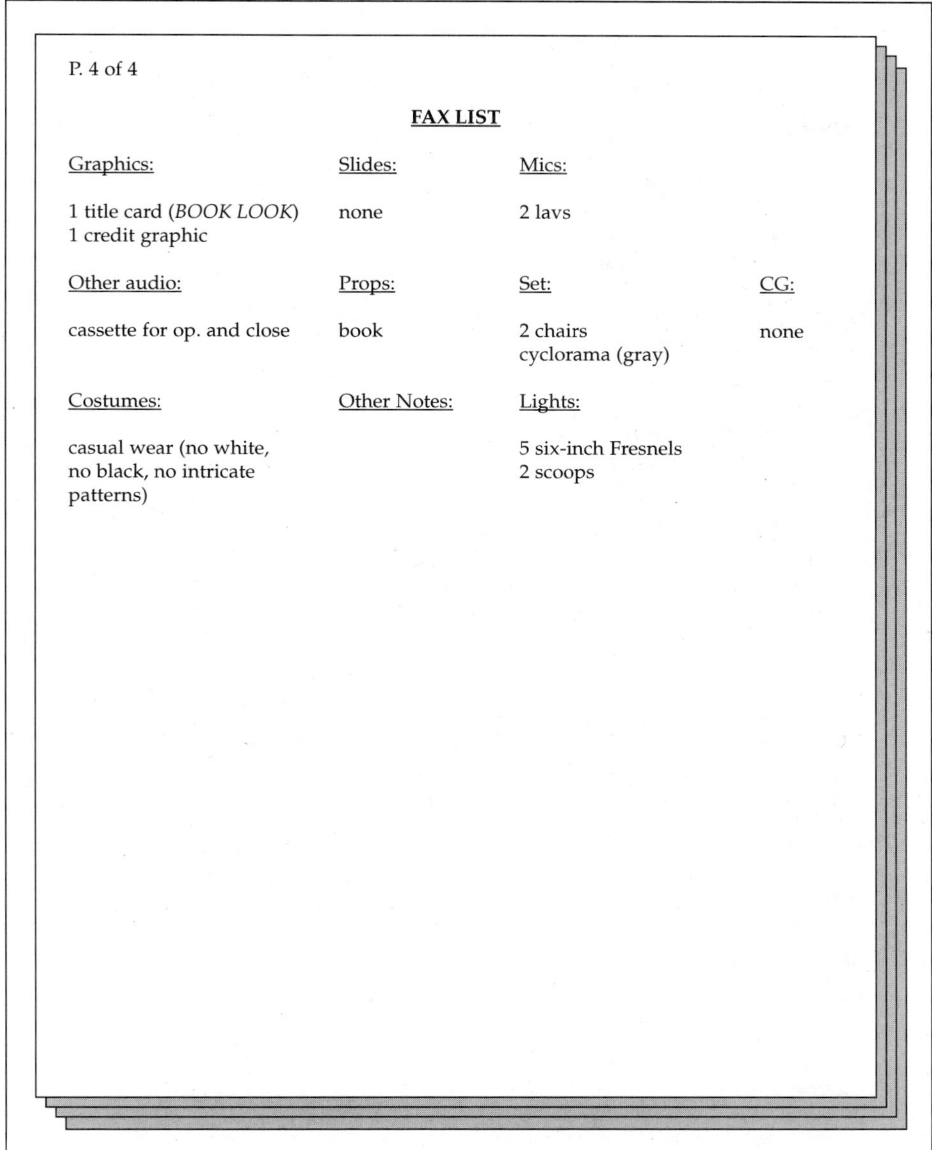

P. 4 of 4

<u>**FAX LIST**</u>

<u>Graphics:</u>	<u>Slides:</u>	<u>Mics:</u>	
1 title card (*BOOK LOOK*) 1 credit graphic	none	2 lavs	
<u>Other audio:</u>	<u>Props:</u>	<u>Set:</u>	<u>CG:</u>
cassette for op. and close	book	2 chairs cyclorama (gray)	none
<u>Costumes:</u>	<u>Other Notes:</u>	<u>Lights:</u>	
casual wear (no white, no black, no intricate patterns)		5 six-inch Fresnels 2 scoops	

Figure 12.2 *Continued*

Newswriting, Ethics, and the Law The newswriter must be acutely aware of legal and ethical concerns relevant to the journalism profession. Beyond First Amendment guarantees of freedom of the press and speech, the newswriter should also be familiar with the rights of privacy of individuals under the Fourth Amendment, as well as protection from libel, slander, defamation, and trespass. Laws related to transgressions in these areas affect the way news is gathered and reported. Table 12.1 defines several of these terms.

In addition to federal law, state laws dictate journalistic practices related to rights of access to restricted areas, the use of surveillance equipment, the

PROFESSIONAL POINTERS

Preparing a Split-Page Script

- You do not need a vertical line between the video and audio columns. Leaving the line out keeps the copy cleaner and reduces the suggestion of a barrier between video and audio elements.
- The speaker of *each and every speech* is identified in the audio column by a name or role.
- Speeches are double- or triple-spaced to allow for easy reading.
- Speeches on the first page begin at the halfway point on the page, or lower down, to leave room for intro or ready cues, music cues, and any other cues needed to start the show.
- To limit confusion for the talent, spoken lines are distinguished from other verbal entries in the audio column by a different typeface, type case, or other attributes. For instance, it's far better to write "HOST: Welcome to the show" than "Host: Welcome to the show."
- The first entry in the video column lists opening shots for every camera. Even if the third camera in a thirty-minute show is not used for the first twenty-five minutes, it still has an opening shot listed at the beginning. With this format, camera operators know which shots to prepare for openers without having to ask the director.
- After the opening shots, the left half of the split page contains all the ready cues, in order, in lowercase letters, in parentheses, flush left in the column. Then it shows the actual commands, in uppercase letters, at the right of the column. This arrangement makes it easy for the director to see both ready cues and commands in order at a glance.
- The video column also contains cues referring to nonvideo elements, including mic, music, and talent cues.
- Horizontal arrows (as shown in Figure 12.2) are often drawn from circled camera commands to speeches in the script where camera changes are meant to occur. This method of connecting camera commands to specific speeches is more accurate than simply writing "Take 1," for example, on the line where you want to use that camera.

recording of telephone conversations without proper notice, and access to juvenile trials. The use of cameras in the courtroom may also vary depending on state law. To avoid costly litigation and to protect yourself and the news group you work for, become thoroughly familiar with the laws in your state. If possible, check with your legal department to minimize problems.

Besides legal restrictions, a newswriter encounters questions of ethics. Many ethical questions involve balancing the need to promote an informed

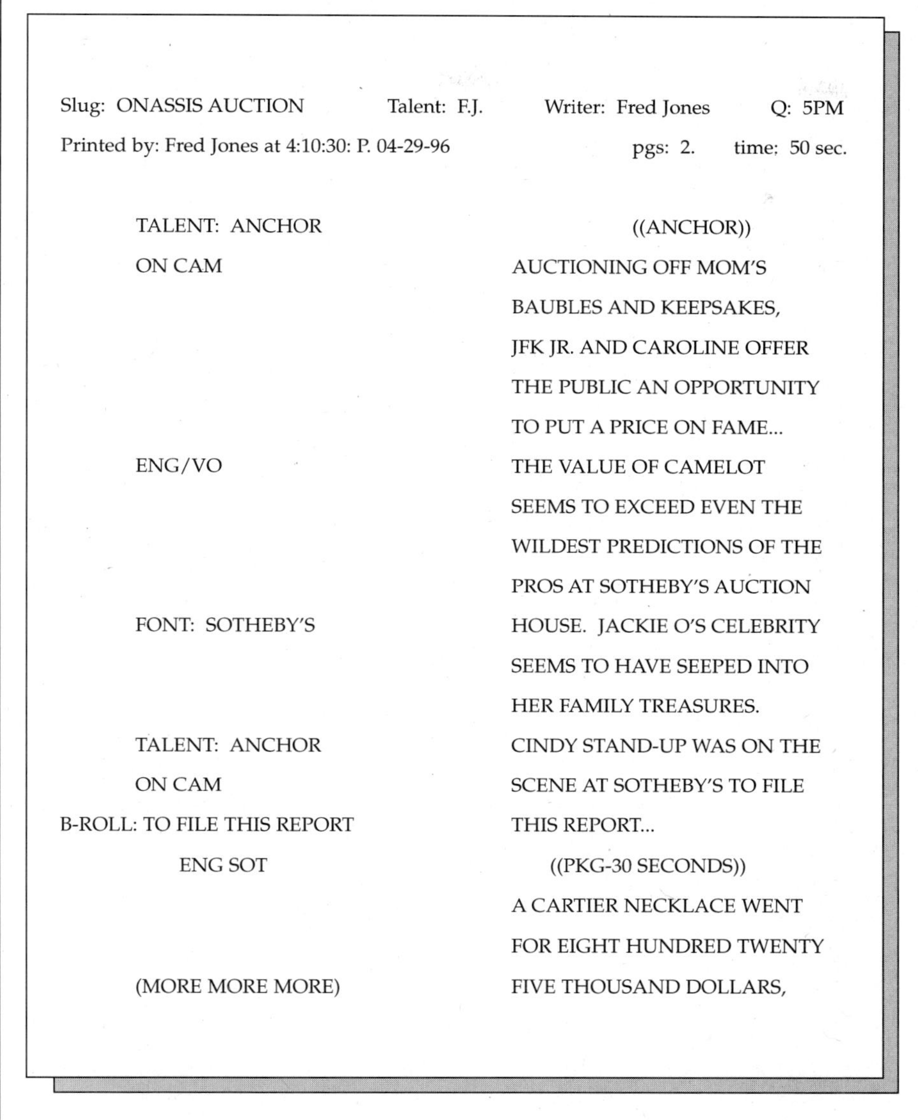

Slug: ONASSIS AUCTION   Talent: F.J.   Writer: Fred Jones   Q: 5PM

Printed by: Fred Jones at 4:10:30: P. 04-29-96   pgs: 2.   time: 50 sec.

TALENT: ANCHOR	((ANCHOR))
ON CAM	AUCTIONING OFF MOM'S
	BAUBLES AND KEEPSAKES,
	JFK JR. AND CAROLINE OFFER
	THE PUBLIC AN OPPORTUNITY
	TO PUT A PRICE ON FAME...
ENG/VO	THE VALUE OF CAMELOT
	SEEMS TO EXCEED EVEN THE
	WILDEST PREDICTIONS OF THE
	PROS AT SOTHEBY'S AUCTION
FONT: SOTHEBY'S	HOUSE. JACKIE O'S CELEBRITY
	SEEMS TO HAVE SEEPED INTO
	HER FAMILY TREASURES.
TALENT: ANCHOR	CINDY STAND-UP WAS ON THE
ON CAM	SCENE AT SOTHEBY'S TO FILE
B-ROLL: TO FILE THIS REPORT	THIS REPORT...
ENG SOT	((PKG-30 SECONDS))
	A CARTIER NECKLACE WENT
	FOR EIGHT HUNDRED TWENTY
(MORE MORE MORE)	FIVE THOUSAND DOLLARS,

Figure 12.3   *A sample news script (continued on the next page).*

citizenry and increase the general welfare against the need to maintain fairness to news sources. Questions of taste, values, and cultural norms may also arise. As a newswriter, you may have to decide whether to use illegally obtained information in preparing a story; whether or how to televise an execution, a suicide, or a hostage crisis; whether a special favor from a news source compromises your objectivity; and whether you can use staged reenactments of crucial events.

The *Code of Broadcast News Ethics* of the Radio-Television News Directors Association, reproduced in Chapter 10, is a useful guide in such decisions. Your own organization may also have its own ethical guidelines. Even so, you

A-ROLL: PRICE OF FAME IS HIGH

      ENG SOT

A HARRY WINSTON PENDANT
FOR EIGHT HUNDRED EIGHTY
THOUSAND. THIS RUBY,
SAPPHIRE, EMERALD, AND
DIAMOND CLIP IN THE SHAPE
OF A FLAMINGO WENT FOR
EIGHT HUNDRED AND SIX
THOUSAND DOLLARS. WHAT
THIS MEANS TO SOME
OBSERVERS IS THAT THE
PRICE OF FAME IS HIGH.
IT ALSO MEANS THE DESIRE
TO CAPTURE A PIECE OF THE
BELOVED KENNEDY PAST IS
PERHAPS STRONGER NOW
THAN IT WAS THIRTY YEARS
AGO. AT SOTHEBY'S

FONT: CINDY STAND-UP

THIS IS CINDY STAND-UP
REPORTING...
      ((ANCHOR))

TALENT: ANCHOR

THANKS FOR THAT REPORT,
CINDY...WELL I WONDER HOW
MUCH I COULD GET FOR MY
EMERALD TIE PIN...BACK WITH
MORE NEWS AFTER THIS...

Figure 12.3   *Continued*

PROFESSIONAL POINTERS

Preparing News Scripts

- Begin each story on a separate page.

- If a story runs longer than one page, use inclusive numbering so that each page shows both the number of the current page and the total number of pages in the script. Also, for multipage stories, write "MORE" at the bottom of each page before the last page.

- To get an accurate idea of running time for each story, establish a standard for margins, type size, and number of lines per page. You may find, for example, that each page equals about thirty seconds of running time.

- For on-air scripts, use paper of a color other than white to reduce light reflection.

- Do not staple on-air script pages together. Unstapled pages can be put aside when they have been read without unnecessary rustling.

TABLE 12.1

Some Journalistic Legal Terms Defined

Defamation	Any statement that harms a person's reputation, name, or character.
Libel	Malicious defamation in written or graphic form. Libel may include any false statement that damages an individual's reputation. A person's business or property can be libeled, as can an institution. To avoid making libelous statements, restrict claims to statements that are true. Further, evaluate the source of a damaging remark to eliminate malice from the motive. Reporters can avoid charges of libel if they express their thoughts in the form of opinions.
Slander	The defamation of a person through oral speech. Guidelines are similar to those for libel. Usually broadcasters are charged with libel rather than slander because their oral presentations generally originate in written notes.
Trespass	Illegal entry into another party's property, land, or premises; or unlawful injury to a person or to a person's rights or property.

will need to reflect on ethical matters yourself, drawing on your own experience and common sense.

Scripts for Commercials

Scripts for television commercials are among the most detailed. Much care is put into them for a number of reasons. First, costs for both production and airtime can be quite expensive. Second, since commercials are usually short messages, sponsors have no time to waste. Finally, since commercials are totally planned in advance, they can be meticulously designed.

Running times for segments within each spot may be specified to the second, as well as the images for each moment of airtime. To convey this information, a storyboard is often used. A **storyboard** is a cartoon panel or sequence of drawings that shows the main shots of a story. Each image in the storyboard sequence may feature corresponding audio material.

Figure 12.4 shows a script with a storyboard for a thirty-second political commercial for a fictitious candidate. The spot fades up from black (F/Blk) onto a text crawl with an announcer's voice-over. Next, we see a dissolve (Dis.) to a medium shot (MS) of the candidate (CAND.) in an interior locale (INT.). The storyboard shows a nice, cozy office, with a desk and fireplace. The storyboard also reveals the candidate speaking directly to the camera, in a business suit, sitting on the edge of his desk. As he reaches the middle of his speech, the camera zooms in (Z-in) to a close-up (CU) of the candidate, as shown in the next storyboard image. As he concludes his speech, the CU dissolves to a closing graphic with the text "BURGERHEAD: LEADERSHIP FOR THE MILLENNIUM." At the end of the thirty seconds, the spot fades to black (FTB).

Notice these other features of the script in Figure 12.4:

- The script lists the name of the production company; the client's name; the date of production; the date of release; the spot number, title, and length; and locations where the spot is to be shown.
- The storyboard provides a visual display of all of the spot's main shots. In addition, the script includes the order and duration of shots and the overall running time.
- From this material, the producer, director, and crew can see what the main camera shots and other visuals (such as the CG text) look like. They can survey the camera angles, shot size, shot content, and so forth.

Full-Page Scripts

Instead of the split-page script, film productions of screenplays often employ a **full-page format,** and this practice is also common in certain types of video. The full-page format is often found in single-camera EFP productions, made-for-TV movies, television dramatic programs, and documentaries. Programs that originate as a series of separate scenes or camera shots—that is, programs whose scenes are shot out of order and then assembled in postproduction—are quite amenable to the full-page script format. This can even include commercials.

POLLY TITIAN ADVERTISING COMPANY

Television script: Title: *"Leadership"*

P. 1 of 1

Client: Linden Burgerhead for President Campaign
Production Date: October 10, 1998
Air Dates: November 1-November 5, 1998
Cities: N.Y., L.A., Chi., Phila., Boston.
Length: 30 seconds
Spot Identification #: x-15

<u>VIDEO</u>	<u>AUDIO</u>
(0:00–0:10) F/Blk text	ANNCR: Some Senate leaders say our country has stopped growing, that we must adapt to a lower standard of living than we've had in the past. Linden Burgerhead disagrees.
(0:10–0:25) Dis. MS CAND. INT.	CAND: We live in the greatest country on earth. Throughout our history, we have managed to solve our education problems, our social strife, and even our economic problems. With our talent, energy, and commitment, all we need is effective leadership.
Z-in CU CAND	
(0:25–0:30) Dis. ID Graphic	
"BURGERHEAD: LEADERSHIP FOR THE MILLENIUM" FTB. F/AUDIO	

Figure 12.4   *A sample script for a commercial, with a storyboard.*

Figure 12.5 offers a sample of a full-page script illustrating many of the characteristics common to the format. Notice the following features:

- A title page bears the writer's name, as well as the producer's name if there is one. It also provides copyright information (optional) and revision dates. Immediately after the title page, there may be pages describing the main characters, their physical attributes, their wardrobes, and so forth.

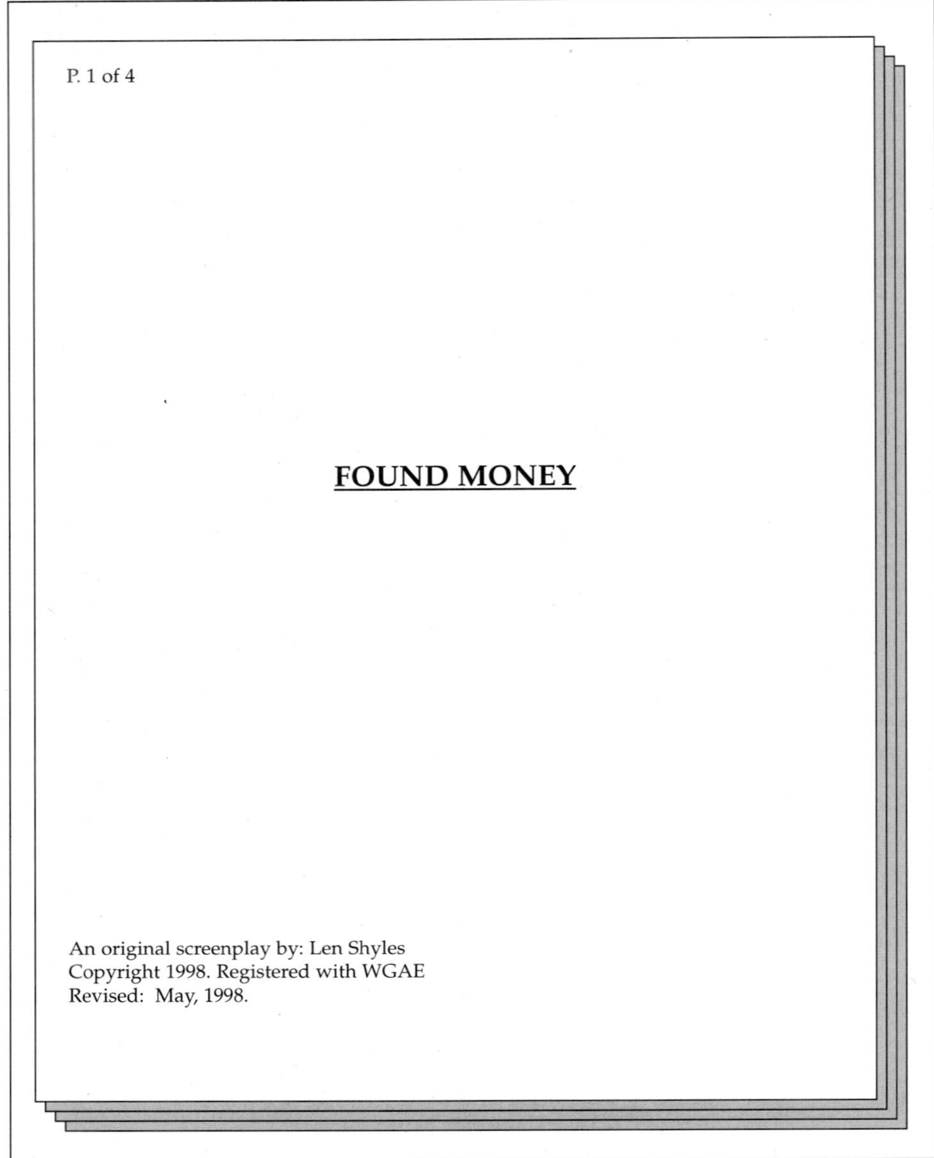

Figure 12.5   *A sample of a full-page script format (continued on the next three pages).*

- All pages are numbered with inclusive numbering, and the date of last revision appears on every page. New pages containing revisions may be on different-colored paper so that they stand out for cast and crew.
- Descriptions of scene locations, time of day, and camera directions are given in capital letters, as are names of characters, actions, sound effects, and music.

P. 2 of 4

<u>Main Characters:</u>

Father McElroy:

A young priest, recently ordained, shy, innocent, and ignorant of the real-world problems that beset him. He is pathetically unprepared for the work of running the AIDS hospice in south central LA to which he has just been assigned, yet run it he must. He is even less able to supervise his newly inherited underling, Sister Mary Elizabeth, who seems to know more than he does about everything. He is daunted by the hospice; the task is simply too big for him to handle. Aside from providing care and counsel to his patients, he must also raise money to keep it open. He is minister, fundraiser, and bookkeeper.

Sister Mary Elizabeth

A middle-aged nun, strong, attractive, serving the community for the last five years by working the AIDS clinic. The Sister is fervently committed to finding ways of making her hospice work more fulfilling by increasing the number of beds available to the community, which is suffering from a lack of funding and an increasing need as the epidemic continues to rise. She has been toiling in the vineyards long and hard, and knows her time is limited.

(5/20/98)

Figure 12.5   *Continued*

- Single spacing is used within paragraphs, double spacing between paragraphs. Double spacing is also used between scenes, between different characters' speeches, and to set off transitions such as "CUT TO" and "FADE OUT."
- Dialogue is typed in a column centered on the page, about 3 inches wide. The character's name is centered in capital letters above each speech. Stage directions, indicating how the talent is to deliver the lines, are presented in parentheses after the character's name.

P. 3 of 4

1   FADE IN.   1

INT. SOUTH CENTRAL COMMUNITY AIDS CLINIC—NIGHT

SISTER MARY ELIZABETH finishes her phone call to the convent, telling
SISTER FRANCINE that she is leaving to come home. She notices the clock
and calendar on her plain metal desk. It's eight P.M., Friday, and she looks tired.
Just as she hangs up, she notices that FATHER McELROY has entered looking
dejected. She has never seen him looking quite this bad.

        SISTER MARY
Good evening Father. You don't look
very happy tonight. What's wrong?

        FR. McELROY
  (with despair)
I feel like I've been brushing back the
waters of the ocean with a broom.
Between trying to comfort dying patients and
their family members, and trying to raise money
to cover the bills due tomorrow, I feel
like I've been torn in two or four major pieces.

        SISTER MARY
  (gentle yet firm)
Well this is just one more normal day...
You need some rest. You'll be fresh
when the relief staff arrives, giving us
both a well-earned weekend off.

        FR. McELROY
Look, don't tell the relief this, but you
see, we're out of money. If we don't pay the bills that
come due tomorrow, especially rent and phone,
this place gets shut down. I hope you have carfare.

        SISTER MARY
  (turning toward him)
You mean we're evicted???

McELROY shakes his head in agreement. He steps aside as SISTER
MARY heads for the door.

                              CONTINUED (5/20/98)

Figure 12.5   *Continued*

Scriptwriting Software

Computers are not yet capable of writing Emmy-winning television programs. They can't even write serviceable flops; people must still do that.
However, word processing programs make script formatting easier than ever.
Software also makes it easy to count words, set margins, use a thesaurus,
check spelling, move copy, and perform other editing functions. For these reasons, word processing programs are especially helpful in news organizations,
where copy must be revised and produced quickly.

P. 4 of 4

                    SISTER MARY
                 (resolutely)
              We'll get the money. We'll get by.

2   EXT. THE BUS STOP ON THE STREET—NIGHT   2

       SISTER MARY steps on board and pays the L.A. bus fare with the last bit of
       change in her modest plastic purse. The bus is empty, since the rush hour has
       ended over two hours ago. The camera outside the bus PANS with her as she
       makes her way to the rear of the bus. As she plops down with a THUD, the
       bus moves away from the curb, the street noise of boom box RAP MUSIC
       quickly fading. As the GRIND of the bus engine begins to invade her brain
       as welcome white noise, her tired eyes soon light upon a soft leather wallet
       lodged between the seats. The twenty thousand dollars in it in large bills shocks
       her to attention.

                                      CUT TO

3   INT. SISTER MARY'S BEDROOM—NIGHT   3

       SISTER MARY tosses the wallet onto her bed, then flops down next to it, flips it
       open, and pulls twenty Gs onto the bedspread. She reaches over to the night
       table for her bedside Bible, opens it up to her favorite Psalms, mouths some
       prayers, and quickly falls asleep.

                                   FADE OUT

                    <u>END ACT 1</u>

                                                         (5/20/98)

Figure 12.5 *Continued*

For screenplays and other scripts that require a standard format, script processing programs such as *Dramatica* and *Scriptor* are popular. They attempt to make the formatting even easier than regular word processing programs do. Check with your software supplier to find the software best suited for the work you intend to do.

A FINAL NOTE

The writing tips and examples of script formats presented in this chapter provide some useful guidelines for preparing scripts for most television productions. However, these ideas are just a start. More in-depth treatment of these topics is available in texts dealing with screenplay writing, news reporting, advertising, and, more generally, broadcast writing and writing for the mass media. If you are at all interested in writing for video, make it a point to seek further information as well as practical experience.[3]

KEY TERMS

liftout *(302)*	semiscripted format *(305)*
telegraphing *(303)*	rundown sheet *(305)*
recapping *(303)*	in-cues *(305)*
proposal *(304)*	out-cues *(305)*
treatment *(304)*	split-page format *(308)*
pilot *(304)*	storyboard *(318)*
cold start *(304)*	full-page format *(318)*

QUESTIONS FOR REVIEW

1. How do characteristics of the television medium, audience, and program genre influence the writer's task?

---

[3]For further examples of scriptwriting formats, see Blum (1995), Haag and Cole (1985), Cole and Haag (1989), Brady and Lee (1988), Walters (1994), and especially Wolff and Cox (1988). For more on storyboards and their use, see Walters (1988), Willis and D'Arienzo (1993), Hickman (1991), and Orlik (1994). For more on screenplay writing, see Armer (1993), Blum (1995), Brady and Lee (1988) and Lucey (1996). For more on writing and reporting the news, see Shook (1996), Stephens and Lanson (1986), and Mencher (1994). For more on writing copy for a variety of media forms, see Walters (1994), Newsom and Wollert (1988), Mencher (1993), Orlik (1994), and Huchison (1996). For more on legal and ethical issues in news reporting, see Shook (1996), Mencher (1994), and Walters (1994). For manuals and handbooks on writing style for broadcasters, see the *UPI Stylebook* (1992), Papper (1995), and MacDonald (1987).

2. What is meant by the idea that content carried in the video and audio track "co-determine" each other? Give some examples.

3. What aspects of the competitive video marketplace influence television writing?

4. What are the differences between fully scripted and semiscripted shows? Between rundown sheets and other script formats? When is it appropriate to use each?

5. What advantages does a split-page format offer the director?

6. What are some of the legal and ethical concerns involved in newswriting?

7. How might a storyboard be used in a thirty-second commercial for an exercise machine?

13 Producing and Directing

Producing and directing for video are highly varied activities. For the most part in this chapter, we will discuss producing and directing as separate roles. In many cases, however, especially early in your career, you may find yourself filling both roles on the same project, becoming a *producer-director.*

It takes a skilled, talented individual with an aptitude for handling details to manage the job of the producer-director. Of course, having broad experience helps also. The producer-director wears many hats, as the cliché goes, coordinating and orchestrating the development of a video project, in many cases bringing it from the concept stage through to the completed program or series, at times even planning the details of its distribution and exhibition. Famous producers and/or directors in television include Norman Lear, Stephen Cannell, Steven Bochco, Aaron Spelling, Angela Lansbury, and Dick Clark, to name but a few.

This chapter describes the responsibilities, roles, and tasks of the producer and the director in the making of video programs. The topics covered include:

PRODUCING

different producer levels • producing in the preproduction stage • producing in the production stage • producing in the postproduction stage • dealing with artistic and technical unions • audience measurement and ratings

DIRECTING

qualities of directors • directing in the preproduction stage • directing in the production stage • directing in the postproduction stage

A FINAL WORD

PRODUCING

Depending on the nature of a given project, the producer's job can vary tremendously. For example, a theatrical program may require the producer to spend a significant amount of time hiring talent, but for the news or sports producer, this task may be quite small, perhaps even nonexistent. Conversely, whereas a news or sports producer may routinely have to rent satellite facilities, a theatrical producer who tapes fiction dramas may spend no time at all dealing with satellite transmission.

Other tasks of the producer may include formulating program proposals and budgets, approving the script, renting or building sets, acquiring studio space, getting copyright clearances and release forms, working with union contracts, and marketing, distributing, and exhibiting the finished program, including arranging distribution rights and syndication deals. Of course, if the producer is acting as a producer-director, the list can be even longer. Table 13.1 lists some of the typical concerns of a producer-director, classified according to the five crucial factors of space, time, material, money, and personnel.

Once personnel are hired, the producer's job involves *coordinating* crew and talent and *delegating* responsibilities, confident that the team members know their jobs better than the producer does. Then mutual trust and respect between the producer and a talented team become the producer's greatest assets.

Different Producer Levels

Just as the producer's job may vary according to the type of program being made, the scope of the job may differ depending on the level of producer you are. In a program of sufficient complexity, producing responsibilities may be differentiated according to a hierarchy. Some of the titles used to define different producer positions include *executive producer, staff producer, agency producer,* and *free-lance producer.* Additional titles include such refinements as *associate producer* and *co-executive producer.* The following paragraphs describe each of these levels of responsibility.

Executive Producer   The executive producer is usually involved in originating project ideas but does not get into specific details. Rather, the executive producer is responsible for funding and budgeting projects and then supervising production. In addition to approving projects for production, the executive producer may check in from time to time to make sure everything is on schedule. To handle the daily details of particular projects, the executive producer often assigns a staff producer.

Staff Producer   The staff producer is responsible for the overall production of a particular project. On that project, the staff producer may be in charge of all production decisions. It is the staff producer's responsibility to ensure the highest possible quality for the project.

Agency Producer   Sometimes an advertising agency assigns a producer (often called the *agency rep*) to a production being done by a station or a produc-

---

TABLE 13.1

Some Concerns of the Producer-Director

Space	Time	Material	Money	Personnel
Studio	Schedule	Facilities	Budget	*Above the line:*
Field locations	Order	Equipment	Salaries	Director
Equipment storage	Calendar	Artwork	Union fees	Writers
Parking	Hours	Cameras	Insurance	Producers
Rehearsal rooms	Overtime	Lights	Parking fees	Talent
Meeting rooms	Rain dates	Microphones	Hotel bills	Art directors
	Rehearsals	Cables	Food costs	Scenic designers
	Shooting	Audio mixers	Overtime	Graphic designers
	Editing	Videotape machines	Equipment rentals	Musicians
	Clock time	Costumes	and purchases	*Below the line:*
	Backtiming	Graphics	Studio rental	Technical director
	Front timing	Script		Floor manager
	Deadlines	Floor plan		CG operator
	Meetings	Props		Videotape technicians
	Appointments	Sets		Engineering
	Program length	Vans		Editors
	Pace	Satellites		Assistant director
	Time slots	Phone lines		Camera operators
	Air dates	Intercoms		Audio technicians
		Music		Utility crew
		Sound effects		Lighting director
		Walkie-talkies		Set construction
		Catering		Props
		Hotels		Wardrobe
		Transportation		
		Clearances		
		Release forms		

tion house. The agency rep works for the advertising agency's client, making sure the program accomplishes the client's objectives. The agency rep may execute the contract, and, depending on the arrangement, the production company or another outside source may provide the facilities and crew. Frequently the agency provides the script and audiovisual materials. The rep may also supply talent and locations. In determining who should produce the program, the rep may entertain several bids from various production houses.

Freelance Producer   Freelance producers work for production companies or agencies on a single-project basis. This arrangement saves the agency or production house a full-time salary and makes it possible to bring in specialists on an as-needed basis. The daily cost for a free-lance producer is usually higher than it would be for a regular employee, but it is worth it because the overall cost is still lower than employing a full-time staff person all year long. The level of work is expected to be of the highest quality, since the free-lancer is brought in as a specialist. In such cases, the producer routinely handles such diverse responsibilities as financing, budgeting, hiring, production scheduling, and, of course, the execution of the production itself.

Associate and Co-executive Producers   As these titles imply, associate and co-executive producers share the burden of the production, often taking on a predetermined set of responsibilities. Large productions such as network sports programs and large-scale theatrical performances may warrant involving several individuals in producer roles to handle various aspects of the production, from financing script development to acquiring talent and crew and from scheduling production tasks to arranging hotels and food. During the postproduction phase, the editing, distribution, exhibition, and syndication rights may also be handled by associate and/or co-executive producers. Of course, in such situations, it is essential that all involved have a clear idea of who is supposed to do what and in what order.

Producing in the Preproduction Stage

One way to view the producer's job is in terms of the different roles producers play in bringing a program idea into final form. These roles include aesthetic judge, technical expert, psychologist, business manager, sales agent, audience analyst, marketing master, legal authority, organization executive, and detail virtuoso. Depending on the type of program being made, each of these roles will demand more or less energy. To get a more detailed idea of how such talents may be used, we will now look at how they come into play during each production phase, beginning in this section with preproduction.

Concept Development   Among the first concerns of the producer in the preproduction stage is to develop a program proposal from a concept deemed to have some audience interest. Although ideas can come at any time from almost anywhere, at the concept stage it is wise to take an inclusive approach. During preliminary brainstorming sessions, write every idea down, no matter how unlikely it sounds at first. As a project develops, it is always easier to eliminate ideas than to invent new ones.

   Next, a winnowing process occurs, selecting the best idea from the large list of possibilities. Your selection should be determined by the quality of the idea, the availability of an audience for it, and an analysis of production feasibility (cost and availability of talent, space, and facilities). During this phase, strategies and compromises are made as you entertain different stylistic approaches. Some approaches may offer advantages in terms of audience appeal, whereas others may be more attractive from the standpoint of scheduling and budget. When the program idea becomes more stable and clearly defined, you are ready for the next step.

The Program Proposal   After making a firm commitment to a particular program idea, the producer usually must develop a full-blown proposal to convince a production company or sponsor to finance the idea. The program proposal should specify needs with respect to the five key factors listed in Table 13.1: space, time, material, money, and personnel. Further, the program's objectives and target audience should be clearly identified. Although formats for program proposals can vary, here are some of the elements your proposal should include:

1. *Title.* As mentioned in the last chapter, all programs should have a title from the outset. The title will identify the program so you can refer to it without ambiguity during production meetings (a real help when you are working on several projects simultaneously). *Most important, titles should capture the subject matter, content, and tone of the program for the sponsor, crew, talent, and eventually, of course, the audience.* (Even tentative titles should do this.)

2. *Program purpose.* The goal(s) of the program should be briefly stated. This can be done in a single sentence, as in the following example: *The purpose of this program is to demonstrate to students at the Culinary Institute how to prepare Filet Mignon.*

3. *Intended audience(s).* This part of the proposal clearly identifies the group(s) of people you wish your program to reach (the *target* audience). The audience can be mentioned elsewhere, such as in the statement of program purpose, but it should be clearly identified in its own right as well. It is regular practice to identify audiences in terms of sex/age categories (women ages eighteen to thirty-four, teenagers, and so on), but that is by no means the only way to do it. Try also to identify the audience in terms of the variables most relevant to the program you are producing: for example, all viewers who have purchased a General Motors car in the last five years.

4. *Program treatment.* As mentioned in Chapter 12, the treatment is a narrative description of a program written to convey a feel for the characters, setting, and action. Though the treatment is sometimes handled as a separate stage, it may be included in the original proposal. If so, it should briefly describe the program's content, subject matter, length, pace, style, and any other aspects that telegraph the nature of the program to the prospective sponsor. The treatment should also tell whether you intend to shoot in studio or on location, using film or tape, in sequence or out of sequence, with one or several cameras. You can include a bit of dialog to convey a sense of the characters to the reader, but in general it is better to leave such detail to the script stage.

5. *Preferred date, time, and channel (if appropriate) for showing the program.* This information tells how you intend to reach the target audience. Obviously, a show can successfully reach its audience only when they are available to see it. For a children's program, for example, you would probably specify an after-school hour on weekdays or perhaps Saturday morning. Identifying the date, time, and channel also forces you to think more clearly about the competition.

6. *Budget.* Include a budget so the client can see how you have computed all expenses. Figure 13.1 shows a sample budget form. Costs can often be determined using **rate cards** (Figure 13.2) that list prices for services, materials, and labor; these are available from various production houses and unions. In your budget, try to show the price for every conceivable item, whether purchased or rented. Consider script preparation; talent and crew salaries; rental of equipment and studio space; fees for location shoots; expenses for cos-

PRODUCTION COST ESTIMATE

Production Co:__________________   Bid by:____________________
Address:______________________   Client:____________________
Telephone Number:______________   Bid Date:__________________
Contact:______________________   Client contact:______________
Director:______________________   Tel:______________________
Producer:______________________   Fax:______________________
Writer:________________________   Project title:______________
preproduction days:____________
build/strike days:_____________
studio shoot days:_____________
Location sites:________________

Summary of estimated costs:

CREW	DAYS	RATE	OVERTIME	TOTAL	ACTUAL
Producer					
Assistant Director					
Camera op.					
Technical Director					
Utility					
Audio					
Boom op.					
Makeup					
Hair					
Playback					
Wardrobe					
Script clerk					
VTR op.					
Teleprompter op.					
Scenic design					
Postproduction coord.					
Other					

Figure 13.1 *A partial budget form (continued on the next three pages).*

tumes, sets, props, and makeup; editing time; and videotape. Include also estimates for food, lodging, transportation, hotel accommodations, insurance, copyright fees, and parking. Allow some leeway to cover extra charges that inevitably result, such as additional equipment rental expenses if a day of location shooting is lost to bad weather. It is generally reasonable for an actual budget to go over the original estimate by 10 percent, but coming in under budget is always preferable from the sponsor's standpoint.

One traditional budgetary practice in television is to arrange costs in terms of above-the-line and below-the-line expenses (discussed in the next section).

Preproduction materials and expenses:

	UNITS	RATE	TOTAL	ACTUAL
Auto rental				
Air fare				
Lodging				
Camera rental				
Taxis				
Telephone and fax				

Location expenses:

	UNITS	RATE	TOTAL	ACTUAL
Permits				
Car rentals				
Parking, tolls and gas				
Air freight				
Food				
Limousines				
Cabs				
Gratuities				
Misc.				

Studio rental and expenses:

	UNITS	RATE	TOTAL	ACTUAL
Rental for build days				
Rental for shoot days				
Rental for strike days				
Meals for crew and talent				
Set guards				
Total power charges and bulbs				
Stage manager				
Misc.				

Equipment rental:

	UNITS	RATE	TOTAL	ACTUAL
Video cameras				

Figure 13.1 *Continued*

Other budgets arrange costs in terms of preproduction, production, and post-production expenses. It is wise to ask the client what is preferred and present the budget accordingly.

7. *Schedule.* The schedule should indicate the time involved for completing the production. Try to show order as well as duration for each production element. If specific dates and times are known in advance, as is often the case with programs about public events, indicate them.

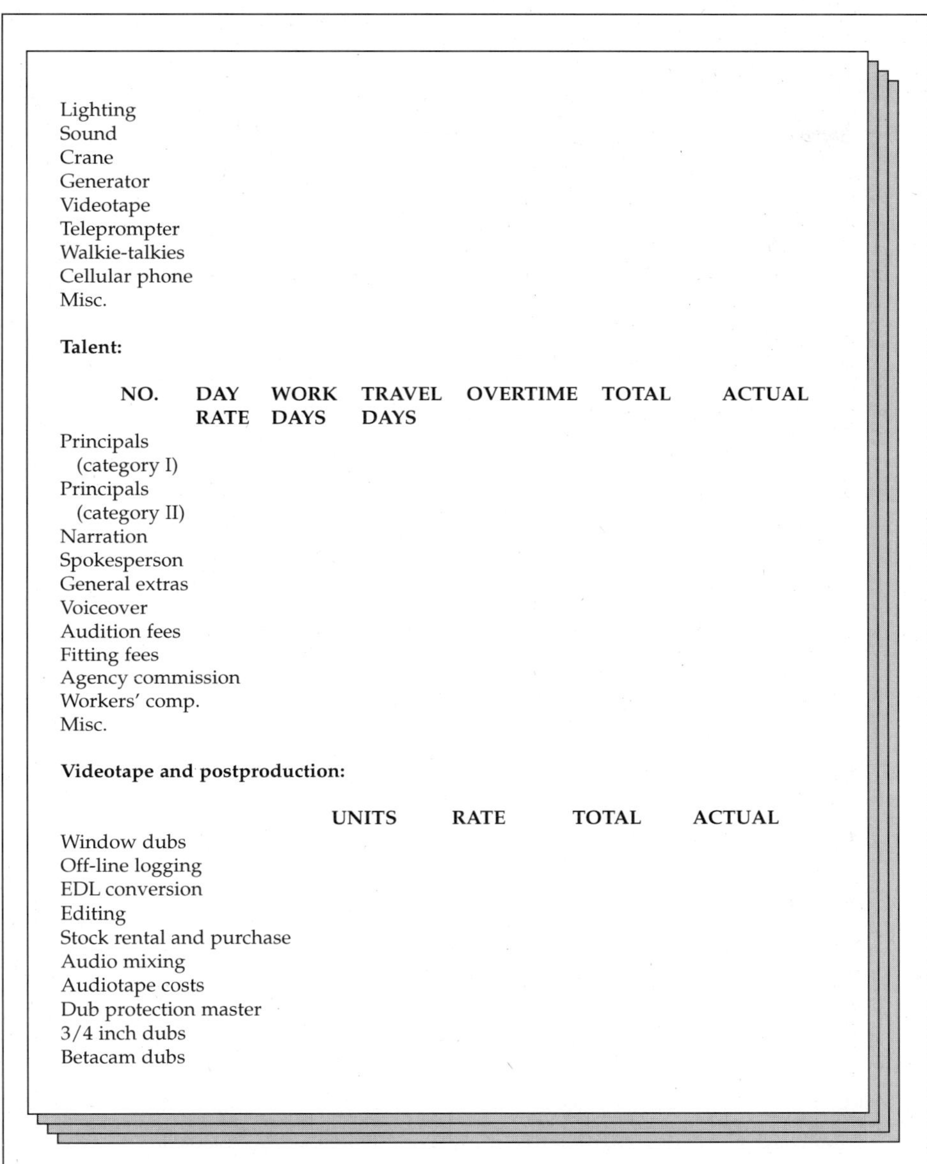

Lighting
Sound
Crane
Generator
Videotape
Teleprompter
Walkie-talkies
Cellular phone
Misc.

Talent:

	NO.	DAY RATE	WORK DAYS	TRAVEL DAYS	OVERTIME	TOTAL	ACTUAL
Principals (category I)							
Principals (category II)							
Narration							
Spokesperson							
General extras							
Voiceover							
Audition fees							
Fitting fees							
Agency commission							
Workers' comp.							
Misc.							

Videotape and postproduction:

	UNITS	RATE	TOTAL	ACTUAL
Window dubs				
Off-line logging				
EDL conversion				
Editing				
Stock rental and purchase				
Audio mixing				
Audiotape costs				
Dub protection master				
3/4 inch dubs				
Betacam dubs				

Figure 13.1 *Continued*

Planning Meetings and Script Preparation   Once the proposal has been approved, you need to plan the practical aspects of the production. This means coordinating all personnel, informing them about the program and how it is to be executed.

The people involved in initial planning meetings should come largely from the creative team of the production—the *above-the-line* personnel, as they are called in budgetary terms. These include writers, directors, talent, art and

Video effects and animation
Misc.

Total Miscellaneous Expenses:

SUMMARY:

<u>Estimated</u>   <u>Actual</u>

GRAND TOTAL:

Comments: ____
____
____

Figure 13.1   *Continued*

graphic designers, executive and associate producers, and music composers and musicians, if any.

Initial planning meetings should be used to develop a script everyone can use to transform the project from paper to the screen. One practical way to prepare the script is to have the writer write the words while the director comes up with accompanying images. Working with the producer, the entire creative team should suggest ways to reflect the essence of the treatment in the script so that it delivers what was promised. At this stage, the producer

RE-DE-PRO PRODUCTION CO., INC.

Equipment	Rate
Videotape machines:	
D-2 VTR	$150.00/hr
D-1 VTR	$100.00/hr
Cameras:	
BetacamSP	$50.00/hr
Camcutter	$100.00/hr
Editing suite (on-line editing):	
2 VTR D-2	$350.00/hr
3 VTR D-2	$500.00/hr
Editing (off-line):	
AVID and EDL	$650.00/day

Figure 13.2 *Sample rate card for equipment and editing services.*

should also work with the art director to plan sets and graphics that fit the tone of the project.

As producer, it is your job to keep the creative team on track. Keep in mind the program's objectives, intended audience, style, length, tone, and pace, and everything else you have promised to the client. After the script is completed or at least brought into working shape, you can hire the crew—the *below-the-line* personnel, including technical directors; camera, lighting, and audio operators; videotape technicians, production assistants, and floor managers; and so forth.

Facilities Requests At this stage, you can also finalize the scheduling of equipment. Use a facilities (fax) request form such as the one shown in Figure 13.3 to list all the equipment you need, including cameras, mics, sets, props, lights, and graphics. The fax sheet should also include the date, time, and place each piece of equipment will be needed. To determine exactly what to order or rent and when, confer with the rest of the staff.

FACILITIES REQUEST

Date:_____________________   Submitted by:_____________________

Production Title:_____________________

Producer:_____________________   Director:_____________________

Recording date:_____________________   Air date:__________ Time:__________

Studio or field:_____________________

Location address:_____________________

EQUIPMENT LIST:

a)

b)

c)

d)

e)

f)

g)

h)

(Include all audio, video, lights, graphics, editing, cables, intercom, transmission, vans, etc.)

Requestor's signature:_____________________

Approved by:_____________________

(producer's name)

Approved by:_____________________

(engineer's name)

Date of approval:_____________________

Figure 13.3 *Sample fax request form.*

Remember to avoid using the latest equipment just for its own sake. Nothing is more unprofessional than wasting space, time, material, money, and personnel in this way. The rule for selecting equipment is to pick only what you need to get the job done, and no more.

Of course, knowing when equipment and personnel are needed and for how long is a function of the schedule you have devised. Be sure to allow some flexibility in case of delays. Communicate the final schedule in writing to all parties to reduce confusion and misunderstandings.

RELEASE FORM

I hereby give permission to _____________________ Production Company and
_____________ , their agents, successors, assigns, clients, and purchasers of their services
and/or products, to use my photograph (whether still, motion, or television) and
recordings of my voice, and my name in any legal manner whatsoever.

(1) I fully and irrevocably release and hold harmless _____________ and
_____________________ Production Company and their respective employees, agents,
and contractors from all liability, loss, claims, demands, and actions arising directly or
indirectly out of my appearance in the Program which was videotaped on ___[date]___.

(2) Further, I irrevocably authorize _____________________ Production Company to use
my voice, likeness, and appearance in the Program for educational purposes and for other
noncommercial purposes.

(3) I irrevocably assign to _____________________ Production Company the following:
        (a) all of my right, title, and interest in the copyright and any other property
interest I may have in the Program; and
        (b) all of my right, title, and interest in any proceeds or royalties from the sale,
rental, loan, publication, license, or other disposition of the Program.

(4) I irrevocably grant to _____________________ Production Company total ownership
of the Program, the right to edit, exhibit, or otherwise communicate the Program, the
right to secure a copyright in the Program, the right to sell, publish, and advertise the
Program, and the right to license or assign to others the foregoing rights enumerated in
this Paragraph 4.

Intending to be legally bound hereby:

Date:_____________________________
Signed:___________________________
Parent or guardian:_______________________
Address:___________________________
City:_____________________________
State:_____________________________
Phone:___________________________

Figure 13.4 *Standard release form.*

Keeping a Production Book As producer, it is essential that you keep an
up-to-date, accurate, and thorough production book listing everyone's phone
number (fax number and e-mail address, too), hours of availability, best times
to reach the person, and any other information needed to get the job done. Besides the cast and crew, include key field contacts, such as the police commissioner's secretary and the caterer. Also note the flight schedules and hotel information for out-of-town talent. Further, the production book should list
possible sources of last-minute equipment. Keep the rehearsal and production

schedules in the book, as well as a set of release forms, clearances, and permits to cover copyright and other legal needs. Being a consummate detail person is invaluable.

Release Forms and Copyright Clearances   Finally, be sure to get release forms signed and any final music clearances approved. This is best done in the preproduction stage to avoid the nightmare of being unable to use the program after investing time, money, and effort in its production. Figure 13.4 offers a sample standard release form.

Both pictures and musical works (including lyrics) may be copyrighted. Therefore, it may be necessary to pay a fee for the use of such materials. In the music industry, two organizations collect fees for music performance rights: the American Society of Composers, Authors, and Publishers (ASCAP) and Broadcast Music Incorporated (BMI). Virtually all music written is registered with one of these organizations. By paying for the rights to use their catalogs, you are legally covered when you use their members' works.

Photographs and motion images may also be copyrighted, making it illegal to use them without permission. You can use such materials in your programs if you pay the fees in advance. Photo libraries are good sources for such materials. Check with the organization from which you solicit the materials to make sure you have proper clearance. In some cases, especially if you can use older images, you may be able to find good noncopyrighted material at the Library of Congress and at state or local historical societies.

Producing in the Production Stage

Once you have developed and approved a script and determined the five key factors listed in Table 13.1 in sufficient detail, it is time to enter the production phase.

Coordinating and Delegating Responsibility   The producer's role during the production stage is to coordinate and delegate responsibility, making especially sure that the production stays on schedule. Check actual progress against the proposed schedule to stay reasonably close to what was originally promised. By keeping careful tabs on each day's work, you can evaluate the progress and make adjustments as you go. Treat the proposal as a template of the actual program, and try to match your production to it.

In delegating work, be as explicit and specific as possible. For example, while it is obvious that the writer is responsible for the writing, it may be less clear whether you want the art director or the floor manager to design the floor plan. Similarly, other duties may not be clearly defined until you tell people what is expected of them.

To answer questions regarding task assignments, hold production meetings. Indicate a chain of responsibility in case a crew member has an emergency and cannot participate. Keep the operation running smoothly by providing phone numbers of every crew member to all involved so everyone can be reached if need be. Communication is crucial. At the end of each rehearsal and production day, conduct a short status meeting to iron out problems and realign tasks and schedules.

Chuck Will

Senior Associate Producer, CBS Sports

Q: Talk to me about your roots. How did you get to CBS?

A: What got me involved was golf. I was a pretty good player. I won the City Amateur championship of Baltimore one time playing on a short course.

I came to CBS to work free-lance because of my golf acumen and because the producer thought I was good with people. In the beginning, we'd tape all the golf for eighteen holes and then play it back after it was edited. I'd be on the course with an earpiece in my ear and a recorder. Then we'd come back to New York, where I'd watch the editing process and learn without getting paid for that. I was paid for the field work.

So I got to watch, and I got interested. Then the producer said, "I think you can help the director down in Augusta to pick up the key shots." That was in 1969. The first Masters Tournament I did was in 1970. I worked for the director from the remote truck, warning the director about where the next shot was coming from. I brought golf expertise to the television crew.

I still tell people, if you want to get ahead, stick around, sink into the woodwork and learn, even if you don't get paid. When someone comes to a remote broadcast and sees what we do in rehearsal and during a broadcast, they are amazed. Golf is supposed to be such a slow game, but when you see what we do with it, it's not slow. I've done many sports as associate producer and producer. In golf, I still see television as a producer's medium. In sports in general, television is still a producer's medium.

Q: What does your job involve?

A: As associate producer of golf for the network, I am the detail person. I make sure the production supervisor gets everyone housed and fed. I make sure that everyone has the right automobile or air ticket and that the production support people who report directly to me know what they have to do. If they are new, I explain what I want them to do and when.

The secret to my job is encompassed in one sentence: I make myself available. If a person in television is full of answers but not available, then he or she shouldn't be in television. If that person is out taking a la-dee-dah look at a camera angle, there's going to be a line of people when she or he gets back and fifteen phone messages with people looking for answers. In my case, whoever calls, I try to get them an answer immediately. The answer has to be there to get the job done.

The thing that turns me on about what I do is that it's live and we can't edit over mistakes. If we take a camera wrong with a couple of reverse angles, the viewer at home may not notice it, but we do. And there is no seven-second delay. The caller who curses the host on a call-in radio show can be deleted by the delay, but we go on live without it.

Q: How come CBS doesn't work with a delay?

A: I don't know. I hope it has to do with the caliber of personnel, with the overall confidence we have.

Q: What happens when things go wrong?

A: The best example of a gaffe in golf coverage happened in 1986 at the PGA Golf Championship at Inverness in Toledo, Ohio. Bob Tway and Greg Norman were locked in a battle all the way down the stretch, tied coming into the eighteenth hole. There were thousands of people at that hole, where you're going to get a terrific reaction because they've been watching all day to see these two guys. The hole is a short par-four. That means the transition between the second shot and the walk to the green is like a snap of the finger.

Tway hits the ball in the bunker, Norman into some short grass near the green. Now they walk to the green. At the moment of the second shot being struck, the producer calls for a commercial. During the commercial, Tway gets in the bunker, and they say to the associate director, "Dump out of commercial! We're goin' live." And he informs the producer that they can't do that because they are in a station break. Tway hits the ball and holes it. When they come back from the break, his arms are raised, and the crowd is cheering.

Now what's the moral of this story? The moral is they should never have gone to commercial even though someone was paying a lot of money for that spot. It's in the interest of the advertiser for the viewers to see the "theater" involved with these two players. If I were in the truck in a similar situation, I'd have knocked the headset off the AD's head so we would stay with the show and not go to commercial.

If we get a breaking news event during golf, we tape the highlights and work them back into the show after the news interruption. In that case, the breaking news event is like a long commercial. In such cases, we miss some commercials and make good on them later.

Q: How do you guard against missed assignments?

A: We give out a schedule each day, and we have the workers check it to make sure everything gets done as planned. In my golf brochure, we suggest procedures to follow to avoid missing assignments. We have a daily schedule at the remote site showing time and location for each person to make themselves available. Every production person is required to consult the office trailer bulletin board on all assignments. It says, "For all clarifications, check with the associate producer. No person should leave an assigned position for any reason unless released by the director."

So I know when everyone is arriving, and I help the production supervisor by reporting all the things he or she needs to know to coordinate the production effort. I designate one of my staff to be the arrivals person so that I know when people are coming in. I also make sure the required credentials are left in the proper place, including parking permits, so that everyone can get in to the sports event.

Q: What do you look for in an intern?

A: I want someone with common sense, someone who can come in out of the lightning, someone who really wants to learn TV production. I want someone intelligent, someone who can pick up the information quickly. I don't want someone who is looking to impress me, who says, "Oh, that's a Grass Valley 10,000." It's just a switcher. I want someone who is willing to go get coffee and not ask why. If we go play golf somewhere, where I'm not the boss, then we can argue. But on the job, I'm the boss. Don't argue with me.

The executive producer of CBS golf today is Frank Chirkinian. He virtually invented TV golf coverage. He taught me a long time ago that in television or any other endeavor, the boss is the boss. It's a producer's medium, and if the producer acts in a despotic way, it's sometimes absolutely necessary. Later we can say, "Forgive me." In our work, the bottom line is: you capture the moment or you die.

Publicity and Promotions   The producer should also notify the public about the upcoming program. To do this, work through your company's promotion and publicity department, or similar channels outside your company, to make the public aware of the show. After all, a program can have no impact if no one sees it.

At the network level, publicity efforts include the production and distribution of press kits for new shows containing glossy promotional brochures that showcase the program and talent. The kits are sent to newspaper, broadcast, and other media outlets. If your publicity budget does not permit you the luxury of a flashy press kit, you can always resort to a short telephone campaign and a less sophisticated press release, such as a one-page announcement mailed in advance to various media outlets in the market where the program is about to air.

Observing and Taking Notes During Shooting   During rehearsals and production, the producer should let the director call the shots. At this point, the producer's job is to observe and take notes. If you have concerns or suggestions, write them down and discuss them with the director and crew during rehearsal breaks. During the broadcast or taping, do not interfere. You as producer have assembled all the elements. Now let the director and crew do their work.

Producing in the Postproduction Stage

Depending on the show you are producing, the postproduction phase can include editing, perhaps both off line and on line. If it does, you will need to arrange editing facilities. You may wish to be involved in the process or leave it to the director and editor. You may also have to show the rough cut (the off-line edited product) to the client before getting final approval to complete the job.

Aside from such production concerns, the postproduction phase involves recordkeeping and accounting tasks that must be handled competently. For example, there may still be bills to pay. Pay them promptly to preserve goodwill with outside agents whom you may need for your next project. Remember also to thank everybody in writing, by phone, or in person and to complete the publicity effort for the debut of the show. Finally, keep a copy of your program as well as the production book on file for archival purposes. You never know when a past effort or some portion of it will become useful in the future.

Dealing with Artistic and Technical Unions

One of the producer's responsibilities at all stages of production is to deal with the appropriate unions. In the media industries, unions have long worked to improve working conditions, salaries, and benefits. For example, in 1933, film actors were lucky to be paid $15 for a day's work ($90 for a six-day week). Unregulated hours and working conditions were often viewed as a greater problem than the pay. In March of that year, producers unilaterally enacted a 50 percent pay cut for all actors under studio contracts. With no organization or collective bargaining power, the actors had no means of maintaining their wages.

As a response to the producers' decree, a small group of Hollywood actors formed a self-governing group called the Screen Actors Guild (SAG) to give actors a stronger voice. A four-year struggle ensued for union recognition and better contracts with producers. In 1937, the actors voted to strike if necessary to win union recognition. More than 95 percent of the major stars involved threatened to walk off the job. Faced for the first time with an effective threat, the producers agreed to improve both working conditions and wages.

The formation of a strong union was a powerful means of protecting actors from unfair labor practices, as well as a potent instrument for promoting their future interests. Today, video producers often deal with unions that represent and protect the interests of television workers. Union rules and regulations can have a significant impact on work schedules, hourly pay rates, overtime, holidays, and many other working conditions that can directly affect budgets, schedules, and therefore programs.

Depending on the type of work people do during a show or rehearsal, they can be classified as performers, guests, or crew members. These distinctions are important, because if a worker's efforts go beyond normal duties, you as the producer may be subject to talent fees and even legal penalties from unions not covered in your budget. To avoid problems and unanticipated expenses, be aware of the various labor boundaries that have become the province of each major union involved in television work. Using nonunion labor in a union setting also requires proper clearance.

In television work, unions exist for creative personnel (such as writers, directors, performers, and musicians) and technical employees (electrical work-

---

TABLE 13.2

Major Unions for Television Workers

Unions for Creative Personnel

American Federation of Television and Radio Artists (AFTRA)	AFTRA was founded in 1937 to represent the interests of professional broadcasters. Today AFTRA ensures fair contracts for radio and television workers, sets minimum pay scales, ensures safe and equitable working conditions, and provides legal representation in disputes. It also offers health and retirement plans, credit unions, scholarships, and other services. AFTRA has more than 77,000 members, including network news anchors and correspondents, announcers, actors, and other on-air talent. Its National Code of Fair Practice lists pay scales for talent working in various programs in different capacities.
Screen Actors Guild (SAG)	SAG represents more than 84,000 actors (including singers, dancers, models, and extras) working for producers of motion pictures and video programs. Its employment contracts govern wage scales, holidays, meal periods, overtime, parking, residuals, rest periods, reuse of photography, script readings, stunt coordinators, safety regulations, travel costs, and more.
Writers Guild of America (WGA)	WGA represents professional writers in motion pictures, television, and radio. Membership is available only through the sale of literary material or through employment for writing services in one of these areas. The Guild is divided into two membership corporations, one on each coast, but for practical purposes it is a single, national organization. The WGA protects its members' rights with Minimum Basic Agreements that cover fees, payments, rights, and credits for authored material. The Guild also offers pension and health benefits and a registration service to protect the rightful ownership of literary material.
Directors Guild of America (DGA)	Formed in 1960 by a merger of the Screen Directors Guild and the Radio and Television Directors Guild and enlarged since then by mergers with other creative organizations, the DGA now represents the interests of directors, associate directors, stage managers, and production assistants working in live and taped television.
American Federation of Musicians (Musicians Union Local 47)	Organized in 1894 as the Los Angeles Musical Association, the AFM today has more than 9,000 members. Its pay scale agreements set the pay rates for musical performances used throughout the television and movie industries. The AFM bargains collectively for members involved in musical performances on network radio and television, videotape, educational television, music videos, electronic transcriptions, commercial advertisements, cable and pay TV, nontheatrical and non-TV documentary and industrial films, and traveling theatrical productions, among others.

TABLE 13.2

Major Unions for Television Workers (*continued*)

Unions for Television Technicians	
National Association of Broadcast Employees and Technicians–Communications Workers of America (NABET–CWA)	Formed in 1994 by the merger of NABET and CWA, the combined organization represents the interests of broadcast engineers, technicians, newswriters, announcers, photographers, and stage service workers, among others. Most of the 10,000-plus members work for two major television networks (ABC and NBC) and more than fifty private radio, TV, film, and videotape companies in the United States and Canada. The areas in which NABET–CWA has improved working conditions for its members include work schedules, compensation for travel time, meal periods, holidays, vacations, on-camera appearances, retirement plans, and sick leave.
International Brotherhood of Electrical Workers (IBEW)	Formed in 1891, the IBEW negotiates contracts for technical staff involved in broadcasting work, including camera operators, videotape and audio personnel, editors, and maintenance workers. IBEW has national agreements with the CBS and Fox television networks, as well as local agreements governing daily hires (per diem contracts) for freelance technicians performing television work. Currently IBEW has more than 800,000 members, only a small proportion of whom work in broadcasting.
International Alliance of Theatrical Stage Employees (IATSE)	IATSE was formed in 1893 to represent behind-the-scenes workers in theater production. Today IATSE includes behind-the-scenes workers in the television industry as well. With more than 85,000 members, including wardrobe and makeup artists, camera operators, and prop, sound, and grip personnel, IATSE represents people involved in television series, commercials, and made-for-TV movie productions.

ers, camera operators, and so on). Table 13.2 lists the most prominent unions, with pertinent information on each.

Audience Measurement and Ratings

Since the producer must convince a client that the program can reach a suitable audience, it is important to know how audiences are measured. Audience measurement is critical for commercial television because a commercial station's revenue, which comes mainly from the sale of commercial time, is determined primarily by audience size.

In measuring audiences, levels of audience attention or involvement are rarely determined. Rather, surveys are used to estimate the number of households or viewers tuned to a program. Two such estimates are often computed: ratings and shares. The **rating** for a program is the proportion of all television households (TVHHs) in a market that are tuned to a particular program. The **share** is the number of homes tuned to a particular program as a proportion of all homes actually *using* television (HUTs) at the time.

Knowing both the TVHHs and HUTs tuned to a program allows you to judge both its overall popularity and its performance against its immediate competition in the same time slot. For example, in a market with 1,000 households (assuming, for simplicity, one set per household), if 20 percent of all TVHHs are tuned to a program, that program gets a 20 rating. If all 1,000 households have their sets turned on, the share will also be 20. However, if half the sets are off, leaving only 500 sets in use, the rating of 20 will mean that 40 percent of all HUTs are tuned to that program, giving that program a 40 share.

A late-night show usually has a relatively low rating because the number of sets in use is generally lower than during prime time. However, the show must still compete with others available at that time, and this is where knowing the share becomes important. In the late-night slot, a program with a 5 rating (5 percent of all TVHHs are viewing it) and a 40 share (40 percent of HUTs are tuned in) may easily win its time slot and therefore remain on the air. In contrast, a program in prime time may get a rating of 10 but only a 13 share and be canceled because it fails to compete well against other shows that air at the same time.

Of course, in addition to learning *quantitative* audience information, you could learn valuable *qualitative* information, including demographic audience characteristics (age, sex, educational level, race, income), psychographic measures (attention level to your program, favorite characters, motives for watching), and lifestyle characteristics (hobbies and interests). As Chapter 1 made clear, such knowledge can provide a solid basis for program planning.

In nonbroadcast settings, audience measurement can also help in making programming decisions. One approach is to use questionnaires designed to tap audience members' reactions to a program they have just watched. Such postprogram evaluations can be used to determine the effectiveness of instructional videos, training tapes, information campaigns, and so forth. There are also informal avenues for audience feedback, including letters, phone calls, e-mail, and computer Web sites and bulletin boards.

DIRECTING

Once the basic conceptualization and planning are done, it is the director's job to execute the production. However, directing is so critical to the success of a video program that it is essential to involve the director from the earliest planning stages, not just when the program is ready to shoot.

Just as the activities of the producer vary depending on the type of program, so may the director's. For example, live broadcasts place different demands on the director than programs shot in the field and edited in postproduction. Program genres may also influence the directing effort: a sitcom, for example, requires different treatment than a musical variety show, a sports program, or an instructional video.

Qualities of Directors

For all directors, communication, preparation, vision, flexibility, and tact are essential. The competent director must be able to communicate clearly what is wanted, solve problems as they arise, and deal effectively with temperamental people. Maintaining group unity and cohesion and promoting a positive attitude among talent and crew can be difficult when the group is a disparate collection of tender egos.

Good directors must also be demanding, have high standards, and know which things are going right, as well as those going wrong. Of course, a vision of the final product must first exist in the mind's eye (and ear) of the director so that it can be conveyed clearly to the talent and crew. Only then can quality be maintained as the script gets translated from paper to the screen. If the concept of the show is not clear to the director, it cannot be communicated to the team, and the chance of making a successful program will be slim.

Another way to think about the director's talents is in terms of the many roles the director has to play. The following sections look at the director as artist and technical expert, psychologist, manager, organizer, coordinator, and leader.

The Director as Artist and Technical Expert   As artist, the director determines the sequence of the shots and their duration. The director must call for every audiovisual change or transition. The director must constantly assess whether what is seen and heard is necessary and appropriate to the goals of the program and whether it will maintain audience interest.

As technical expert, the director must be familiar with the capabilities of each piece of equipment used in the production. The director should be able to judge whether the equipment selected for a particular job is the best available for the task in light of the budget.

The Director as Psychologist   As psychologist, the director must be sensitive to the needs of the crew and talent, recognizing that human beings deserve dignity and respect. (Translation: remember the Golden Rule, and treat others as you would like to be treated.) In addition, the director must recognize when talent are performing to the limits of their abilities, when they are about to burn out, and when they need rest. The director should try to create an esprit de corps, conveying to the crew the importance of teamwork and the value of putting the goals of the production first.

The Director as Manager, Organizer, and Coordinator   As manager, organizer, and coordinator, the director schedules the tasks to be performed and delegates some of them to others. The director's immediate support staff includes the assistant or associate director (AD), the technical director (TD), the floor manager, a production assistant (PA), and perhaps a lighting director (LD). By delegating tasks to these key people, the director maximizes efficiency.

In practice, appropriate delegation means the director shouldn't climb a ladder to adjust lights if the lighting director can take care of it. Similarly, the director shouldn't stop a rehearsal to pin a mic on a talent if the audio operator (or a utility person) can do it. Of course, in a union shop, such actions could be in violation of the union contract, in which case the director should not be doing the work in the first place. The key point is this: the director must delegate specific tasks to others so that she or he can perform the more general task of running the production.

The Director as Leader   Finally, as leader, it is the director's job to supervise the production, making sure all program elements, crew, and talent are brought into play when and where they are needed. In this way, the director is like the conductor of an orchestra and the crew and talent are like the musicians.

Directing in the Preproduction Stage

The preproduction stage is unarguably the most important part of the television production process, since it is here that the most time is spent formulating and perfecting the program concept, working out details, and solving production problems. Without a fruitful preproduction stage, the production stage has virtually no chance to succeed. And with a bad production stage, no amount of postproduction manipulation will, as the saying goes, make a silk purse from a sow's ear.

One main goal of the preproduction stage is to finalize the script and floor plan, since these documents anchor the entire production. These documents are to the program what a recipe is to the food a chef prepares. The director's other preproduction objectives include conducting production meetings and read-throughs, refining and marking the script, preparing storyboards and shot sheets if needed, working out talent movements, arranging camera shots and camera movements, conducting rehearsals, and making the final adjustments and corrections clear to talent and crew. At this stage, the director's job differs from that of the producer because the director focuses on bringing the program into final form rather than gathering and arranging production elements and personnel. The following sections look at these preproduction tasks in greater detail. Though some tasks must precede others, many can be done simultaneously.

Conducting Production Meetings   Prior to actual rehearsals, it is useful to conduct production meetings on an as-needed basis to communicate with crew and talent, answer their questions, iron out difficulties, and communicate last-minute changes. Depending on the type of show being produced and how it will be shot, the amount and type of preparation can vary tremendously, and thus the detail handled at the meetings will also vary. Even for simple productions, however, production meetings should be conducted as needed throughout all the production stages.

Refining the Script and Floor Plan   When you are directing, it is helpful to visualize the shot sequence as you read through the script. Read through several times to finalize your camera shots, camera angles, and microphone and talent positions. Place the floor plan beside the script as you read to confirm that everything you want to do is physically possible. For your purposes as director, the script should be, insofar as possible, a written sequence or time line of every audiovisual element that will be used, in order, from the first fade-in of a program's opening shot to the last fade-out of its closing credits. Similarly, the floor plan should provide a complete map of the program, showing the positions of all set pieces and talent, as well as all relevant off-air production elements such as crew members, cameras, lights, and booms.

Also, mentally listen to the dialog in the script to "hear" what it sounds like. Are there some lines that could be improved? Do they say everything that needs to be said? Are the speeches each talent delivers consistent with his or her character? Can some speeches be deleted without sacrificing the overall meaning of the program?

Think through your rationale for each production decision, justifying why you have decided to shoot things the way you have. Does each decision fit with the overall concept of the show? Is there a better way to do it? If you think of a better way to go, is it feasible to make the change, or will your late inspiration require such radical revision that is better to leave well enough alone?

Script Marking   After the script has been finalized, mark up a fresh, final copy that is legible and accurate. Set off all speeches and dialog clearly. Include all transitions and descriptions of shot content, music, and sound effects. Be as complete as possible. See Figure 12.2 on page 309 for an example of a director's working script. Above all, this copy should be so clear and easy to read that you can tell what's happening at a glance without having to search for any information.

Preparing Storyboards and Shot Sheets   As mentioned in Chapter 12, a storyboard is often used in fully scripted shows, especially in commercials, to precisely convey the content, appearance, order, and screen time of a program's main shots. If a storyboard is being prepared, the director will make certain it is complete and accurate. In other cases, though, a shot sheet (Figure 13.5) can serve a similar function. The **shot sheet** provides a list of shot descriptions for each camera. The shot sheet conveys to each camera operator which shots he or she will be responsible for and the order in which they will be called. One advantage of a shot sheet is that it enables the director to call for a particular shot by number so that ready cues are faster than they would be if they had to be described. For example, saying "ready camera 2, shot 12," is faster than saying "ready camera 2, close-up head shot of host, please."

Talent and Camera Blocking   **Blocking** is the process of coordinating the movement of talent and cameras through space to achieve the desired effect on screen. Blocking rehearsals are crucial to cover talent movement on air, because without practice it may be impossible for camera operators to carry tal-

CAMERA THREE	
<u>Shot number</u>	<u>Description</u>
1	Close-up host and guest
3	Extreme close-up product shot
6	Bust-shot host holding product
8	Two-shot host and announcer tasting food

Figure 13.5   *Sample section of a shot sheet.*

ent smoothly from one point to another, especially if they don't know where or when talent will be moving.

Before an actual rehearsal, you can use the floor plan to help you visualize and plan the moves you want the talent and camera operators to make. The floor plan will help you confirm that certain movements are possible within the space constraints.

In the blocking rehearsal, at first let the talent perform their moves as they would do them normally. This will give you an idea of how to block them more naturally. After seeing how they perform and taking time to consider how the blocking can best serve the program, make the talent aware of any adjustments or modifications you need.

As they perform the adjusted moves, make talent aware of the framing, shot size, and camera angles you intend to use for each camera. If talent must cross from one set area to another, make them aware of which camera will carry them and whether they should address the camera directly. If they are to handle props, let them know which camera will be used for prop shots so that they can maneuver the props without hiding them from view and without losing focus or framing.

To rehearse blocking in studio, use the floor plan, a program monitor, and headsets if needed to communicate with talent and crew. With the floor manager and relevant crew members on headset, you can verbally cue talent and cameras to move to desired locations. Alert talent and crew that you may need them to perform certain movements repeatedly until key shots are mastered. If need be, mark the floor with tiny pieces of masking tape to indicate places where talent *and* cameras need to go. During the blocking rehearsal, note the amount of time needed to perform transitions so that you know which changes are reasonable to ask for and which are not.

Depending on the time available, the complexity of the moves, and your personal style of directing, you may wish to conduct blocking rehearsals from the studio floor rather than from the control room. Doing it from the studio is usually faster because it eliminates the need to work through the floor manager. If you do it this way, have the technical director operate the video switcher for you in the control room. Set up a line monitor in the studio so that you can see the shots you are calling and a live studio mic to allow control room personnel to hear your commands. After ironing out the major problems, conduct the last blocking rehearsals from the control room through the floor manager.

Early Rehearsals   The rehearsal stage is the time when all production elements are integrated into the finished product, a video program. Rehearsals are often known as **run-throughs,** though the early, rough passes can look more like tumble-throughs.

To avoid wasting valuable studio time and reduce crew salaries, equipment costs, and rental fees, preliminary rehearsals are often conducted outside the studio facility or field location. For the early stage, often called a **dry run,** the director and talent may assemble in a meeting room to read through the script and discuss the approach they will take. Questions are answered and problems are worked out. Such preliminary meetings are often conducted for dramas, comedies, and musical variety programs. For blocking, the dimensions of the space where the actual program will be shot may be simulated in the meeting room with chalk or masking tape. Sets may also be simulated with available furniture and materials.

When the rehearsals are held in the actual studio or field location, some directors prefer to split them into phases, each focusing on a different level of activity. For example, **primary movement** includes the motion of the talent and other elements in front of the camera. A primary rehearsal, therefore, could just concentrate on directing talent to move to their various locations as they say their lines. After such a practice of "marks and lines," the director could hold another primary rehearsal to layer in the timing and props.

When the talent can perform these elements successfully, the director can move to the next level of rehearsal, dealing with secondary movement. **Secondary movement** includes camera movement, framing, focus, zooms, and so forth. When this phase is mastered sufficiently, the director may tackle **tertiary movement,** the control room activities such as cuts, wipes, keys, special effects, and other transitions.

The advantage of working in phases is that the crew as a whole will need fewer full rehearsals. This method can cut expenses, including unnecessary wear and tear on personnel. It also increases the director's chance to work personally with each individual in the production, and it boosts the crew's confidence that the production is being built on one solid foundation after another.

Full Studio Rehearsals   Depending on the type of show being produced, the director might choose to perfect and then tape a single segment before going on to the next. Some EFP single-camera programs lend themselves to this approach. Alternatively, the director may decide to rehearse the entire show from beginning to end several times, bringing it up to speed and polishing it further with each pass. Live-to-tape studio productions might reasonably be done this way.

In **interrupted run-throughs,** the participants may spend most of the rehearsal time practicing complicated transitions between segments and very little time rehearsing the program segments themselves. For example, if a live shot in a ten-minute interview is to carry the host from the interview set to a separate area for a commercial, the director may spend only a few seconds rehearsing the interview itself and use most of the time perfecting the more challenging move from one set to the other. Interrupted run-throughs cut to the chase, so to speak, using rehearsal time to solve the most pressing and complex production problems.

The seasoned director distinguishes between major problems, which may require calling a cut and bringing everything to a screeching halt, and minor errors that can be fixed later. Not every mistake is worth killing the rhythm of a rehearsal to fix. Sometimes it is better to let the talent and crew go for awhile to get a feel for the project and give you a sense of the pace and timing of the show. If you work this way, have your production assistant or the assistant director write notes as the run-through progresses so that you can later recall what you wanted to fix. On the other hand, you must not accept garbage, nor should you allow the cast and crew to memorize a string of mistakes that will be hard to change. Deciding how to treat problems in rehearsals is a critical judgment call the director must make.

Dress Rehearsals   The final studio rehearsal should be a **dress rehearsal,** a replication of the actual show, including every production element: every sound effect, music cue, and credit, down to the last detail. Dress rehearsals are sometimes taped in case segments go better than in the actual program. Then it may be possible to cull the best bits from the dress rehearsal and combine them with the best from the actual show to make the final product even stronger.

Before airtime or final tape, the director should communicate any last-minute suggestions, corrections, or changes. Any such items the director thinks of (as well as noteworthy contributions by others) should be noted during the rehearsal period and communicated to the crew and talent before showtime.

Directing in the Production Stage

During the production stage, the director orders every audiovisual change that occurs by issuing commands to crew members. To do this, the director usually delivers verbal cues through a headset from the control room to the crew in studio or field. The director may also speak to the master control room to let technicians know when to begin rolling tape before the show begins.

Techniques for Delivering Director's Cues   Several techniques can help you, as director, deliver cues effectively. First, speak clearly and succinctly to keep confusion to a minimum. Second, to improve crew members' poise in executing the commands, deliver each cue in two stages: first give a **ready cue,** then give a **command:**

> *"Ready take camera 1. Take 1."*
> *"Floor manager, ready pull card. Pull."*

Third, since many people may hear your voice, it helps to specify *who* is being addressed as close to the beginning of the speech as possible:

> *"Camera 1, when we get off you, please break to commercial set for product shot. OK camera 1, break."*

This technique alerts the proper crew member to listen to the command, and it frees up others from having to listen to the entire cue.

Finally, keep verbiage to the bare minimum. For example, it is better to say "Ready take camera three. Take three" than to say "Ready *to* take camera three. Take three" because the word *to* may be misunderstood as a camera number.

The Importance of Anticipation   In white-water rafting, it is critical to anticipate the rocks that lie ahead of you to avoid being bumped or tossed from your craft. Rocks already passing by are of no concern at all. Similarly, for the director, it is important to concentrate on what lies ahead, not on what has just been broadcast. What has already been broadcast can never be called back; therefore, from the director's standpoint, it is useless to worry about it.

To deal effectively with both planned and unplanned situations, you as the director should become adept at anticipating future conditions. In terms of planned situations, anticipation involves the ability to deal with normal, moment-to-moment program changes, including setting up the next shot, knowing where the script is headed, and being aware of the preview and off-air monitors. Familiarity with the script and thorough preparation will equip you to deliver a professional performance.

In terms of unplanned situations, the director must be a contingency thinker, considering in advance what to do in case of disaster. Much of this thinking should be done during the preproduction stage so that if disaster strikes during production, you will have already prepared ample fall-back positions. Ask yourself plenty of questions and work out the answers:

Question: What if camera one goes down while it is on air during the interview?
Answer: Cut immediately to the cover shot on camera two while we reposition camera three.
Question: What if the talent gets sick and fails to appear at the last minute?
Answer: Shoot other segments that do not require the absent talent.

At times such as this, the poise you show as director will go a long way toward building confidence among the cast and crew. In contrast, if you lose your cool, the entire production can grind to a halt.

The Director's Use of Time Time pressure is inherent in directing. Since facilities are in demand, often rented for a fee, their use is always limited by budget constraints. Time is money, and making good use of production time is essential. Further, program time is often determined in advance, making it necessary to time each program segment precisely so the show comes out to the required length. To use production time wisely, set up a production schedule and *stick to it*. Figure 13.6 shows a simple production schedule set up in blocks of time.

The major advantage of a production schedule is that it provides a blueprint for getting the job done in the allotted time. Of course, the schedule is useful only if it is followed. Therefore, avoid the temptation to sacrifice time blocks to perfect the work you are doing at the moment. It is generally better to move along and stay on schedule.

One reason strict timing is so important in commercial television is that programs and commercials that a station runs often originate from different sources, making it necessary for the station to switch program feeds repeatedly throughout the broadcast day. If programs were not of a set length, synchronous switching between sources would be impossible. In contrast, with precise timing, stations can switch from one source to another without losing part of either message, thus increasing their revenue.

In controlling program time, several concepts are helpful. **Clock timing** refers to the use of a separate clock in the control room (usually a digital clock or a stopwatch) to time a program or program segment during run-throughs and actual taping. This tells the director exactly how long each segment takes. For every run-through, the clock can be reset.

Backtiming is a means of determining how much time is left in a show by subtracting the program's present time from its total length. As we saw in Chapter 7, backtiming can be especially useful when you need to figure out when an audio or a video source should begin so that it ends precisely at a desired moment. For example, imagine you have a forty-eight-second piece of music that you would like to use to close your half-hour show. If you have the AD cue the audio operator to roll it at precisely the twenty-nine-minute, twelve-second mark, then regardless of when you fade that music in, it will end exactly as your clock reaches thirty minutes. This use of backtiming can give a professional, polished quality to your program.

With **front timing,** you add the running times for specific audio or video segments to the starting time of a program so that you can compute when to run certain inserts. Front timing is especially helpful in news programs, in which prepared packages of known lengths are frequently inserted. If you find that an insert comes up later than it was supposed to, you know immediately that the show is behind schedule.

PRODUCTION SCHEDULE

8:00–9:00 A.M.	Production Meeting
9:00–11:00 A.M.	Equipment setup and lighting
11:00–11:30 A.M.	Technical meeting
11:30–12:30 P.M.	Talent and camera blocking rehearsals
12:30–1:00 P.M.	Notes and reset
1:00–1:30 P.M.	Lunch
1:30–3:00 P.M.	Dress rehearsal
3:00–5:00 P.M.	Tape
5:00–5:30 P.M.	Strike

Figure 13.6   *Sample production schedule.*

Directing in the Postproduction Stage

Immediately after the appropriate thank-you's, the director often calls a "post mortem" meeting with the crew to discuss problems that arose during the production and how they might be solved in the future. Then, during postproduction, the director must supervise editing, if needed. This process (described in detail in Chapter 11) begins with viewing all unedited tape and deciding what is usable. To temper your judgment, it may be wise, if time permits, to distance yourself from the project for awhile, at least for a few days.

When you finally do sit down to watch, remember that the editing process lets you shuffle reality like a deck of cards. You may be able to use some sub-segments of shots in ways that were unanticipated. Listen to the audio with a critical ear, and don't forget that visuals can be salvaged even if the accompanying audio must be discarded, and vice versa.

A FINAL WORD

As the beginning of this chapter pointed out, today's video industry often allows people to work as producer-directors, combining the two roles into one. But even in a large, traditional setting where the two jobs are distinct, they are inextricably bound together. The director fundamentally depends on the producer to gather and arrange all production elements so the project can be put into final form. Conversely, the producer depends on the director to do justice to all of the elements placed at the director's disposal. Only through making a mutual professional effort can both parties hope to achieve their shared objective: the creation of a successful video program.

KEY TERMS

rate card *(330)*
rating *(343)*
share *(343)*
shot sheet *(347)*
blocking *(347)*
run-through *(348)*
dry run *(348)*
primary movement *(349)*
secondary movement *(349)*

tertiary movement *(349)*
interrupted run-through *(349)*
dress rehearsal *(349)*
ready cue *(350)*
command *(350)*
clock timing *(351)*
backtiming *(351)*
front timing *(351)*

QUESTIONS FOR REVIEW

1. In what activities does the producer get involved during the preproduction, production, and postproduction stages of a video project?

2. What should you include in a program proposal to maximize the chances of getting your project approved?

3. What information should a production book include to make the producer's job run smoothly?

4. Name the major unions for television workers, and discuss their impact on video production. What should producers know about union rules to enable them to stay on schedule and budget during a production?

5. Define *ratings* and *shares,* and explain why it is useful to compute both. Besides ratings and shares, what qualitative measures can help you assess the success of a television program?

6. What main roles does the director play? How does the director work with the script and the floor plan?

7. What are the usual stages of rehearsals, and what typically happens in each stage?

8. How should director's cues be delivered to minimize confusion?

14 Performing

Performing as video talent covers a wide range of activities. But whether you are acting in the role of a fictional character, performing as yourself as a sports commentator or an interview guest, or working with talent as a producer-director, you should know certain basic guidelines.

If you appear as a video talent, you should consider how technical, social, and aesthetic aspects of video production will affect the way you look and sound as you communicate with your audience. Since television presents live human action within certain technical limitations, it is important to craft your presentation accordingly. After all, what good is a key speech if it is delivered off mic or a product demonstration if done out of frame and out of focus? Further, since video production is a collaborative effort, you need to understand the social aspects of performing: your best performance may be lost if you miss the floor manager's cues. Finally, it's important to consider the aesthetic dimensions of the program, including its content, meaning, tone, and purpose. As an obvious example, your favorite comedic expression may have little place in a serious drama. Moreover, to make the best possible appearance, you need to know how to select the appropriate wardrobe and use makeup properly.

This chapter explains how technical, social, and aesthetic aspects of video production influence the look, sound, and performance of television talent. The topics covered include:

TECHNICAL ASPECTS OF VIDEO PERFORMING

scanning system • color, contrast ratio, and lighting • depth of field and framing • orientation and movement of talent • using microphones

SOCIAL AND AESTHETIC ASPECTS OF VIDEO PERFORMING

auditioning • performer and director • performer and floor manager • performing with other talent • performer, audience, and program content • video makeup

355

TECHNICAL ASPECTS OF VIDEO PERFORMING

Since video presents material composed of sound, images, color, and motion, it is important to consider each of these elements in performing. The following sections examine several technical characteristics of the medium that affect the way your performance appears on the screen.

Scanning System

Remember from Chapter 2 that, at least until digital television takes over, the NTSC standard for American television produces a relatively low-resolution image. For the video performer, this means wardrobe must be selected not on the basis of how it looks to the naked eye but based on how it appears on screen after being rendered by the camera. Unless you are performing as a clown or a space alien, it is best to wear clothing free of intricate patterns that challenge the camera's imaging capability.

Small and intricate clothing patterns such as herringbone, hound's tooth, and narrow stripes are generally poor choices for video because they can create a distracting *moiré* effect, or color vibration, that results from the interaction of the clothing design and the scanning lines of the television system. Solid colors avoid this problem and thus are generally better choices.

Color, Contrast Ratio, and Lighting

Wearing clothes similar to or the same color as the sets and backgrounds in which you appear makes it difficult for viewers to see where your form begins and ends. Worse, wearing clothing the same color as the chroma key effect can result in losing portions of the image of your clothing. To present yourself clearly, find out beforehand what the set looks like in terms of color, and then try to wear something different that fits the situation. Further, highly saturated colors, such as deep reds and other primary colors, can command the viewers' attention and perhaps distract them from your performance. For that reason, seasoned performers often avoid highly saturated colors unless the program justifies them.

Clothing choices should also be conditioned by the contrast ratio of the video camera. *Contrast ratio,* as we saw in Chapter 3, refers to the difference between the brightest and darkest portions of the picture that the camera can accurately reproduce. In video a contrast range of about 30:1 (40:1 in some CCD cameras) is common, so the brightest portion of the picture is at most thirty or forty times brighter than the darkest portion. In practical terms, this means that if you wear a dark blue hat with black lettering on it, the lettering may be invisible on screen. Further, a completely white outfit can cause tonal compression or reduction (*clipping*) in the contrast range of the television image, resulting in an unrealistic darkening of the face. Conversely, a pure black velour suit can cause the iris to open wide, making the face appear washed out and overexposed. To avoid contrast range problems caused by your

---

P R O F E S S I O N A L　　P O I N T E R S

Choosing a Video Wardrobe

- In the absence of compelling reasons to do otherwise, choose fabrics in solid colors.
- Wear mid-range colors rather than highly saturated ones, bright whites, or pure blacks.
- Choose colors that do not match the set, the background, or the planned chroma key effect.
- Do not try for wardrobe effects that rely on subtle contrasts in tone.
- Avoid horizontal stripe patterns (unless you want to look chubbier).
- Avoid shiny jewelry and other accessories, such as large belt buckles, that could cause strong reflections.

---

wardrobe, the best color choices are those that avoid these extremes, such as mid-range colors and muted pastels.

Jewelry can also be a problem in the presence of bright studio lights. Shiny jewelry can cause excessive reflections that can be quite distracting. To reduce or eliminate reflections, apply dulling spray to the jewelry's surface or rub it with a bar of soap.

Finally, most experts agree that the television image tends to add weight to the performer. This may be due in part to the camera angle used to picture talent, as well as to the fact that video renders three-dimensional objects on a flat screen. Or perhaps the illusion of added weight arises because the body is enclosed in a tight visual frame. Whatever the reason, you can offset the illusion of added weight by wearing well-tailored clothes that fit properly and avoiding horizontal stripe patterns.

Depth of Field and Framing

Television's technical limitations with respect to framing and shot composition directly affect what talent can and cannot do. The fact that video reduces three dimensions to two makes it critical to block positions precisely. If you are standing farther back in the set than another talent, the camera's rendering of the distance between you and the other person may differ greatly from your own perception. Moreover, depending on the camera's depth of field, a small movement toward or away from the camera can take you out of focus. Sometimes, if you are not the center of attention, it may be proper to move out of focus, but at other times it obviously is not appropriate.

In terms of framing, as you already know, a camera shot can vary from an extreme close-up to an extreme long shot. For television talent, it is important to know the framing at any particular time to be able to judge how much lateral movement is possible. If, while you are being carried in a tight head shot, you extend your arms to show the size of the fish you recently caught, the gesture will be lost (along with the information about the size of the fish) when your arms go out of frame. If a shot is tight enough, merely shifting weight slightly from one foot to the other can move you completely out of frame. Therefore, it is often necessary to remain stiller than you would in real life. Also, if you are going to move or gesture, it helps to telegraph that fact to the director in the control room. For instance, you might use a verbal cue ("Let me show you how big this fish was") as you slide to the edge of your chair. The verbal cue and the preliminary movement signal that you are about to do something different and give the director time to request a wider shot.

Framing also influences the way you handle and move props. For example, if you are cued to show an item featured in a tight shot—say, a dish of pineapple cubes to be used in a drink recipe—it may be better to point to the dish while discussing it rather than pick it up from the table. Picking it up can move it out of frame, nullifying the shot the camera operator has just set up.

Though tight framing is a constraint, it also has advantages. It enables you to convey a wide range of emotions with very little movement. For example, in a close-up, a mere tilt of the head or a shifting eyebrow can speak volumes about what you are thinking. Remember that in close-up, you usually do not need exaggerated movements and expressions; instead, rely on subtler indicators.

Orientation and Movement of Talent

When you make eye contact with the camera lens, your orientation to the camera is called **direct orientation**. Many performers use a direct orientation, including newscasters, announcers, hosts, and speechmakers. The direct orientation is appropriate for addressing the home viewer in the most intimate manner, with no intermediaries.

In many cases, though, it is more appropriate to avoid direct eye contact with the camera lens. Such an **indirect orientation** to the camera is usually preferable in a sitcom or fiction drama, where the characters are supposedly unaware of viewers. Likewise, in an interview, it is more natural for the host and guest to look at each other than at the camera. The indirect orientation conveys the sense that talent are primarily involved in action on the set. Some notable exceptions to this rule include the pioneering "George Burns and Gracie Allen Show" and the "Gary Shandling Show," in which the title characters occasionally break through the mythical *fourth wall*, as it is known in theater, to address the audience directly.

For the director, deciding whether the talent's camera orientation should be direct or indirect helps determine where to place prompting devices or cue cards. For example, if the director wanted to use an indirect orientation, a teleprompter attached to the front of the camera would be useless. The prompting device would have to be placed in some other off-camera location so that the talent's performance would appear as natural as possible.

Camera Switching Newscasters often must deal with on-air camera cuts. Imagine that you are a sports reporter appearing in a two-shot with the news anchor. The anchor tosses you the story with an "intro" line, and as you begin your report, the director cuts to your one-shot on another camera. When this happens, it is important to know exactly when the cut occurs so that you can maintain eye contact with viewers.

To make smooth transitions, you need to follow the floor manager's cue or watch for the tally light change indicating when the cut occurs. A common trick is to glance down for a beat as though you are checking your notes and, when you look up, reorient yourself to the on-air camera. This method of changing cameras appears more natural and fluid than simply turning your head on air from one camera to the other. If many cuts are planned, you may wish to mark your script for them in advance.

Blocking and Pace As noted earlier, an unexpected movement, even a relatively small one, can take you out of focus or out of frame. Moreover, shots intended to be combined with graphic inserts in postproduction may be spoiled if you fail to hit your marks precisely and at the right moment.

To turn in the most professional performance in terms of blocking and pace, work hard to match your actual performance to what was rehearsed. Do not deviate from what was agreed on in rehearsal, and be sharply aware of running times for each segment so that the show times out properly. On the other hand, in the back of your mind, always consider the places in the script where lines, business (a bit of action), or movement can be padded or shortened to accommodate last-minute changes. In other words, if you are performing, be precise and consistent in your moves and delivery. Don't surprise the director, but do prepare to be flexible if the need arises.

Cheating to the Camera To accommodate the camera, it is often necessary to position talent differently from the way people might be positioned in real life. For example, when two people are conversing, they normally face each other. To get the best coverage of their faces on video, though, it is often better to twist each talent's position slightly toward the camera. Shifting talent to enable the camera to capture a fuller view of their faces is called **cheating** talent to the camera. This often occurs in soap operas: two actors engaged in dialog with each other are positioned with both faces toward the camera, the shorter person in the foreground and the taller one behind.

Sometimes people speak of "cheating" props and display items to the camera, too. To determine whether and how much talent (or a cereal box) should be cheated to the camera, assess their appearance on the line monitor.

Controlling Dead Space Since television yields low-resolution images, it is often necessary to use close-ups and tight framing to provide greater picture detail. For video talent, this means that performers shown together in the same shot may need to be closer together than they normally would be in real life. Under daily circumstances, when you talk to a friend in a face-to-face conversation, you may stand several feet apart, but on television the result would be too much dead or wasted space between you. To get the desired level of picture detail in an attractively composed two-shot, talent are sometimes placed just a few inches apart. At first, this kind of invasion of talents' personal space may be discomforting, but as a performer you need to adapt to accommodate the camera.

Performing in Edited Productions Frequently video acting requires performing in scenes that are shot out of sequence and edited later in postproduction. When this style of production is used, it is difficult to maintain continuity in both your appearance and your intensity level. If you shoot a crying scene in which you are being led away from a hospital room and the next scene shows you, still weeping, being helped into a car to be driven home, you need to maintain the same demeanor even if the scenes are shot days apart.

In such cases, it helps to take a snapshot or shoot some video of the final moments in the first scene and write notes about your acting during that scene. Studying these materials will help you guard against obvious discontinuities in terms of props, clothing, and hair and make it easier to recapture a similar level of intensity when the follow-up scene is shot.

Using Microphones

Whether your voice is carried on a lavaliere, desk, boom, hand-held, or other mic, visible or concealed, you must be close enough to maintain the quality of being "on mic" as you deliver speeches. To turn in an air-quality performance, be aware of where your microphone is when you speak, and speak clearly.

In rehearsals, speak at the same intensity you intend to use for your opening speech. Then trust that the audio technician has taken proper levels during rehearsals and has placed the mics used to cover you in the proper position. The technician will "ride the gain" to accommodate varying intensity levels as your performance continues.

If you must handle or move your microphone, do so gently. Also, do not pull a mic cable by the microphone; pull the cable itself to reduce the danger of damaging the connections. If you must move from one set to another with a hand mic and cable, make sure you have enough cable to make the transition. Practice making transitions and crossovers without getting tangled in furniture and other talent.

If you must move from one area to another with a wireless mic, confirm in rehearsal that the equipment can capture the audio at each location. To do this, the audio technician may ask for a voice check at each location.

It is also important to control unnecessary incidental noise resulting from props, costumes, and jewelry. For example, if you are handed stapled copy to read on air, remove the staple in advance and then place pages gently to the side as you finish reading them. Noise from other paper props, such as newspaper and construction paper, can be reduced by spraying them lightly with a water mist shortly before airtime.

As for costumes, assess the fabrics you select not only for their visual compatibility with the camera but also for their noise level. Crinoline and other coarse, stiff fabrics may look all right on the line monitor but may sound terrible. Remember that lavaliere mics are especially susceptible to rustling noise from clothing.

Finally, the best test for jewelry is to wear it in rehearsal to determine whether it causes an audio problem. If you detect a problem, you may be able to correct it with some bits of masking tape in a few strategic places. If you can't keep the jewelry from rattling or clanging, eliminate it from your wardrobe.

SOCIAL AND AESTHETIC ASPECTS OF VIDEO PERFORMING

Auditioning

If you wish to be considered as talent for a video production, the first step will often be an audition for the job. Audition calls are posted in talent agencies, advertised in trade papers or newspapers such as *Variety*, and advertised on radio or television.

How should you prepare for an audition? To make the best showing, bring a portfolio of your past work, especially videotapes and photos (*head shots*). Next, be ready to perform, either acting, reading, singing, or dancing. If possible, prepare in advance a piece similar to the one for which you will be auditioning. (Some auditions, though, are nonspecific *cattle calls*, making it impossible for you to prepare anything in advance.) Finally, leave your name, address, and phone number so that you can be called back if you are judged to be right for the job.

If you are called back, remember that you may not be the only one still under consideration. You may be asked to demonstrate your skills once again. Be prepared to delve further into the part, and, if asked, be ready with suggestions about how the job might be done better.

Those who assess your performance will likely look for several characteristics, including the way you look and sound, your **range** (ability to handle different types of material), your pace and delivery, your **flexibility** (ability to give more than one type of performance), your ability to take direction, and your **consistency** (ability to repeat key parts of your performance exactly the same way each time).

Performer and Director

Aside from being involved with technical matters, scheduling and logistics, and content issues, the director must work to get the most out of the talent. It is the director's job (and sometimes the producer's) to assess the performance, note weaknesses, and make improvements.

However, television directors commonly have little time during production to work with talent. In many cases, the director is too busy in the control room with other aspects of the production. For these reasons, the production period is not always the best time to meet with the director. Even during rehearsals, the talent generally works with the floor manager and crew. It is early in preproduction that the director is best able to give you the attention you may need to answer questions about your upcoming performance.

If you are employed as talent, work hard in preproduction to prepare your presentation. If you have lines to memorize or ask questions about, the earlier you do it, the better off you will be. If you are supposed to demonstrate a product, practice beforehand to perfect your pitch. Practice with the props you will have to use, and note any problems that need the attention of the director. Then, when you do have a chance to meet with the director and discuss your role, have your questions ready.

Performer and Floor Manager

In general, the crew member with whom you will interact most is the floor manager. If you are not able to hear the director's voice through an earpiece, the floor manager may be your only link to the director. Therefore, it is critical to understand the silent hand signals the floor manager will use to cue your performance. During preproduction, learn that set of hand signals so that you have a clear understanding of what each gesture means. The floor manager should be able to cue you to slow down or speed up, go to a commercial, wrap up a segment, speak more loudly or more softly, get closer to a microphone, and many other things. Figure 14.1 displays a number of hand signals floor managers commonly use to communicate with on-air talent.

In acting assignments, when you are playing a fictional character in a fully scripted and blocked production, your communication with the floor manager may be more limited than if you are performing as yourself in the role of host or announcer. Nevertheless, some cueing may still be necessary, so it is wise to come to an understanding with the floor manager even in those instances.

Performing with Other Talent

As television talent, you may find yourself working with others from all parts of the performing spectrum, including experienced professionals, newcomers, major stars, and prima donnas. The people you work with may come from drama, comedy, talk shows, news, the music industry, and numerous other parts of the entertainment and nonentertainment worlds. How can you craft your performance to make working with others most rewarding?

First, be dependable. Show up on time, since a great number of individuals may be relying on you for getting the show produced. Also, be sensitive to time constraints. There is usually a limited time available to get the program made.

Second, know your assignment. Be familiar with the script. Remember, there is no substitute for thorough preparation. Remember also that at times, such as during rehearsals, it is better to listen than perform.

Third, don't throw curve balls. Be consistent. Do not deviate from the rehearsal and script so much that it throws off your co-workers. Though ad-libbing may appear to be the bread and butter of a Robin Williams, most of the time a polished ensemble performance is carefully plotted out in advance. Stick to the script.

Try also to get a sense of the overall pace, tone, and meaning of the production. Tune your delivery and intensity level to the program, and permit others' performances to come through as well.

Finally, be sensitive to the vast range of skill levels and temperaments of your co-workers. Some may be consummate professionals with years of experience, whereas others may be the greenest first-time guests with no clue about how to be television talent. Some may detest small talk; others may make you detest it. Getting to know the skill levels, backgrounds, and personalities of your co-performers can help you understand production needs and simplify the entire process.

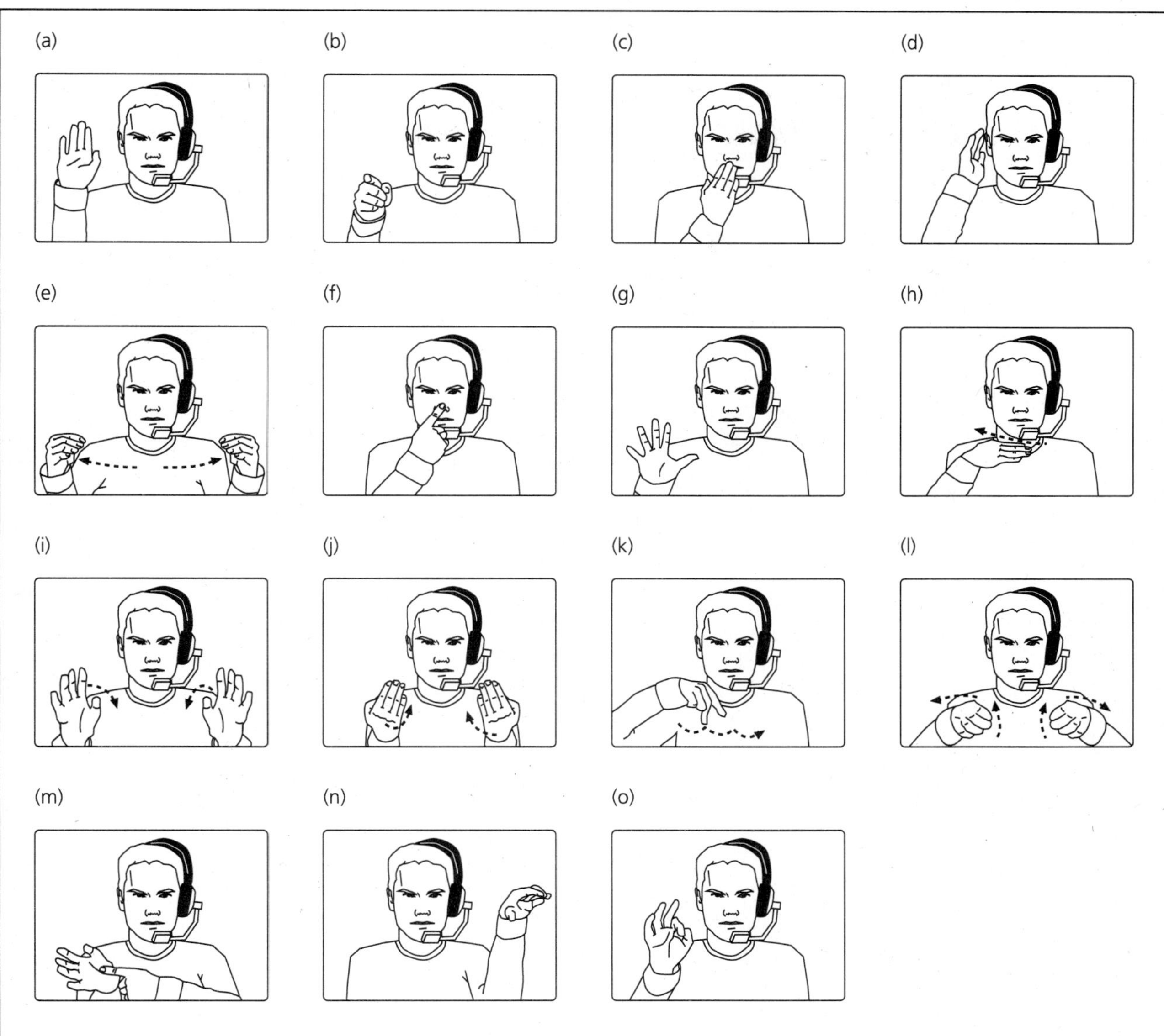

Figure 14.1   *Hand signals typically used by floor managers to cue talent. Although the ones shown here are fairly common, anything that communicates successfully is acceptable: (a) stand by; (b) cue (you're on); (c) speak more softly; (d) speak louder; (e) stretch (slow down to fill time); (f) we're on schedule; (g) five minutes left; (h) cut; (i) move back; (j) come closer; (k) walk (move to next performing area); (l) go to station break; (m) go to commercial; (n) go to commercial; (o) everything is okay.*

In general, try to make the experience of performing together fun. Be respectful of others, for aside from the intrinsic social rewards gained from enjoying people enjoying themselves, you never know when you will have to work with someone again in the future. The goodwill you cultivate is invaluable.

Performer, Audience, and Program Content

Ultimately, your objective in being a video performer is to communicate some message to an audience. However, the vast majority of the target audience is usually not present while you are performing. While there is always some crew around, and sometimes a studio audience, the home viewer is not there to react to you, making it difficult to assess your effectiveness. This means relying on your own concept of the audience and your sensitivity to the material you are working with.

One way to surmount the difficulty of the absent audience is to imagine the camera lens not as a proxy for potentially millions of faceless viewers but as a small group of close friends or even a single best friend. The quality of intimacy this approach achieves may be exactly what you need to be most effective.

Finally, remember that viewers' expectations are often strongly shaped by program genre. Try to attain a strong sense of the program's genre and its main objectives. Steep yourself in prior examples of the genre, and then pay attention as well to the particular objectives of your program. Some programs are made to entertain, whereas others are intended to educate, inform, or persuade. Some programs have multiple goals. The purpose behind the program should shape the tone, pace, and other aspects of your performance.

Video Makeup

Makeup can improve, control, or change the way you look on camera. If used properly and in keeping with your program purpose, makeup can be one of the cheapest and fastest ways to enhance your video appearance.

The video camera unflinchingly captures faces, often in intense close-up and under bright lights. Under such conditions, the television system may reveal and even emphasize blemishes, blotches, and other minor skin imperfections. Although minor defects may not be noticeable under normal conditions, they can detract from your television appearance.

In male talent, the darker appearance of the face in the beard area even after a recent shave can be quite distracting. Some people's skin is more translucent than others, resulting in uneven lighting effects when two or more people are pictured together in the same shot. Being subjected to hot lights causes some performers to perspire and release oils to the surface of the skin, resulting in distracting shine when photographed by the camera. Skin color also varies across subjects, sometimes making it necessary to use makeup to offset unflattering color distortion or contrast range problems resulting from the limitations of the television system.

Of course, if talent looks good in tight close-up (the acid test) without any makeup at all, none may be needed. However, if close-ups reveal unwanted or distracting effects such as those just described, the proper application of makeup is often the solution. Figure 14.2 illustrates many of the makeup procedures described in the following sections.

Using Makeup to Control Shine and Uneven Skin Tone   To control shine caused by perspiration and oil on the skin and to hide minor blemishes, apply a **foundation** or **base makeup** to cover the skin and give it a more even appearance. Foundation makeup is also used as a background for all other makeup applications.

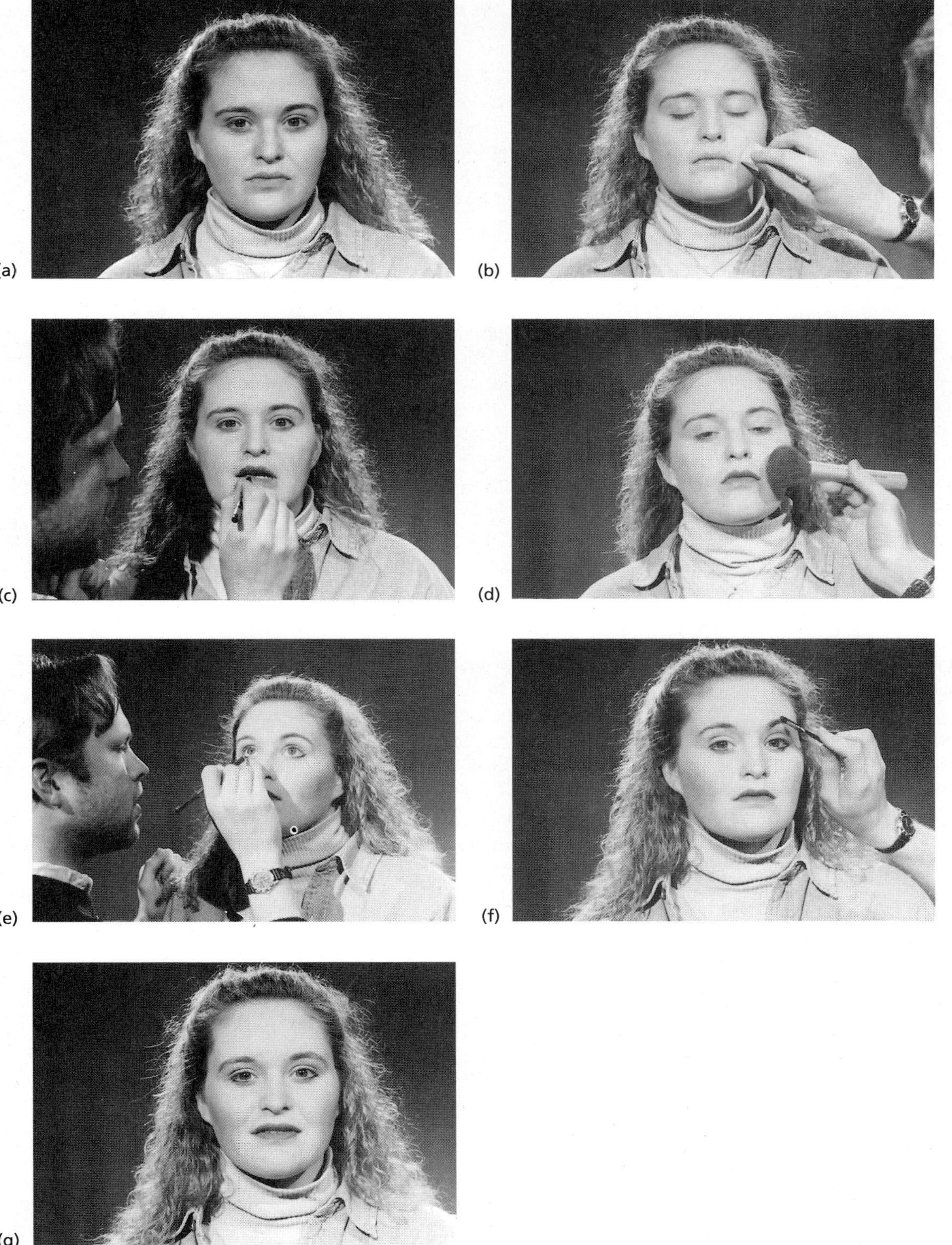

Figure 14.2 *Some of the steps in applying makeup for video talent. (a) The talent before any makeup is applied. Notice how the lights cause shine on her nose, chin, and forehead. (b) Applying fluid foundation to control the shine; powder will then be added to "set" the foundation. (c) Lip liner, which will be followed by lipstick. (d) Rouge. (e) Mascara for the eyelashes (in this photo, eye shadow has already been applied). (f) Eyebrow pencil to darken the brows. (g) The talent after a final application of powder.*

Joyce Simonelli
Makeup Artist, Fox Television, New York City

Q: How long have you been in this business?

A: About nine years. I originally got interested at age twelve. I would see fashion magazines and try to make up my sister and myself to look like the models on the covers of the magazines. I'd also apply makeup to my friends.

Q: How did your career start?

A: I worked for a dancing school; that's how I came to New York. At the school, I took some of the dancers and applied makeup to them. I knew nothing of the makeup career when I first came, but some dancers recommended I try a career in makeup and they bought me a kit.

I started as a makeup artist in print work. This was a good background because it is the most precise work and you really see your mistakes. I moved from print work to video by calling around at some independent stations. I gradually got called for television work as a fill-in to replace regular house makeup artists who would take a day off for a commercial or some other freelance job.

I found out that there is really no set pattern or training program you follow to become a makeup artist. What you must do is find teachers who've been artists for years. I studied with two teachers. One was Burt Roth. He's head makeup artist at ABC, and he still teaches a class. Another is Jim Cola, who was the head makeup artist at Fox.

Q: What did they teach you?

A: There are things that you really must know. Number one is your beauty makeup. Within that category you have to understand how to contour, shape, shadow, enhance, and highlight the face. You also have to understand lighting. How will the light affect the face you are working with? If you are working outside in natural light as opposed to studio light, you need to know the impact. There's also soft light and hard light, and colored gels used within a set. You have to be able to compensate for the lighting with the tones of makeup.

Another style of makeup is certain special effects, including how to age a face, either with prosthetics or strictly with color foundations [highlighting and shading]. Depending on how far the talent will be from the camera and how close the shots are, you use different techniques to get the job done. Other special effects include under-eye bags, scars, bruises, bullet wounds, burns, black eyes, and the use of blood. For example, say you are working on a soap and a character gets punched in the eye. You have to know how to make that person look as though they were punched in the eye. If a character is shot, you have to give them a bullet wound. But when it comes to heavier-duty things like blood exploding out of the chest, that is a different area of special effects. In those cases, there is usually a team of people that do it.

In extensive prosthetic work, you take an actor, make a mold of his or her face, then make several heads or busts from it. Then you create pieces for the actor to use. You do not sit down with the actor for each scene; you use the molds and work from them. Artists who do such work are mainly out of California, and they work in teams. Shows like "Saturday Night Live" do it also. It's a lot of very time-consuming work. That's why it's usually done by teams.

Q: What kind of money can you make starting out these days?

A: If you work for a television station, you can work union or nonunion. The union I belong to is IATSE Local 798 in New York. Depending on the show and what you're doing, rates can differ. Right now, a basic union rate is $160 a day, which is very low. What you do is work a lot of overtime to get the rate up. On talk shows, the base rate is about $350 per day. On commercials, rates can be $350 to $1,000 a day. If you work on a soap, you can make several hundred a day. One of the nice things about working a union job is that the rates are stable.

Q: On a soap, do you make up the whole cast?

A: On a soap or large-scale production, you usually have a first, second, and third makeup person doing the work.

Q: Are there any major differences in applying makeup to males versus females?

A: Obviously there is a lot more makeup applied to females than to males, but you can really well-groom and enhance a man's face with makeup. It's necessary.

Q: Then should everyone on TV be made up?

A: Anybody that appears on TV should be made up.

Q: Why? Don't some people look good without makeup?

A: The lighting on TV makes it necessary. Natural oils reflect poorly, so you need to cover that. Without makeup, you'll see beard shadow and capillaries on the skin. The makeup can be as minimal as a little powder, or it can be very thick deck foundation, but something must be applied.

Q: Do you do hair?

A: In a union job, no. There's a separate hair person. In a nonunion job, artists may be required to do hair. And that means styling as opposed to cutting. If you can learn to style hair nonunion, you can get more work.

Q: How have things changed over the last ten years?

A: Technology has changed. Digital cameras are now able to manipulate the look of the face of the on-camera talent. You can actually remove facial lines without distorting the clarity of the image and without changing the focus. You can now electronically soften the look of the skin. And it makes the makeup job a lot easier. So when you're dealing with someone with serious under-eye bags, wrinkles, or bad skin that's very marked, you can even out the skin tones. You can't cover the indentations on the skin, though. The light on someone with very bad acne will still reveal those marks. What the camera can do is smooth out the skin somewhat, making it look more attractive.

Another big change has been that, with cable, we have many more television stations. The field is really open right now. There's twenty-four-hour news, there are stations like HBO making their own films. It's still very competitive, but things are opening up.

Q: From a practical viewpoint, how does working in television differ from print?

A: In TV you can't take the time you can in photography. You have to have the talent ready when the director is ready to go to air. You have to work quickly. You may have ten faces to prepare in an hour, and you must be able to look at each face, know the products you're going to use, apply them, and get the talent out on the set. The show must go on!

Q: Does any single philosophy guide your work?

A: You have to enjoy people and realize that the subject is *not* a canvas but a human being. There is a human element that must not be overlooked. Most talent are concerned with how they look. Sometimes you see an underlying concern or even an insecurity or vulnerability that no one else sees. Therefore, you need to be very gentle with their egos and listen to what they are saying. You also need to be able to take charge of the situation and not let them run your creative side. There is a fine line when dealing with talent. They are nervous about going in front of the camera. They are thinking about it. You need to make the makeup room calm. Make it a peaceful environment and help them feel ready to go in front of that camera.

Artists who are caught up in their own egos need to remember they are dealing with human beings. You are sometimes also dealing with novices, like guests on talk shows, who may be sweating. Nervous sweat can be a problem, so it's good to keep the room cool and keep a sweater available.

Q: What if someone gets out of the chair, looks at the mirror, and hates it? Has that ever happened?

A: Well, they probably wouldn't get out of the chair [laughs]. You correct the problem. You clear away whatever is wrong and start again. Sometimes you correct it by applying over it. You have a responsibility to make the face right. Remember, your talent is your best friend. If they like your work, they're going to want to keep you around.

Foundation makeup comes in fluid form, cream, or cake. If you use foundation, it is generally considered best to match the color to the skin you are covering. For men, though, it is sometimes desirable to use a foundation one or two shades darker than the skin you are covering to maintain a masculine appearance. For women, lighter colors should be used only for very light-skinned subjects.

Fluid foundations are best for video work, and they are available in oil-free formulas for oily skin. Cream foundations are available in stick form, compact, or case. Among the brands most widely available are Bob Kelly Creme stick and Kryolan Paint stick. Product lines by Mirage and Joe Blasco have also been developed for television. Water-based cake foundations have been developed for face and body makeup by Max Factor, Bob Kelly, and Kryolan. A basic beauty kit, used for both male and female talent, is illustrated in Figure C.7 in the section of color plates.

Dealing with Color Distortion Skin tones differ from person to person across a wide range of colors, from very light to very dark. In addition, television viewers judge the accuracy of their color reception using skin tone as a reference point. Therefore, it is important to render skin tones accurately.

With darker faces, the television system may overemphasize cooler hues (blues and greens). You can offset this with foundation makeup that contains more of the warmer colors. Similarly, with light faces, when the television system overemphasizes the warmer hues (reds and oranges), you can counteract the problem with foundation of a cooler color. Judge the makeup on the basis of how it looks on screen and under the actual lighting conditions that will be used during the production.

Powdering All cream foundations require powdering after they have been applied. Without powdering, the skin will appear shiny, especially under bright lights. Powdering provides a matte finish that helps show off the features of the face. Powder also sets and holds the makeup during the performance. The most useful powder is loose, translucent powder with no color value. To apply it, press it into the foundation using a cotton ball or smooth, flat sponge. If applied properly, there will be no extra to brush away.

Shading and Highlighting After applying foundation, **shading** and **highlighting** makeup can be used to define or de-emphasize facial features. Shading makeup darkens areas of the face that would fall into shadow under natural lighting conditions. Highlighting makeup helps emphasize natural highlights of the face.

Shading is done first with darker colors using powders and creams, applied with brushes and worked in with the fingers. *Grease liners* that resemble pencils may be used for more precise shading work. Highlighting with lighter colors is done last, also using creams and grease liners. Use the fingertips to blend in the makeup to eliminate noticeable edges or abrupt color differences. After making the shading and highlighting additions, use powder to set the work.

Shading and highlighting may be used to thin or broaden the face, change the appearance of the nose, reduce tired-looking shadows around the eyes, and emphasize facial structures such as cheekbones and eyes. In general, follow the principles of naturalism discussed in Chapter 4. Remember that we are accustomed to seeing human faces the way they look when lit from above by a single main source of light, namely the sun. Therefore, when using

<table>
<tr><td colspan="2">

TABLE 14.1

Shading and Highlighting Applications and the Effects They Achieve

</td></tr>
<tr><td>

Application

</td><td>

Effect

</td></tr>
<tr><td>

Shading
Shade under the eyebrows

</td><td>

Emphasizes the eyes

</td></tr>
<tr><td>Shade under the cheekbones</td><td>Thins the face, deepens facial structure</td></tr>
<tr><td>Shade on the point of the chin</td><td>Shortens long faces</td></tr>
<tr><td>

Highlighting
Highlight eyelids

</td><td>

Emphasizes the eyes

</td></tr>
<tr><td>Highlight cheekbones</td><td>Emphasizes natural facial highlights</td></tr>
<tr><td>Highlight middle of nose</td><td>Emphasizes natural highlights of nose</td></tr>
</table>

makeup to achieve a natural effect, it makes sense to sculpt the facial features by adding shading where natural light is most likely to cast shadows: under eyebrows and under cheekbones. Table 14.1 lists the most widely used shading and highlighting applications along with the effects they produce.

Rouge   *Rouge* is a powder or blush of red, peach, or rust color used to offset the uniform appearance of foundation makeup and for emphasis on the face. Female talent usually use lighter colors of rouge and male talent darker colors. For fairer skins, lighter, peachy tones of rouge generally work best, especially on women and children. For darker skins, deeper, rust colors work best.

For video, cream rouges are better than powders, since creams provide more of a glow to the skin. The rouge should be applied to the area where the

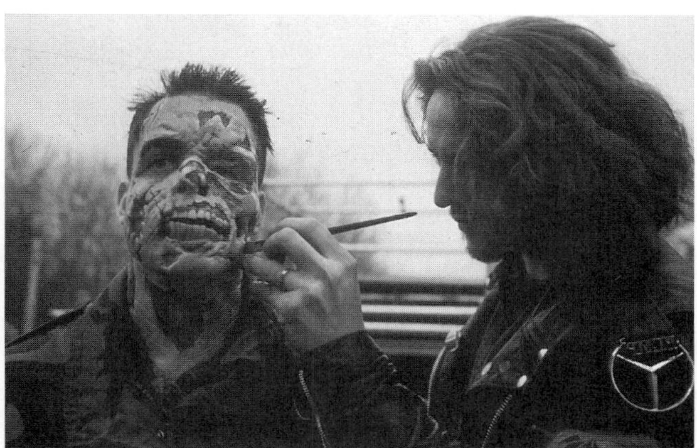

Figure 14.3   *An example of what can be done with special-effects makeup.*

Makeup Procedures

- *Makeup area:* Set aside a separate, well-organized makeup area with a chair, table, and large mirror. The area should be well lit so that you can see the work as it takes shape. If possible, to accommodate the color temperatures that will be used in the production, match the lighting conditions in the makeup area to the actual studio or field lighting. To further reduce color temperature problems, decorate the makeup area with neutral colors.

- *Washing and drying the face:* Before applying makeup, wash the face thoroughly with soap and water to remove surface dirt, oil, and perspiration. Then dry thoroughly with a towel.

- *Protecting your clothes:* Use an apron to keep stray smudges of makeup from getting on your clothes or costume.

- *Applying foundation:* As the first step in applying makeup, apply foundation with a sponge to the face, neck, and ears. On bald men, apply to the head also. After applying, gently sponge off excess foundation to avoid streaks and smears.

- *Applying powder:* Always set the work with a coating of colorless, loose powder. Use a cotton ball to press on the powder gently and lightly; do not rub.

- *Applying rouge:* To break up the flat look of the foundation and powder, use rouge to bring out the natural glow of the skin. Use a brush to apply rouge to the cheekbones, center of forehead, and nose.

- *Defining the eyes:* Use mascara and eyeliner to define the eyes clearly. Use just enough to reveal the eyes to the camera. Remember that in tight closeup, only a little is needed to get the job done. In general, especially for men, it is better to err on the side of too little rather than too much. In women, more eye makeup is acceptable because many women use such cosmetics in everyday life. If eye shadow is used, stick to the medium browns rather than blues and greens, which look less attractive on television.

- *Lipstick for women:* Have a selection of natural colors. Use lip liners to improve the definition of the lips. Top off with lip gloss to add a moister appearance, if desired.

- *Hair care:* After preparing a careful makeup job, it helps to keep the hair out of the way so that the camera and audience can see the face. To keep unwanted hair from flopping in front of the face, use hair spray or some strategically placed pins.

- *Makeup removal:* Remove makeup with cold cream and a soft cloth. Soap and water are less effective and can be harsh on the skin.

cheeks appear full when smiling. A touch can also be applied to the center of the forehead and nose. After applying the rouge, use a soft brush to fade the edges of the color into the surrounding area.

Eye Makeup *Eyeliner, shadow,* and *mascara* provide emphasis and definition to the eyes and should be used for both male and female talent. Normal street makeup may be used, especially brown tones, which tend to provide the most natural appearance on television.

Advanced Makeup Materials So far, our discussion of makeup for television has focused on enhancing the talent's natural appearance. However, more extensive makeup may also be used to alter appearance entirely, as is often done on such television programs as "The X Files" and "Star Trek: The Next Generation" (Figure 14.3).

Makeup is commonly used for aging, providing special effects such as wrinkles, cuts, stitches, scars, and black eyes, and rendering characterizations in period and fantasy productions. Among the most widely used materials for such makeup jobs are *latex* for age wrinkles, *collodion wax* for scars, *spirit gum* for securing hair pieces (wigs, mustaches), *mortician's wax* or *putty* with red liners for cuts and stitches, and dark liner colors for black-eye effects. For more on using makeup for special effects, see Swinfield (1994).

Makeup Equipment Most makeup equipment is available from cosmetics counters in department stores, but some of the more exotic equipment may be available only from theatrical suppliers. Here are the basics:

- *Tissue and cotton balls:* 100 percent cotton is recommended for applying powder and foundation.
- *Brushes:* A variety of sizes are useful for different jobs, from applying rouge to shading to highlighting.
- *Sponges:* Useful for powdering.
- *Pencils and grease liners:* Useful for eye makeup and for providing detail.
- *Cleansers:* To maintain good skin condition, do not use soap and water to remove makeup. They will not do a thorough job, and they can leave old material on the skin that can eventually clog skin pores. Instead, use a good cleansing cream or lotion, available at most cosmetics counters.
- *Additional musts:* The basic makeup kit should also include combs and brushes, scissors, hair spray, a razor, and false eyelashes.

KEY TERMS

direct orientation *(358)*   consistency *(361)*
indirect orientation *(358)*   foundation (base) makeup *(364)*
cheating *(359)*   shading *(368)*
range *(361)*   highlighting *(368)*
flexibility *(361)*

QUESTIONS FOR REVIEW

1. What technical characteristics of video affect talent's on-air appearance? Describe how the television system influences the selection of wardrobe in particular.

2. What technical characteristics need to be considered when working out blocking for on-air talent?

3. When is it advisable for talent to use direct and indirect orientations to the camera? Describe several situations where it is appropriate to have talent "cheat" to the camera.

4. How do framing and depth of field affect the locations of talent with respect to one another?

5. What special performing problems does acting in edited productions involve?

6. What are some good points to remember about using microphones?

7. What are the main purposes of makeup in video performing?

8. Describe the various types of makeup available and their functions.

Adams, Michael H. (1992). *Single camera video: The creative challenge.* Dubuque, IA: Wm. C. Brown.

Alten, Stanley A. (1994). *Audio in media.* 4th ed. Belmont, CA: Wadsworth.

Andrews, Bart (1985). *The I Love Lucy book.* New York: Doubleday/Dolphin.

Armer, Alan A. (1993). *Writing the screenplay.* Belmont, CA: Wadsworth.

Bazin, Andre (1967). *What is cinema?* Berkeley, CA: University of California Press.

Beck, C., and Andrews, H. (ca. 1903). *Photographic lenses.* London: Lund Humphries.

Bellman, W. F. (1974). *Lighting the stage.* San Francisco: Harper & Row.

Bentham, Frederick (1980). *The art of stage lighting.* 3rd ed. London: Pitman.

Bettinghaus, Erwin P. (1973). *Persuasive communication.* New York: Holt, Rinehart and Winston.

Blum, Richard A. (1995). *Television and screen writing: From concept to contract.* Boston: Focal Press.

Brady, Ben, and Lee, Lance (1988). *The understruture of writing for film and television.* Austin: University of Texas Press.

Brandt, H. (1968). *The photographic lens.* Garden City, NY: Focal Press.

Browne, Steven E. (1993). *Videotape editing: A postproduction primer.* 2nd ed. Boston: Focal Press.

Catsis, John R. (1996). *Sports broadcasting.* Chicago: Nelson-Hall.

Clark, Charles N. (1966, July). Characteristics of incandescent lamps for theatre stages, television, and film studios. *Illuminating Engineering,* 61, 7.

Cohler, David K. (1985). *Broadcast journalism: A guide for the presentation of radio and television news.* Englewood Cliffs, NJ: Prentice-Hall.

Cole, Hillis R., Jr., and Haag, Judith H. (1989). *The complete guide to standard script formats. Part I: The screenplay.* North Hollywood, CA.: CMC.

Compesi, Ronald J., and Sherriffs, Ronald E. (1994). *Video field production and editing.* 3rd ed. Boston: Allyn and Bacon.

Cooper, Lane (1960). *The rhetoric of Aristotle.* Englewood Cliffs, NJ: Prentice-Hall.

Cox, Arthur (1974). *Photographic optics: A system of optical design.* New York: Focal Press.

Cremer, Charles F.; Keirstead, Phillip O.; and Yoakam, Richard D. (1996). *ENG: Television news.* New York: McGraw-Hill.

Cronkhite, Gary (1969). *Persuasion: Speech and behavioral change.* New York: Bobbs-Merrill.

Ditchburn, R. W. (1953). *Light.* New York: Interscience Publishers.

Drude, P. (1922). *The theory of optics.* New York and London: Longmans Green.

Dmytryk, Edward (1984). *On film editing.* Stoneham, MA: Focal Press.

——————, and Dmytryk, Jean Porter (1984). *On screen acting.* Stoneham, MA: Focal Press.

Evans, Ralph M. (1948). *An introduction to color.* New York: Wiley.

Faigin, Gary (1990). *The artist's complete guide to facial expression.* New York: Watson-Guptill Publications.

Fink, Donald J. (1957). *Television engineering handbook.* New York: McGraw-Hill.

——————, and Lutyens, David (1960). *The physics of television.* New York: Anchor Books.

Fishbein, Martin, and Ajzen, Icek (1975). *Belief, attitude, intention and behavior: An introduction to theory and research.* Reading, MA: Addison-Wesley.

Franke, G. (1966). *Physical optics in photography.* New York and London: Focal Press.

Gaskill, Arthur L., and Englander, David A. (1967). *How to shoot a movie story.* Hastings-on-Hudson, NY: Morgan Press.

Gaunt, Leonard (1981). *Zoom and special lenses.* New York: Focal Press.

Giannetti, Louis (1993). *Understanding movies.* Englewood Cliffs, NJ: Prentice-Hall.

Greenleaf, A. R. (1950). *Photographic optics.* New York: Macmillan.

Haag, Judith H. (1988). *The complete guide to standard script formats. Part II: Taped formats for television.* North Hollywood, CA.: CMC.

Hawken, William R. (1981). *Zoom lens photography.* Somerville, MA: Curtin and London.

Head, Sydney W.; Sterling, Christopher H.; and Schofield, Lemuel B. (1994). *Broadcasting in America.* 7th ed. Boston: Houghton Mifflin.

Hedgecoe, John (1992). *The book of photography.* New York: Alfred A. Knopf.

Hewitt, John (1995). *Air words: Writing for broadcast news.* Mountain View, CA: Mayfield Publishing.

Hickman, Harold R. (1991). *Television directing.* Santa Rosa, CA: Cole.

BIBLIOGRAPHY

Hutchison, Earl R., Sr. (1996). *Writing for mass communication*. 2nd ed. New York: Longman.

Inoue, Shinya (1987). *Video microscopy*. New York: Plenum Press.

Karlins, Marvin, and Abelson, Herbert I. (1970). *Persuasion: How opinions and attitudes are changed*. New York: Springer.

Kehoe, Vincent J-R. (1985). *The technique of the professional make-up artist for film, television, and stage*. Boston: Focal Press.

Kiesler, Charles A.; Collins, Barry E.; and Miller, Norman (1969). *Attitude change: A critical analysis of theoretical approaches*. New York: Wiley.

Kingslake, Rudolf (1978). *Lens design fundamentals*. San Diego: Academic Press.

_____________ (1989). *A history of the photographic lens*. San Diego: Academic Press.

Lee, H. W. (1927). The development of the telephotographic objective. *Proceedings of the Optical Convention, 869*.

Levin, Robert E. (1984). *Sylvania GTE lighting handbook for television, theatre, and professional photography*. 7th ed. Danvers, MA: GTE Products Corporation, Sylvania Lighting Center.

Longhurst, R. S. (1957). *Geometrical and physical optics*. London: Longmans.

Lucey, Paul (1996). *Story sense: Writing story and script for feature films and television*. New York: McGraw-Hill.

Luckiesh, Matthew (1948). *Light, vision and seeing*. New York: D. Van Nostrand.

Luckiesh, Matthew, and Moss, Frank K. (1943). *The science of seeing*. New York: D. Van Nostrand.

MacDonald, R. H. (1987). *A broadcast news manual of style*. New York: Longman.

McAdams, Katherine C., and Elliott, Jan Johnson (1996). *Reaching audiences: A guide to media writing*. Boston: Allyn and Bacon.

McCain, Thomas A., and Shyles, Leonard, eds. (1994). *The thousand hour war: Communication in the gulf*. Westport, CT: Greenwood Press.

McCandless, Stanley R. (1958). *A method of lighting the stage*. 4th ed. New York: Theater Arts.

Mencher, Melvin (1993). *Basic media writing*. 4th ed. Madison, WI: Brown and Benchmark.

_____________ (1994). *News reporting and writing*. 6th ed. Madison, WI: Brown and Benchmark.

Merritt, Douglas (1993). *Graphic design in television*. Oxford, UK: Focal Press.

Minnick, Wayne C. (1957). *The art of persuasion*. Boston: Houghton Mifflin.

Mott, Robert L. (1993). *Radio sound effects*. Jefferson, NC: McFarland & Co.

Myleaf, Harry, ed. (1966). *Electricity one-seven*. Rochelle Park, NJ: Hayden.

_____________ (1976). *Electronics one-seven*. Rochelle Park, NJ: Hayden.

Neblette, C. B. (1965). *Photographic lenses*. Hastings-on-Hudson, NY: Morgan and Morgan.

_____________, and Murray, A. E. (1973). *Photographic lenses*. 2nd ed. Dobbs Ferry, NY: Morgan and Morgan.

Newsom, Doug, and Wollert, James A. (1988). *Media writing: Preparing information for the mass media*. Belmont, CA: Wadsworth.

Nisbett, Alec (1979). *The technique of the sound studio*. London: Focal Press.

Noll, A. Michael (1988). *Television technology: Fundamentals and future prospects*. Norwood, MA: Artech House.

Oenslager, Donald (1978). *The theatre of Donald Oenslager*. Middletown, CT: Wesleyan University Press.

Orlik, Peter B. (1994). *Broadcast copywriting*. 3rd ed. Boston: Allyn and Bacon.

Osborne, B. W. (1968). *Colour television reception and decoding techniques*. New York: Hart.

Oringel, Robert S. (1972). *Audio control handbook for radio and television broadcasting*. New York: Hastings House.

Parker, Sybil P. (1988). *Optics source book*. New York: McGraw-Hill.

Parker, W. Oren, and Harvey K. Smith (1979). *Scene design and stage lighting*. 4th ed. New York: Holt, Rinehart and Winston.

Penzel, Frederick (1978). *Theatre lighting before electricity*. Middletown, CT: Wesleyan University Press.

Pudovkin, Vsevolod (1970). *Film technique and film acting*. New York: Grove Press.

Ray, S. (1979). *The photographic lens*. New York and London: Focal Press.

Rees, Terence (1978). *Theatre lighting in the age of gas*. London: Society for Theatre Research.

Reisz, Karel (1955). *The technique of film editing (basic principles for T.V.)*. New York: Visual Arts Books.

Richards, Stanley (1971). *Best mystery and suspense plays of modern theatre*. New York: Dodd, Mead.

Rosen, Marvin J., and DeVries, David L. (1993). *Introduction to photography.* 4th ed. Belmont, CA: Wadsworth.

Ross, Raymond S., and Ross, Mark G. (1981). *Understanding persuasion.* Englewood Cliffs, NJ: Prentice-Hall.

Rubin, Joel E., and Leland H. Watson (1954). *Theatrical lighting practice.* New York: Theatre Arts.

Russell, Bertrand (1945). *A history of western philosophy.* New York: Simon and Schuster.

Ryan, Dr. Rod, ed. (1993). *American cinematographer manual.* 7th ed. Hollywood, CA: ASC Press.

Sellman, Hunton D., and Merrill Lessley (1982). *Essentials of stage lighting.* 2nd ed. Englewood Cliffs, NJ: Prentice-Hall.

Settel, Irving (1983). *A pictorial history of television.* 2nd ed. New York: Frederick Ungar.

Severin, Weerner J., and Tankard, James W., Jr. (1988). *Communication theories: Origins, methods and uses.* New York: Longman.

Shetter, Michael D. (1982). *Videotape editing: Communicating with pictures and sound.* Elk Grove Village, IL: Swiderski Electronics.

Shiers, George (1977). *Technical development of television.* New York: Arno Press.

Shipley, Joseph T., ed. (1956). *The crown guide to the world's great plays from ancient Greece to modern times.* New York: Crown.

Shook, Frederick (1996). *Television field production and reporting.* White Plains, NY: Longman.

Spottiswoode, Raymond (1965). *Film and its techniques.* Berkeley and Los Angeles: University of California Press.

——————— (1969). *A grammar of the film: An analysis of film technique.* Berkeley and Los Angeles: University of California Press.

Staley, Karl A. (1960). *Fundamentals of light and lighting. G. E. Bulletin LD2.* General Electric Co.

Stephens, Mitchell, and Lanson, Gerald (1986). *Writing and reporting.* New York: Holt, Rinehart and Winston.

Swinfield, Rosemarie (1994). *Stage makeup step by step.* Cincinnati: Betterway Books.

Taylor, H. D. (1906). *A system of applied optics.* New York: Macmillan.

Taylor, J. Traill (1898). *The optics of photography and photographic lenses.* 2nd ed. New York: Macmillan.

Thompson, Roy (1993). *Grammar of the edit.* Jordan Hill, Oxford: Focal Press.

United press international stylebook: The authoritative handbook for writers, editors, and news directors (1992). Lincolnwood, IL: National Textbook Co.

Verna, Tony (1993). *Global television: How to create effective television for the future.* Boston: Focal Press.

Walters, Roger L. (1994). *Broadcast writing.* 2nd ed. New York: McGraw-Hill.

Welford, W. T. (1962). *Geometrical optics.* Amsterdam: North-Holland.

Wentworth, John W. (1955). *Color television engineering.* New York: McGraw-Hill.

Whittaker, Ron (1992). *Television production.* Mountain View, CA: Mayfield.

Wilkie, Bernard (1979). *The technique of special effects in television.* London: Focal Press.

Willis, Edgar E., and D'Arienzo, Camille. (1993). *Writing scripts for television, radio, and film.* New York: Harcourt Brace Jovanovich.

Wood, R. W. (1967). *Physical optics.* New York: Dover.

Wolff, Jurgen, and Cox, Kerry (1988). *Successful scriptwriting.* Cincinnati: Writer's Digest Books.

Wurtzel, Alan, and Rosenbaum, John (1995). *Television production.* 4th ed. New York: McGraw-Hill.

Zettl, Herbert (1992). *Television production handbook.* 5th ed. Belmont, CA: Wadsworth.

Zworykin, V. K., and Morton, G. A. (1954). *Television.* 2nd ed. New York. Wiley.

———————, and Ramberg, E. G. (1949). *Photoelectricity.* New York: Wiley.

A-B roll editing   Editing two tape sources through an edit controller.

A-B rolling   Running two synchronized tape sources through a video switcher and treating them as if they were two live sources, editing them on the fly.

ac (alternating current)   Electric current that does not always flow in the same direction but alternates, flowing in one direction and then reversing and flowing in the other.

AD (assistant director)   Crew member who assists the director in all phases of a production.

AFM (American Federation of Musicians)   Union formed in 1894 that today represents more than 9,000 musicians performing on network radio, television, music videos, commercials, cable, and pay TV outlets.

AFTRA (American Federation of Television and Radio Artists)   Broadcasting union founded in 1937 that currently represents the interests of more than 77,000 radio and television workers, including news anchors, reporters, correspondents, and announcers.

AM (amplitude modulation)   Using sound to vary the amplitude of a radio carrier wave.

amplitude   Maximum value reached by a current or voltage in an electric circuit.

antialiasing   In character generators and other digital text displays, a function that reduces or eliminates the stairstep or jaggy appearance of the curved parts of letter forms.

aperture   Opening through which light passes in a camera.

arc   (1) To move the camera along a curved dolly or truck. The move can be tight or wide, toward or away, or to the right or left of the subject. (2) A type of lighting instrument filament. (3) A story line in a script.

aspect ratio   Width-to-height proportion of the television screen, currently set at four units to three, soon to be changed to sixteen units by nine for high-definition television.

assemble editing   Transferring all tracks of audio and video from a source tape to an edit tape, including the control track, in order. Assemble editing is most often done when raw footage must be put into final form quickly, as is often the case in ENG production.

assistant director   See *AD*.

asynchronous sound   Sound featured in the audio track that does not coincide with the image presented on screen.

attack phase   The initial part of the sound envelope, when a sound first begins. In stringed instruments, the attack phase of a note usually occurs by plucking or bowing.

attenuation   The weakening and eventual dying out of a sound or radio signal as distance from the source increases.

axis-of-action rule   Imaginary line along which the main action of a scene takes place. As a rule, to avoid false reversals of action, keep all shots on the same side of the line throughout your shoot.

back light Light that illuminates the back of a subject.

backtiming Determining how much time is left in a show by subtracting its running time from its total length.

bandwidth The frequency range of a radio signal between its lowest and highest frequencies.

barn doors External adjustable black metal flaps on the front of a lighting instrument used to control the spread of the light beam.

base light Diffuse, even light level required for optimal operation of a television camera.

bass roll-off Control on a microphone that attenuates bass frequencies.

bidirectional microphone Microphone that picks up sound in front and back but not on its sides.

blanking signal Part of the complex video waveform that turns off the electron gun during retrace.

blocking Rehearsing and marking the movements of talent and cameras through space.

bounce light (1) Light reflected off a wall, card, or other flat surface behind talent to create backlight. (2) Light reflected as above to soften lighting on talent.

bust shot Medium shot of a talent from head to upper chest.

C-clamp A piece of hardware, shaped like the letter C, used to hold lighting instruments to a lighting grid.

C/K (chroma-key insert) Script notation indicating the location of a chroma-key effect.

CAD (computer-aided design) Computer software programs that simplify and speed up the process of creating floor plans, light plots, and other drafting materials.

cameo lighting Lighting that illuminates the subject but leaves the background dark.

camera control unit (CCU) Equipment that allows a video technician to adjust the video signal during a program or taping, including contrast and brightness, color rendition, and picture registration.

capacitor (condenser) microphone Microphone featuring a plate that vibrates along with sound impinging on it, generating a varying electrostatic field with another plate in back of it.

cardioid pickup pattern Heart-shaped pickup pattern of a unidirectional microphone.

CCD (charge-coupled device) Solid-state electronic circuit containing thousands of discrete picture elements (pixels) used to transform light into electrical energy. CCD cameras are also called *chip* cameras.

CCU See *camera control unit.*

CG See *character generator.*

character generator (CG) Computer with keyboard used to type electronic letters and numbers for text graphics.

charge-coupled device   See *CCD*.

chip chart   Card display of a television gray scale used by technicians to set contrast and brightness levels on a television camera.

chroma key   Special effect that replaces a video signal of a preselected hue (usually blue) with a signal from another source.

chrominance   Color or hue information of a color television signal.

classical editing   Editing to emphasize and vary the intensity of a narrative. Dramatic, physical, and psychological story aspects may be manipulated with classical editing techniques.

clip   (1) To close someone's microphone. (2) To limit the black and/or white levels of a video signal.

clock timing   Using a separate clock in the control room (e.g., digital or stopwatch) to time a program segment during rehearsal or taping.

close-up (CU)   Camera shot picturing the subject at close range.

cold start   Opening a program with the first scene, even before credits and theme song. Cold starts have become more common in recent years in the television industry as a means of adapting to fiercer competition from cable channels and independent program suppliers.

color bars   Color reference standard used by the television industry to align cameras to achieve proper color rendition. Color bars may be internally generated by a video switcher or by a field camera.

color temperature   The relative amount of a given color (red, blue, etc.) exhibited by a light source, measured in degrees Kelvin. Television studio lighting is in the 3,200-degree range, and normal daylight is in the 5,600-degree range.

component recording   In a Y/C format, a recording system in which the luminance and chrominance parts of the signal are recorded separately using two separate wires; in an RGB format, the red, green, and blue portions of the signal are recorded separately using three separate wires. The advantages of separating out signal components is that many more dubs are possible without appreciable signal loss. However, the disadvantage is that more expensive circuitry is required to process the separate signals.

composite recording   Recording system in which both color and brightness information are recorded together. Composite recording is the original recording method standardized by the NTSC.

compositing   Electronically overlaying parts of video images from different sources.

constructive interference   Occurs when sound waves received in phase with one another reinforce one another, resulting in a louder sound.

continuity   Preserving a smooth flow of uninterrupted action from shot to shot in an edited scene or program.

continuity editing   Editing to maintain the smooth, uninterrupted flow of action.

contrast ratio   Difference in brightness between the darkest and lightest parts of a television picture, measured in foot-candles of reflected light. For most cameras, the contrast range of light to dark is roughly 30 or 40 to 1.

control track   Part of the videotape used to record sync pulse information and keep a count of each frame of video; also called *pulse track*.

convergence   Term describing the growing intercompatibility of different mass media technologies. For example, the ability to use a cell phone to transmit computer data over a modem combines the once autonomous technology of the telephone with the computer.

cover shot   Camera shot that includes the widest area of the set for a scene or program.

crane   Camera mount featuring a counterweighted arm on a heavy, wheeled dolly; some cranes are capable of automated movements.

cue cards   Poster cards with lines of copy printed on them used to deliver scripted copy to on-air talent.

cut   An instant change from one video image to another; also called *take*.

cutaway   Shot that leads the viewer's attention away from one scene and links it to another. Cutaways provide bridges or transitions between unrelated scenes. A common cutaway uses a wide exterior shot of a building in which the next scene is to take place.

cut-in (insert)   Close-up that captures a key moment of visual business to emphasize a story's main point.

cutting on action   Going from one shot of an action to another before the action is completed to provide a smooth, uninterrupted flow. *Cutting on action* generally holds audience interest more than cutting to the next bit after the first action is finished.

cyclorama   Continuous stretched curtain running from the floor to the top of the set area in a studio; used to provide a backdrop for sets and talent. Also called a *cyc* (rhymes with *bike*) for short.

dead space   Parts of the screen area that include unnecessary content or are left unused.

decay phase   Part of a sound after it begins to diminish in intensity until it attenuates completely.

decibel (dB)   A value that expresses the ratio of two sound or voltage intensities.

degrees Kelvin   Color temperature scale used to express the amount of color (red, blue, etc.) present in a light source.

demodulation   Process of recovering signal intelligence from a modulated carrier wave.

depth of field   In a camera, the range of distance within which objects appear to be in focus.

destructive interference   Occurs when sound waves received out of phase with one another diminish one another, resulting in a softer sound.

DGA (Directors Guild of America)   Union formed in 1960 to represent the interests of directors, associate directors, stage managers, and production assistants.

diffuser   A spun glass filter put in front of a lighting instrument to soften light intensity.

digital video disk   See *DVD*.

dimmer board   Lighting console that enables an operator to control intensities of individual lighting instruments with a series of faders. Many dimmer boards are now automated.

dissolve   Gradual transition from one nonblack video source to another nonblack video source.

dolly   To move the entire camera and camera mount either toward (dolly in) or away (dolly out) from the subject.

downstream key   Component of the special-effects generator of a video switcher that permits the TD to key in a video source (usually the CG) without using a key bus.

drop-frame time code   Time code method used to adjust for discrepancies caused by the need to change from the original standard frame rate of 30 to 29.97 frames per second to accommodate the addition of a chrominance signal for color television. To keep time code in step with real time, drop-frame time code ignores frames 00 and 01 each minute except at every tenth minute, amounting to a net correction in time code display of 3.6 seconds per hour.

DVD (digital video disk)   Video format that uses a device resembling a CD to store and play back video programs.

dynamic editing   (thematic montage)   Using editing to stress the association of ideas, but without regard to preserving unity in space and time.

dynamic (moving-coil) microphone   Microphone with a pickup element consisting of a diaphragm and a coil moving through a magnetic field.

edit decision list   See *EDL*.

EDL (edit decision list)   Written log of proposed edit points for constructing a program by time code numbers in sequence. The EDL can also specify shot content and desired transitional devices to be used (e.g., cut, wipe, dissolve).

EFP (electronic field production)   A mode of television production using a single field camera with multiple setups, with care given to such production values as talent rehearsals, camera and subject placement, blocking, lighting, sound, continuity, scriptwriting, and editing.

electromagnetic waves   Radio energy transmitted through space at the speed of light, making radio and television broadcasting possible.

electromotive force   Electrical force that results when two charges have a difference of potential.

electronic field production   See *EFP*.

electronic news gathering (ENG)   A mode of television field production characterized by rapid deployment of equipment to the site of a fast-breaking news story. ENG coverage is usually done with a minimum of preproduction planning.

electronic still store (ESS)   Digital device for storing individual video images.

ellipsoidal spotlight   Spotlight that produces a well-defined beam that can be shaped with metal pattern inserts.

ENG   See *electronic news gathering*.

ESS   See *electronic still store*.

essential area   Part of the television screen in the center of the scanning area that is most likely to be seen by the home viewer; also called the *safe title area*.

establishing shot   Shot used to establish setting, talent relationships, and/or situation in a scene or program.

ethos   Source credibility; the degree of believability attached to a message source (e.g., a speaker) by the audience.

fade   Gradual transition from black to a nonblack video source or from a nonblack video source to black.

fiber optics   Technology that uses uniform strands of thin glass to transmit audio and video information with light. Some advantages of fiber optics technology over conventional coaxial copper cable is that there is less signal attenuation and greater capacity or bandwidth.

fill light   Light directed at the subject from the opposite side of the key light to add illumination to shaded areas, reducing shadows. Soft lights are most often used as fills.

filter   Pieces of clear or colored transparent material (either glass or plastic) for eliminating unwanted wavelengths of light while permitting others to pass.

filter factor   Measure of the quantity of light lost when a filter is added to a lens or lens system. Each factor of two reduces the amount of light entering the lens by half.

fishpole boom   Microphone fixed to the end of a stick that is extended into a scene and held by a crew member, usually out of view of the cameras.

flags   Opaque cards used to shield the subject from direct sunlight; often used in field production.

flat lighting   Lighting that provides uniform illumination for talent throughout the acting or performing area. Flat lighting minimizes jarring discrepancies in light levels when cutting between cameras shooting the same subject from different angles. Flat lighting is often used in sitcoms.

flat   Standard set piece used to form backgrounds for interior sets, made of either canvas on wood frames (*softwall* flat) or particle board (*hardwall* flat).

flatbed scanner   Machine used to digitize two-dimensional graphics.

flicker   Perception of static moments in the presentation of motion visuals. Flicker can be a source of severe eye fatigue. In the human perceptual system, flicker can be defeated by making the rate of presentation exceed forty-five presentations per second.

FM (frequency modulation)   Using sound to vary the frequency of a radio carrier wave.

f/number   (1) Standard numerical value expressing the amount of light admitted by a lens at a given aperture setting. (2) The ratio of aperture diameter to the focal length of a lens, in millimeters. Also called *f-stop number*.

focal length   Distance in millimeters between the optical center of a lens or lens system and the plane on which the image of an object at infinity (or greater than fifty feet, practically speaking) is in sharpest focus.

focal plane   Plane on which a lens forms the sharpest image of objects at infinity (or greater than fifty feet).

focus   Adjusting a camera's lens system to achieve a sharp, clear image of the subject.

foot-candle (fc)   Unit of measurement of the amount of light falling on a subject. One fc is the amount of light from a candle falling on a 1-foot square at a distance of 1 foot. A *lux* is a European standard for measuring light intensity. One fc equals 10.75 lux.

footprint   Coverage pattern of a satellite signal on the earth.

formalism   Production style that emphasizes selected elements of scenes to heighten audience attention to particular parts of the story.

frame   (1) A single picture in a motion picture. (2) One complete scanning cycle in a video presentation, occurring every thirtieth of a second. Two fields equal one frame of video.

frame grabber   Device used to capture and store single frames of video.

frame synchronizer   Device used to synchronize video signals from various video sources.

framing   Selecting the perimeter around the picture to limit the field of view. Extreme close-ups of a subject are said to be tightly framed.

frequency   The number of cycles per second exhibited by a radio wave. Frequency is inversely related to wavelength; that is, the higher the frequency of a given wave, the shorter its wavelength.

Fresnel spotlight   Pronounced "fra-nell," the most commonly used light in television production; features a lens with stepped concentric rings enabling it to throw a directional light onto the subject. The spread of the beam may be adjusted with a focusing knob.

front timing   Adding the running times for specific program segments to the starting time of the program to compute when to begin certain segments.

gain   (1) Volume level of an audio source. (2) Level of amplification of a video signal.

gel   Colored filter that may be fixed to the front of a light to give it a desired hue.

genlock   Device that coordinates signals from the synchronization generators of various video sources to provide smooth transitions among them. Genlocking permits smooth transitions between studio and nonstudio sources and between tape and nontape sources.

genre   A program type (e.g., sitcom, drama, western, cop show).

giraffe boom   Microphone boom used to place a static microphone into a performance area; also called a *tripod* boom.

gobo   Short for *go between*, a foreground set piece through which a camera can shoot a scene. A gobo might consist of a window pane to give the impression that the camera is peering into a dining room from outdoors.

gray scale   Chart, usually of seven or nine steps, displaying a transition in light reflectance progressing from TV black to TV white.

ground row   Set piece placed on the studio floor to provide a seamless appearance between the floor and the cyclorama or backdrop.

guard band Part of the bandwidth of a radio or television channel used as a buffer to protect against adjacent channel interference.

HDTV See *resolution*.

head-on shot Shot featuring the main action of the scene coming directly toward the camera.

headroom Screen space above the talent's head.

head shot Shot of subject from the neck up.

helical scanning Diagonal arrangement of video frames on a videotape to increase the surface area of the tape as it appears to record or playback heads of a videotape recorder. Helical scanning makes it possible to use narrower tape formats to store video information with less signal loss or deterioration. Also called *slant-track* video.

HMI lamp Highly efficient, high-intensity light that produces hard shadows; used to simulate the effect of sunlight in field productions. HMI lights operate in the 5,600-degree Kelvin range, using an additional piece of equipment called a *ballast*. The letters HMI stand for hydrargyrum medium-arc iodide.

hue Another word for *chrominance*; refers to the color of a particular light.

Hz Abbreviation for *cycles per second (cps)* named after Heinrich Hertz, who discovered some of the unique properties of radio waves propagated at stable frequencies. For example, 1,000 cps may be rewritten 1 kHz.

IATSE (International Alliance of Theatrical Stage Employees) Union formed in 1893 that today represents the interests of more than 85,000 behind-the-scenes television workers, including makeup artists, camera operators, and prop, sound, and grip personnel.

IBEW (International Brotherhood of Electrical Workers) Union formed in 1891 that today represents technical staff involved in broadcasting, including camera operators, videotape and audio personnel, editors, and maintenance workers.

IFB (interruptible foldback) Audio system that allows voice communication (usually instructions from the director or other production personnel) with an on-air talent through an earpiece. Also called *interruptible feedback*.

impedance Total opposition to current flow presented to a radio circuit by electronic components.

in-cue Notation on a script indicating the phrase or moment where a new segment is to begin. In-cues help maintain continuity and reduce dead air and choppiness.

incident light Light that reaches an object directly from a light source. The amount of incident light from a light source may be read using an incident light meter, aiming the meter directly at the light source from a selected position (e.g., a subject or camera position).

insert See *cut-in*.

insert editing Transferring selected tracks of audio and/or video from a source tape to a preblacked edit tape. *Preblacked* means the edit tape has a continuous control track laid down in advance. Insert editing is done when extensive postproduction work is needed (e.g., when adding additional audio or B-roll "beauty" shots to raw footage).

intelligence   Audio or video information contained in a radio wave. Signal intelligence is provided by modulating a carrier, transmitting it, and demodulating it at the receiver.

interactivity   The ability of a communication system to permit two-way communication among relevant parties in real time.

intercoms   Short for *intercommunication systems*, intercoms provide audio communication among production personnel during a production. Telephone headsets are commonly used as intercoms.

interlaced scanning   method of tracing all odd-numbered lines, then all even-numbered lines in succession by the electron gun in producing television pictures. The original purpose of using interlaced scanning was to eliminate flicker problems. However, interlaced scanning will soon be superseded by a progressive scanning method adopted by the broadcasting industry for the coming all-digital age. In progressive scanning, each line is scanned in succession without skipping any lines, thus making broadcasting more compatible with the computer industry.

interruptible foldback   See *IFB*.

inverse square law   The principle that light intensity decreases as distance from the source increases. The rate of fall-off is inversely proportional to the square of the distance.

jib arm   Camera support that permits a single operator on the ground to raise and operate the camera at heights of about 25 feet.

jump cut   Abrupt transition between shots that violates smooth flow of uninterrupted action, often involving the same subject presented in slightly different screen locations.

key   (1) Key light; the main source of illumination. (2) Electronic effect performed by the technical director on the video switcher, involving the placement of video material from one source (e.g., lettering) over another at full picture strength for each.

key light   The main source of light used to illuminate a subject. Usually the key light is a "hard" light with a directional throw.

keystone   Appearance that a graphic, usually lettering, is deviating from the horizontal. To correct unwanted uphill or downhill appearance, either the right or left edge of the card may be pitched either closer or farther from the camera.

lavaliere mic   Small microphone worn on the lapel or other clothing on the front of talent below the chin; also called *lapel mic*.

leadroom   Screen space located in front of talent moving across the screen from right to left or vice versa. Usually leadroom should be greater than the space behind the subject, unless it is desirable to create a sense of entrapment for the subject. Such framing decisions should be determined from the subject matter of the program and from consideration of the effect the director wishes to create in the audience.

liftout   Portion of video copy that may be used as a stand-alone, shorter version of a message. For example, if properly written, a ten-second spot may be *lifted out* of a thirty-second spot and used by itself without any alteration.

light meter   Device used to measure the light intensity falling on a scene; usually measured in *lux*.

light plot   Two-dimensional plan, similar to a blueprint or floor plan, depicting the location, wattage, and aim or direction of all of the lighting instruments used in a television production.

lighting grid   Support pipes used to secure lighting instruments from the ceiling.

limbo lighting   Illumination of a subject with plain light background to create the feeling of endless space behind the talent.

logos   Pronounced with an *s* rather than a *z* sound, logos refers to rational appeals a message source offers to an audience.

long shot (LS)   Camera shot featuring the subject occupying a small portion of screen space.

longitudinal time code   See *LTC*.

lower-third   Presentation of text material in the bottom one-third of the screen. Lower-thirds frequently present the names and affiliations of interviewees.

LS   See *long shot*.

LTC (longitudinal time code)   Type of SMPTE time code that records frame numbers lengthwise on either the cue track or an audio track of the videotape.

luminance   The amount of brightness information presented in a video signal; usually measured by its location on a gray scale.

lux   European unit of light intensity. One foot-candle (fc) equals 10.75 lux.

M-S microphone   Short for *middle-side microphone*, a stereophonic mic with two mic capsules housed in a single case. One capsule (usually cardioid) covers the middle of a performance area; the other (usually bidirectional) covers the sides.

master shot   Widest shot used in a scene or program. It is useful to see the master shot on the line monitor during setup and rehearsal to plan the locations of lights and booms. Also called *cover shot*.

MCR (multicamera remote) production   Type of field production done in the style of a studio production from a complete control room facility and a number of live camera feeds, usually from the site of a major sporting or public event scheduled in advance (e.g., a political convention). Often involves a truck with either satellite or microwave transmitters.

medium shot (MS)   Camera shot devoting roughly one-half of the screen space to the subject.

MIDI (Musical Instrument Digital Interface)   Digital device that integrates audio signals from a variety of sources.

mise en scène   Pronounced "meez-on sen," the arrangement of objects in the frame.

modulation   Variation of a carrier wave with signal intelligence.

morphing   Using digital graphics to gradually change one image into another.

motivating light   Illumination conveying the presence of a light source in a scene or program (the sun, a flashlight, a lamp, etc.).

mounting head   Device used to support a camera on a tripod or dolly.

MS   See *medium shot*.

multicamera remote   See *MCR*.

multisource editing   Editing using two or more sources.

Musical Instrument Digital Interface   See *MIDI*.

NABET–CWA (National Association of Broadcast Employees and Technicians–Communications Workers of America)   Union formed in 1994 to represent the interests of more than 10,000 broadcast engineers, technicians, newswriters, announcers, photographers, and stage workers.

National Television System Committee   See *NTSC*.

naturalism   Lighting approach advocating that in the absence of compelling reasons to do otherwise, one should imitate the effect of sunlight by providing lighting that simulates a single main source of light from above.

non-drop-frame time code   Time code method that displays one hour of time for each 1 hour and 3.6 seconds of real time. See *drop-frame time code*.

nonsynchronous sources   Video sources that are not in phase with one another. When signals are out of phase, transitions among them may exhibit glitches. To eliminate glitches, a time base corrector or frame synchronizer is used to stabilize transitions.

noseroom   Amount of screen space in front of the subject when turned to the side. Usually noseroom is greater than the space behind the head. This norm is often violated to convey a sense of entrapment of the subject.

NTSC   (1) Acronym for National Television System Committee, the group that developed the American standard for television broadcasting in 1941. The technical standards adopted by the NTSC include bandwidth allocation (6 MHz), number of scanning lines (525), field rate (60 per second), frame rate (30 per second), and a method for transmitting color (chrominance) and brightness (luminance) information. (2) The composite television signal used in American broadcasting. The move to an all-digital television system will soon change these standards.

off-line editing   Editing preparation phase after raw footage has been shot, including viewing master tapes to prepare logs and/or EDLs and searching for and selecting edit points before entering the editing suite to cut the final copy of a scene or program.

omnidirectional (nondirectional) microphone   Microphone with audio sensitivity that is equal in all directions.

on-line editing   (1) Editing a live broadcast or a live-to-tape program by calling shots among multiple cameras in a studio or remote truck. (2) Editing a program into final form in postproduction using high-end editing equipment.

one-shot (1-S)   Camera shot featuring a single subject. Similarly, a 2-S features two subjects, a 3-S three subjects, and so on.

operating light level   Minimum amount of light required for the camera to function.

out-cue   Notation on a script indicating a phrase or moment where a segment is about to end. Out-cues help maintain continuity and reduce dead air and choppiness.

over-the-shoulder (OS) shot   Camera shot from a position behind and above a subject that includes one of the subject's shoulders; often used in interview programs to establish the relationship of the host (interviewer) to the guest (interviewee).

overlapping action   Action in one shot repeated in another shot to provide a logical and continuous edit point.

overtones   Harmonics of musical tones that are not multiples of a fundamental frequency.

PAL (Phase Alternation by Line)   Alternative technical standard to the NTSC television system used in North America, the PAL standard uses 625 lines, 50 fields, and 25 frames per second for signal transmission. The PAL system is most prevalent in Great Britain, Europe, Africa, the Middle East, and the Far East.

pan   To turn the camera from side to side on the mounting head while keeping the tripod or pedestal stationary. You can pan right (or left) by moving the lens to the right (or left) as you look at the camera from the operator's point of view.

pantograph   Device for adjusting the height of hanging lighting instruments.

parabolic microphone   Microphone that uses a dish or bowl-shaped reflector to direct sound from distant sources to a pickup element located at its focus.

pathos   Emotional appeals a message source offers to an audience.

ped   To raise or lower the camera on its pedestal.

pedestal   (1) Vertical column on a camera tripod or dolly that supports the camera. (2) Director's command to either raise or lower the camera during a rehearsal or production. (3) Black level on a camera control unit that permits the camera technician to set levels by comparing it to a standard on a waveform monitor.

periaktos   Column with a triangular shape and swivel base, often colored differently on each side, making it possible to present different faces to the camera quickly.

persistence of vision   Tendency for the human perceptual system to retain an image for a short period of time after it is no longer present to the eye.

Phase Alternation by Line   See *PAL*.

phase   Indicates the time relationship between two waves with alternating voltages and currents. Two waves are said to be *in phase* with each other when they reach maximum and minimum values at the same time.

photoconductivity   See *photoelectric effect*.

photoelectric effect   Effect exhibited by materials that emit electrons when light strikes them; also called *photoemissive effect*.

photoemissive effect   See *photoelectric effect.*

pilot   Prototype of a series episode. A *pilot story* is an outline of all the visual sequences of the initial episode.

pixel   (1) Short for *picture element,* the light-emitting points arranged on a grid of a CCD used to compose video images. (2) Discrete points arranged on the face of a television receiver or computer monitor to register video images when scanned with an electron gun.

point-of-view (POV) shot   Camera shot from the angle or perspective of the subject, showing the audience a scene as it would be seen from the subject's perspective.

polecat   Spring-loaded, retractable stick, affixed between floor and ceiling, used to secure a set piece.

practical light   (1) A light that can actually be operated by the talent (e.g., a lamp on a desk). (2) A lighting instrument used to provide talent with illumination necessary for a bit of action (e.g., a light clipped to a music stand for following music during a musical performance).

preblacked tape   Tape that already contains a control track; essential for performing insert editing.

preroll   Starting a videotape machine for a few seconds until it comes up to speed. Prerolling allows the video signal to stabilize, and is needed in editing as well as viewing tapes.

pressure zone microphone (PZM)   Microphone equipped with a sound-reflecting surface designed to receive sound from varying distances at the same time. Used in round-table discussions and to cover audience sound. Also called a *boundary microphone.*

primary movement   Movement that takes place in front of the camera (e.g., talent walking through a set area).

program bus   Row of buttons on the video switcher used to put pictures from various video sources on the program or line monitor.

propagation   Generation of a radio wave from a transmitter.

proximity effect   Increased level of bass frequencies relative to treble frequencies as mic-to-source distance decreases. Lavaliere microphones are often equipped with a bass roll-off switch to help cancel such effects.

pulse track   Control track of the video tape; the area of the videotape where synchronization information is located. The control or pulse track enables editing machines to count each frame of video as it passes the video heads.

racking focus   Manipulating a lens or lens system to shift focus from foreground to background, or vice versa, to direct the audience's attention to important parts of the picture.

raincoat   Cover for a field camera or other production equipment used to protect it from moisture.

range   Refers to the different types of information and media that can be shared in a communication system (e.g., voice, image, data, motion visuals)

rating   Proportion of all television households in a given market tuned to a particular program.

reach The universe of relevant parties that can be connected to one another by a communication system.

realism Production style that advocates the use of deep focus, long takes, wide shots, and camera movement rather than editing to offer the audience more choice in what to look at in the video image.

rectilinear propagation Property of light rays that permits them to travel in straight lines.

reflected light Light bounced from an illuminated object.

repeater station Ground station that receives, amplifies, and retransmits radio signals.

resolution Level of picture detail possible with a television system. Currently, picture resolution in the American television system is limited by the 525-line standard set by the NTSC. In the near future, that standard will be replaced by a 1,080-line standard (now called HDTV or high-definition television). Picture resolution is also affected by the lens system used and the presence or absence of filters.

reveal Camera shot that gets progressively wider, showing more and more of a scene. Reveals should increase audience interest by giving them more relevant information.

reverse-angle shot Footage of a scene from an opposing perspective, useful for editing purposes. Reverse-angle shots are commonly used in interviews, where footage of the interviewer from over the interviewee's shoulder provides transitions between different portions of the interview.

rf carrier Short for *radio frequency carrier,* the radio wave that is modulated by an audio and/or a video signal prior to being transmitted.

RGB recording See *component recording system.*

ribbon microphone Microphone with a pickup device made of a ribbon-shaped diaphragm suspended in a magnetic field.

rippling In nonlinear editing, the term describing the ability to shift material around to make room for late additions or deletions without having to re-edit material already laid down.

riser Platform used to bring seated talent to eye level with the camera.

rundown sheet List of each program segment in order with approximate running times. Rundown sheets can take the place of full or semiscripts in productions where the program format is well known to a seasoned crew, such as in late-night talk shows, where the show is roughly the same night after night.

safety chain Cable or chain used to secure lighting instruments to the lighting grid if the C-clamp fails.

SAG (Screen Actors Guild) Major union for television actors, singers, dancers, models, and extras, currently with more than 84,000 members.

satellite news gathering See *SNG.*

saturation Degree of purity of a color, free from white light.

scanning area Screen area scanned by the camera's pickup tube or CCD seen in the viewfinder.

scrim   (1) Spun-glass material placed in front of a lighting instrument to diffuse and soften the light falling on a scene. (2) Gauzy material hung in front of a cyclorama to provide a soft, diffusely lit background.

SECAM (Système Electronique Couleur avec Memoire)   Alternative technical standard to the NTSC television system used in North America, the SECAM standard uses 625 lines at 25 frames per second. It is in place in much of the rest of the world not serviced by either the NTSC or PAL systems.

secondary movement   Movement of the camera itself (panning, tilting, zooming, trucking, etc.).

SEG (special-effects generator)   Electronic component of a video switcher that permits keying, wiping, and other effects involving two or more sources, where the picture strength of the sources used is 100 percent.

semiscript   Script that features some but not all of the words to be spoken during the program. News shows that have late-breaking news stories often use semiscripts.

shading   Adjusting the electronic levels of the camera to provide the desired contrast range and color reproduction. Shading is done by a technician in the studio at the camera control unit (CCU).

shadow mask   Metal plate pierced with more than 200,000 small holes and fixed to the inside of a television monitor or receiver between the electron gun and the phosphor coating of the picture tube; used to direct electrons to the phosphor dots of the proper color.

share   Proportion of all television households with sets in use in a given market tuned to a particular program.

shotgun microphone   Microphone with a long tube casing designed to pick up sounds from a long distance.

shoulder mount   Camera mount designed to allow the camera to be supported on the operator's shoulder.

SMPTE/EBU (Society of Motion Pictures and Television Engineers/European Broadcasting Union) time code   Frame-accurate system for coding videotaped footage. SMPTE/EBU time code gives each video frame a permanent "address" throughout the entire editing process. Pronounced "simpty time code" for short.

SNG (satellite news gathering)   Use of a satellite to transmit live and/or recorded audio and video material from a remote location to a home base (e.g., a news station).

SOT (sound on tape)   Program segment featuring audio as it appears on the tape itself, without a voice-over. News scripts often specify whether the audio track of a piece of tape will be used by indicating *SOT* on the script.

sound envelope   Changes in the loudness of a sound through time.

special light   Light set up for only one purpose (e.g., to illuminate a box of cereal).

speed light   Small lighting instrument clipped to the top of a camera to provide illumination of on-air talent in ENG productions.

staging   Having on-air talent perform action they might or might not perform if the camera were not present in order to capture it for a news story. Depending on how staging is done and presented to the audience, ethical issues of misrepresentation may arise. Different news organizations have different policies regarding staging, and should be consulted to follow accepted guidelines of journalistic integrity.

storyboard   Panel of drawings showing the main shots of a scene, story, or program. Storyboards are especially useful in television advertising, where precise picture content and running time are crucial.

superimposition (super)   Mixing two video sources on the same screen at the same time, where the picture strength of each source is less than 100 percent.

surround sound   (1) Sound created through speakers set up at the front, sides, and back of a room. (2) Term referring to Dolby Stereo.

sustain phase   Amount of time a sound remains at a stable level after it begins.

sweetening   Enhancing and/or adding additional audio material in postproduction.

synchronous sound   Sound featured in the audio track that coincides with the visual action seen on screen.

synchronous sources   Signals from different video sources that are in phase with one another. To keep different video feeds in phase, a time base corrector or frame synchronizer may be used. Synchronizing video sources permits transitions among them to be stable and glitch free.

Système Electronique Couleur avec Memoire   See *SECAM*.

tail-away shot   Shot featuring action of the scene moving away from the camera.

take   See *cut*.

tally light   Red light on the front and/or back of a studio camera or on the inside of a field camera's viewfinder that indicates when that camera is on the air.

TBC   See *time base corrector*.

TD (technical director)   Crew member who operates the video switcher. The TD may also be responsible for the overall quality of the video images during the production.

technical director   See *TD*.

telecine   Film-to-tape transfer machine found in the control room of a television studio or postproduction facility; also called *film island*.

telegraphing   Informing the audience in advance about upcoming program segments.

teleprompter   Electronic device for feeding script material to on-air talent; located on the front end of a television camera.

tertiary movement   Movement on screen created by control room manipulation (e.g., wipes, keys).

thematic montage   See *dynamic editing*.

three-point lighting   Most commonly used lighting scheme for illuminating stationary talent. Three-point lighting features a key light, a fill light, and a back light to provide an attractive, sculpted look to the talent while adding depth to the video image.

tilt   To tip the camera up or down on the mounting head while keeping the tripod or pedestal stationary. You tilt up (or down) by moving the lens upward (or downward).

timbre   Unique tonal quality of a sound.

time base corrector (TBC)   Device used to stabilize video signals from video tape recorders.

transducer   Device that changes one form of energy into another.

transponder   A receiver/transmitter aboard a satellite.

treatment   Narrative version of a proposed story, presented as a means of generating interest in a producer to produce a pilot of the program.

trim   (1) Camera shot featuring a zoom-in to reduce the area captured in the frame. Trim shots should be motivated and should direct the viewer's attention to important picture information relevant to the subject. (2) Device on an audio console that reduces volume levels at the mic or line input.

tripod   Camera support with three legs that can be placed on a dolly for easy movement.

truck   To move the camera laterally with respect to the subject. The truck can be to the left (truck left) or right (truck right).

TV black   Darkest part of the television picture with a light reflectance level at about 3 percent.

TV white   Brightest portion of the television picture with a light reflectance level at about 70 percent.

two-shot (2-S)   Camera shot featuring two subjects in the frame.

variable focal length (zoom) lens   Lens system that can change from a narrow angle of view to a wide angle of view with continuous gradual motion between two extremes while remaining in focus throughout the movement.

vectorscope   Device used by a video technician to adjust color levels for studio and field cameras during rehearsal and production.

vertical interval time code   See *VITC*.

vertical retrace   Part of the scanning process when the electron gun is turned off to return the electron beam from the bottom right to the top left of the screen.

video switcher   Device that routes, controls, and shapes video images during a television production

viewfinder   TV monitor on the back of the camera used by the operator to see the image captured by the camera.

VITC (vertical interval time code)   Pronounced "vitsee," a type of SMPTE time code that records frame numbers vertically in each frame of the video track of a videotape. One advantage of VITC time code is that it cannot be lost when audio tracks are added in postproduction.

voice-over (VO)   Script notation indicating that a piece of tape will be accompanied by an announcer's voice-over.

volume unit (VU)   Measure of the relative loudness of a sound, or the percentage of modulation of an audio signal.

waist shot   Camera shot picturing the subject from the top of the head to the waist.

waveform monitor   Device that displays a graph of a television signal electronically; used for setting optimal video levels from black to white.

wavelength   The distance a radio wave travels in the time it takes to complete one full cycle. Since light travels at about 300 million meters per second and radio waves travel at the speed of light, the wavelength in meters of a radio wave of a given frequency equals 300,000,000/frequency.

WGA (Writers Guild of America)   Union representing professional writers in motion pictures, television, and radio.

white balance   Adjustment of a television camera to produce an accurate rendition of white when the camera is focused on a white field. White balancing is necessary to avoid inaccurate color reproduction.

window dub   Time code display "burned in" to a video dub of raw footage, which may then be viewed on a regular videotape player; useful for preparing logs, EDLs, and off-line editing. Window dubs cannot be used for on-line editing because the time code display cannot be erased.

wipe   Video effect that looks as though one video source is pushing another off the screen. Wipes can have a variety of patterns (e.g., checkerboard, circle).

X-Y miking   Miking a performance area with two directional mics positioned horizontally on the same plane, at right angles to each other, and a few inches apart to produce a stereo effect.

Y/C recording   See *component recording system*.

z-axis   Depth dimension of the video image going from foreground to background. Action along the z-axis may keep talent in a tight stationary frame and in focus throughout the entire depth of field.

zoom   To change the focal length of a variable focal length lens. *Zooming in* means moving the lens elements to a narrow angle of view, making the scene appear to move closer to the viewer. *Zooming out* means moving the lens elements to a wide angle of view, making the scene appear to move farther away.

394